KS3

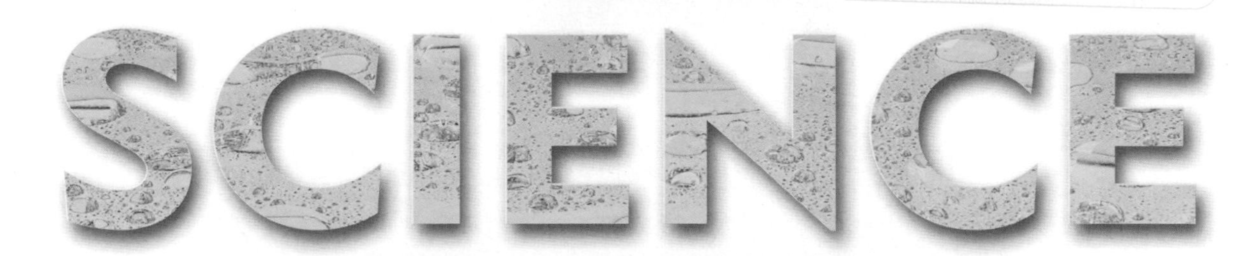

SUMMARY BOOK
SECOND EDITION

Brian Arnold

Hodder & Stoughton

A MEMBER OF THE HODDER HEADLINE GROUP

Photo acknowledgements

The publishers would like to thank the following individuals, institutions and companies for permission to reproduce photographs in this book. Every effort has been made to trace ownership of copyright. The publishers would be happy to make arrangements with any copyright holder whom it has not been possible to contact:

Action Plus (26 bottom, 148 top); Andrew Lambert (126 bottom); Bruce Coleman/Hans Reinhard (34 bottom)/Jane Burton (14 left and top right)/Jeff Foott Productions (32 left)/Staffan Widstrand (39); Colin Taylor Productions (60); Corbis/Alain de Garsmeur (49 left)/Alison Wright (21)/Bettmann (26 top)/Chris McLaughlin (176)/ Craig Lovell (62 right)/David Lees (102 top)/Gary Houlder (48)/ John Calcalosi (57)/John Heseltine (42 bottom right, 49 right)/ Michael Freeman (102 bottom)/Richard Hamilton Smith (148 bottom)/Robert Pickett (35)/Rosemary Greenwood (114); GSF Picture Library (90 top left and bottom left, 91 both, 93, 94 all, 95 both, 108 all, 109 left); Hodder & Stoughton (122); Imperial War Museum (43 left); Life File/Emma Lee (34 top two, 56 giraffes, 62 left, 76 top left and bottom left, 125 both, 170)/Jeremy Hoare (38, 56 snake, 68 left, 150, 155 left)/Jon James (14 bottom right)/Liz Beech (20 right, 126 top)/Massimo Listri (84)/Mike Evans (32 right, 56 shark)/Nigel Shuttleworth (56 puffins)/Owaki-Kulla (112)/ Richard Powers (90 right)/Sally-Anne Fison (155 right)/Ted Spiegel (76 right); PA Photos (20 left); Ping Golf Equipment (159); RD Battersby/Bo'sun Media Services (56 right); Science Photo Library (10 bottom left, 42 top right)/Alfred Pasieka (42 left, 68 right)/ Andrew Syred (10 bottom right)/CNRI (10 top right)/Bruce Iverson (66)/Chris Priest & Mark Clarke (126 middle)/Dr Fred Espenak (188 bottom right)/David Nunuk (188 left)/David Taylor (69)/ Edelmann (16)/Eye of Science (10 top left and middle right)/JC Teyssier (43 right)/PLI (182)/Joe Tucciarone (188 top right)/NASA (189)/Rosenfeld Images Ltd (110 right)/Saturn Stills (47)/Simon Fraser (110 left)/Spencer Grant (109 right).

Orders: please contact Bookpoint Ltd, 130 Milton Park, Abingdon, Oxon OX14 4SB. Telephone: (44) 01235 827720. Fax: (44) 01235 400454. Lines are open from 9.00 – 6.00, Monday to Saturday, with a 24-hour message answering service. You can also order through our website www.hodderheadline.co.uk.

British Library Cataloguing in Publication Data
A catalogue record for this title is available from the British Library

ISBN 0 340 87174 1

First published 1999
Second edition published 2003

Impression number	10 9 8 7 6 5 4 3 2 1
Year	2009 2008 2007 2006 2005 2004 2003

Cover photo from Science Photo Library.
Typeset by Fakenham Photosetting Ltd, Fakenham, Norfolk.
Printed in Italy for Hodder & Stoughton Educational,
a division of Hodder Headline Ltd, 338 Euston Road, London
NW1 3BH.

Contents

STUDY SKILLS AND TEST PREPARATION 2

Study skills 2
Exam revision 5
The day of the exam 7

SCIENTIFIC INVESTIGATIONS 8

LIFE PROCESSES AND LIVING THINGS 10

Cells and reproduction

1.1 Cells 10
1.2 Reproduction 14

Food, digestion and respiration

2.1 Food 20
2.2 Digestion 24
2.3 Respiration 26

Living and feeding

3.1 Living things in their environment 32
3.2 Feeding relationships 36
3.3 Competition and survival 38

Illness and health

4.1 Microbes and disease 42
4.2 Fit and healthy 46

Variation, classification and inheritance

5.1 Classification and keys 54
5.2 Variation and inheritance 56

Plants

6.1 Photosynthesis and respiration 60
6.2 Structure and reproduction 62

MATERIALS AND THEIR PROPERTIES 66

The particle model

7.1 Solids, liquids and gases 66
7.2 Changing state 68
7.3 Solutions 70

Acids, alkalis and simple chemical reactions

8.1 Acids and alkalis 76
8.2 Simple chemical reactions 78

Atoms, elements, compounds and mixtures

9.1 Atoms, elements and compounds 84
9.2 Compounds and mixtures 86

Rocks

10.1 Rocks and weathering 90
10.2 Rocks and the rock cycle 94

Reactions and reactivity

11.1 Reactions of metals and non-metals 98
11.2 Patterns of reactivity 102

Importance of chemistry

12.1 Environmental chemistry 108
12.2 Using chemistry 112

PHYSICAL PROCESSES 118

Energy and energy resources

13.1 Energy and energy changes 118
13.2 Energy resources 120
13.3 Renewable sources of energy 122
13.4 Generating electricity 124
13.5 Temperature and heat energy 126
13.6 Heat transfer 128

Electricity

14.1 Simple circuits 132
14.2 Series and parallel circuits 134
14.3 Current and resistance 136

Magnets and electromagnets

15.1 Magnets and magnetism 140
15.2 Electromagnetism 144

Forces and motion

16.1 Speed 148
16.2 Effects of forces 150
16.3 Balanced and unbalanced forces 152
16.4 Friction 154

Light

17.1 Rays of light and reflection 160
17.2 Refraction and colour 162

Sound and hearing

18.1 Hearing sounds 168
18.2 Different kinds of sounds 170

Pressure and moments

19.1 Forces and pressure 174
19.2 Pressure in liquids and gases 176
19.3 Turning forces or moments 178

The Earth in space

20.1 The Earth in space 182
20.2 The Solar System 184
20.3 Stars and satellites 188

Study skills

Figure 1 Some people find it easy to study and actually enjoy learning! For others studying is not so easy. It is important that you find and use ways of studying that work for you

Listening

Listening is one of the most important study skills and yet it is one we often do badly. You can improve your listening skills by

★ making a real effort to concentrate on every word when someone is explaining something.

★ not getting distracted or letting your mind wander – this isn't easy but you will get better with practice.

★ getting involved – asking questions, jotting down notes etc.

Making notes

It is likely that you will learn from your notes when it comes to revision time. It is, therefore, vital that you are able to read and understand them. It will help if you write them

★ clearly

★ in your own words

★ in small manageable chunks.

Use lots of headings and short paragraphs as these kind of notes are easy to dip into for information.

Figure 2 One of the quickest ways to learn is to listen but it is easy to be distracted

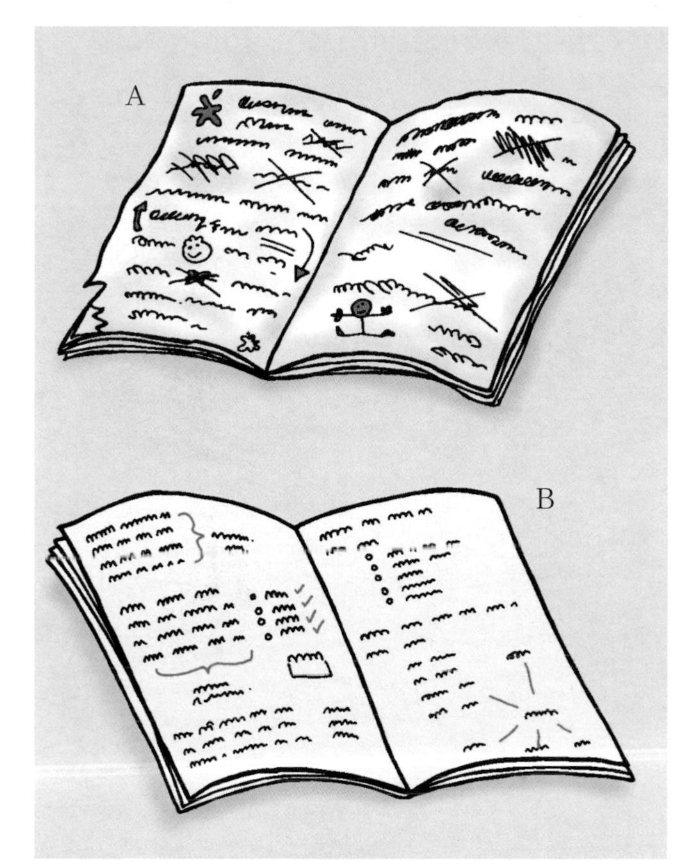

Figure 3 The extra effort student B put into his notes will no doubt be worthwhile when he comes to revise

There are three different types of notes, each of which have their own advantages. You can choose which to use depending upon the type of information to be presented.

1. Structured notes

The information is presented in brief statements with lots of headings, subheadings and numbered points. There are no large sentences or paragraphs.

> ENERGY
>
> 1. Types of energy
>   a) light
>   b) sound
>   c) chemical
>   d) electrical
>
> When energy is changed work is done.
>
> 2. Sources of energy
>   a) fossil fuels
>   b) tidal
>   c) wind
>   d) solar
>
> Fossil fuels are non-renewable sources of energy. They should be conserved if possible.

2. Spider diagrams

The main idea is placed in the centre and all other connected thoughts and ideas are added around it. These should be extremely brief, perhaps just one or two words. More detailed notes could be written on the opposite page. The spider diagram shows an overall view of connected ideas while the notes provide detail. Spider diagrams are very useful for planning essays or making brief notes when revising.

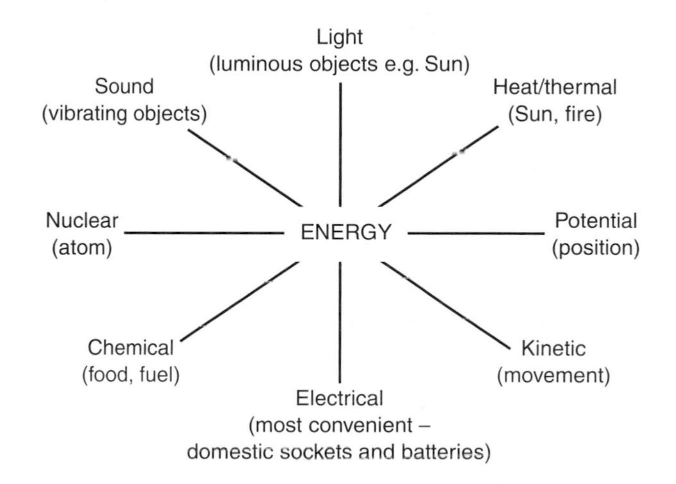

3. Pictures and diagrams

A picture is worth a thousand words. It is often a quick way of presenting, understanding and retrieving certain kinds of information. Here are a few different types of picture that you might find useful. Select those that best suit your needs.

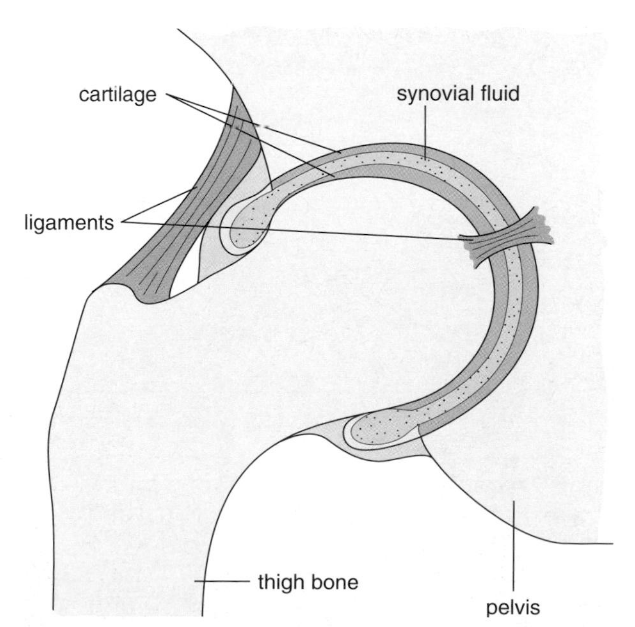

Figure 4 Diagrams like this give a good visual impression which reinforces text

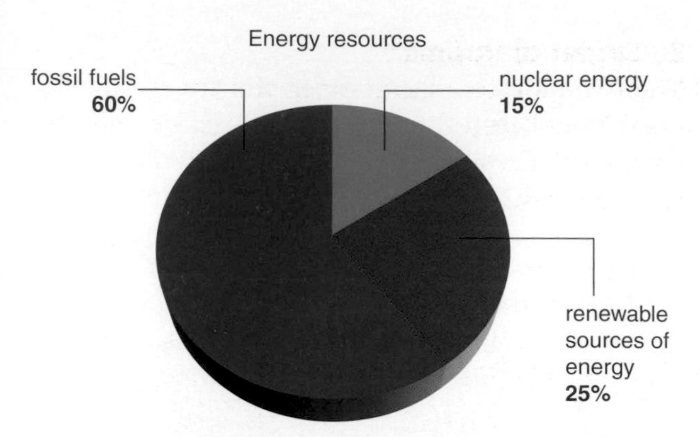

Figure 5 Pie charts are good for comparing several quantities at once

Planet	Time to orbit the Sun (years)
Mercury	0.2
Venus	0.6
Earth	1

Figure 6 Tables allow you to collect and record a lot of information in a very small space

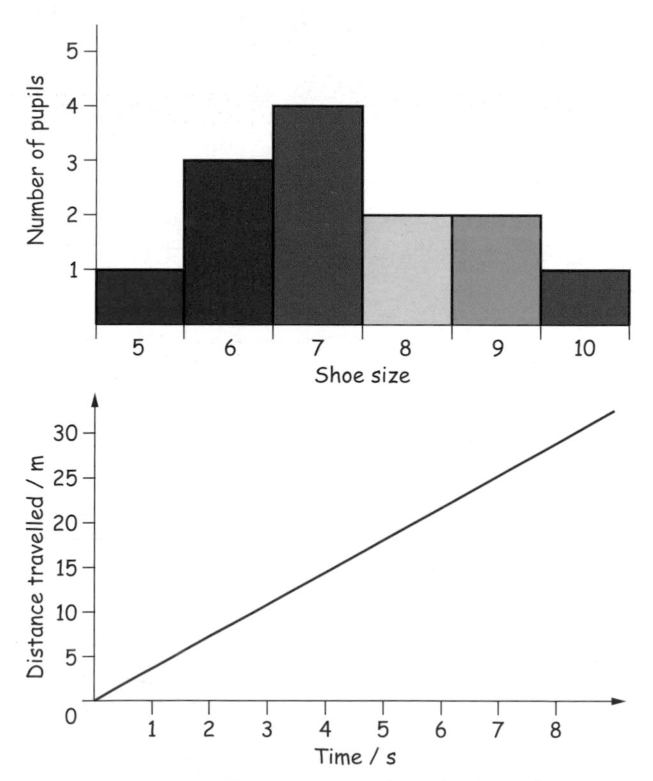

Figure 7 Graphs allow you to record a lot of information in a way which is very visual and compact. They enable you to make quick comparisons and see trends

Deciding where you are going to study

A large amount of your studying is going to take place in school but there will also be times when you need to work at home. Finding the right place with the conditions that suit you is important. Here are some ideas you might like to consider.

Deciding where, when and how to revise

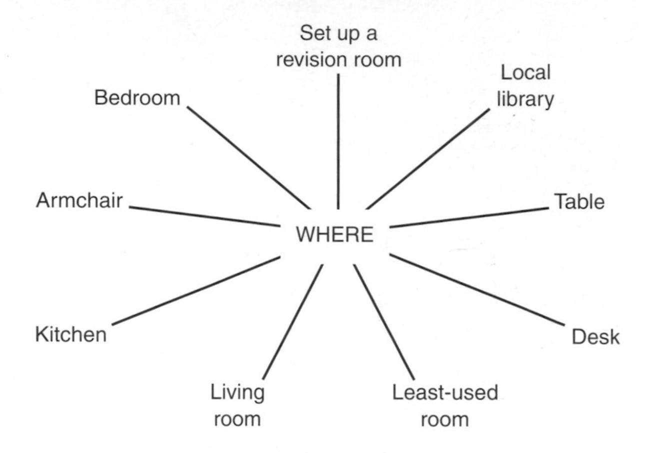

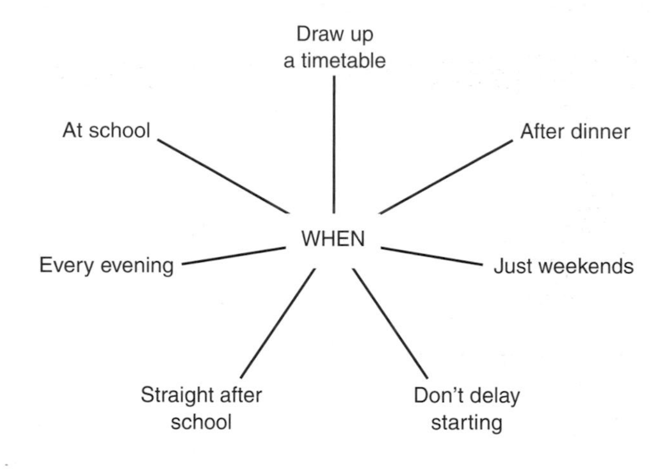

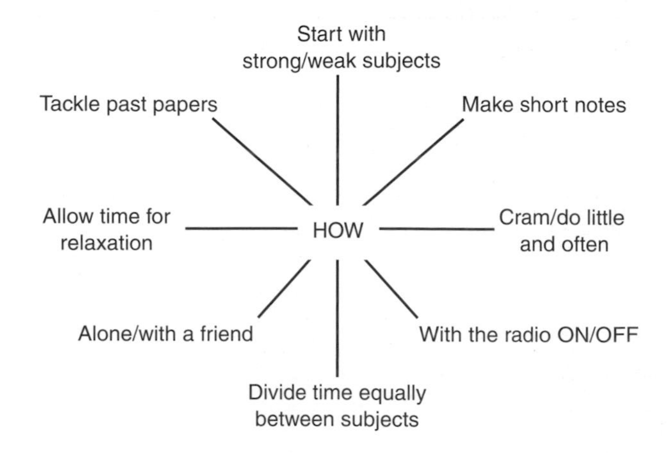

Exam revision

Towards the end of Year 9 you will take your National Tests in Science. A good revision programme and good revision techniques will help you to prepare.

1. Start early

2. Organise yourself

Collect together all the books you are going to need from Years 7, 8 and 9. Draw up a timetable for your revision like the one at the bottom of this page and stick to it.

Figure 8 By starting early you will go into your exams feeling confident and well-prepared. Don't follow the example of the student on the right who thought he could do all his revision the night before

Date	Topic 1	Topic 2	Topic 3
Monday 1 May	Make notes on 'Forces and pressure'	Check Key Terms for 'Forces and pressure'	
Tuesday 2 May	Do questions at end of 'Forces and pressure' topic	Make notes on 'Pressure in liquids and solids'	Check Key Terms for 'Pressure in liquids and solids'
Wednesday 3 May	Do questions at end of 'Pressure in liquids and solids' topic	Make notes on 'Turning forces or moments'	Check Key Terms for 'Turning forces or moments'
Thursday 4 May	Do questions at end of 'Turning forces or moments' topic	Check out 'What you need to know' at end of section	
Friday 5 May	Do 'How much do you know?' questions at end of Pressure and moments section		

Guides to drawing up your revision timetable

1 Don't try to tackle too much in your revision slots. Try working for an hour, then reward yourself with a break of 15–30 minutes and then go back to your work for another hour.

2 Include 'time-out' sessions in your table. It is important that you don't become stale by working too hard and not relaxing. Keep a sensible balance between work and play.

3 Plan the order in which you want to tackle the topics. On days when you have lots of other things to do, such as homework or going out with your friends, choose an easy, straightforward topic. On days when you have a little more time, tackle more demanding areas.

4 Tick off the topics you have covered.

5 Try not to get behind with your revision once you have drawn up your timetable. You will revise much more efficiently if you know you are on top of things.

6 Make several copies of your timetable. Keep one copy in your school bag and stick one up in the kitchen or next to the TV set. This will mean that you won't miss a revision session and will enable your family to see how well your revision is going.

3. Tips on how to revise

a) As you read through your notes use a highlighter pen to mark key ideas and key words. The **Key terms** in this book have already been highlighted and listed for you.

b) Try to understand what you are reading. Don't try to memorise things parrot-fashion. The questions at the end of each topic give you the opportunity to make notes in your own words.

c) Write out key facts on separate cards. This will help to emphasise them.

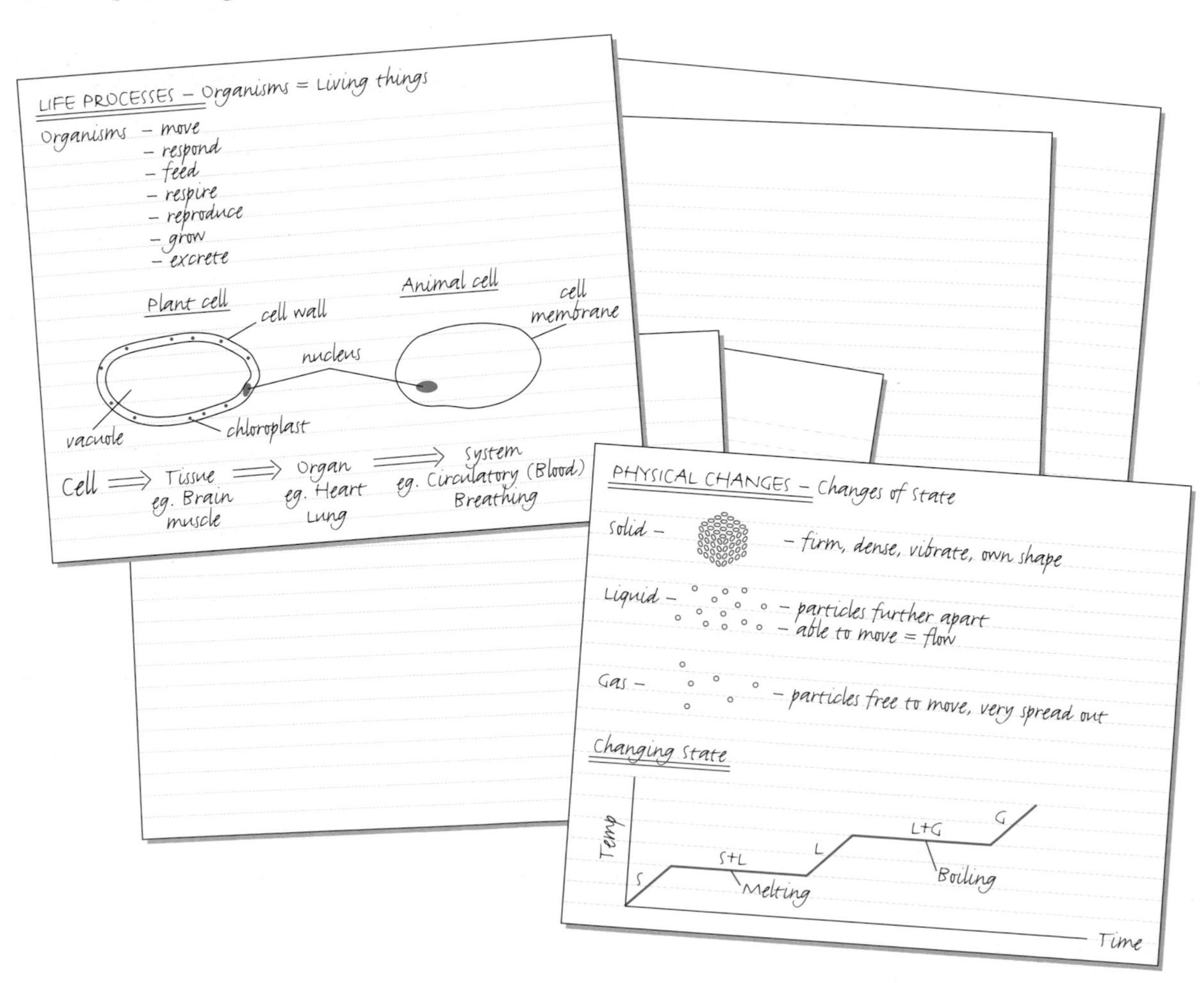

Figure 9 Examples of some revision cards

d) After you have read a piece of work (no more than one page at a time) try to explain to yourself what it is you have just learned or draw a spider diagram of the topic.

e) After you have finished a longer topic, perhaps 10 pages, go back over it glancing only at the highlighted headings and key terms. Check that you understand what they mean and why they are important. A summary of the key ideas, **What you need to know**, is given in this book at the end of each group of topics.

f) When you have finished several topics, try to tackle some questions from past papers. In this book you will find at the end of each group of topics examples of the different types of question you will meet in your exam.

Mnemonics and sayings

These can be useful memory joggers. Some examples are given here:

Red, **O**range, **Y**ellow, **G**reen, **B**lue, **I**ndigo, **V**iolet.

These are the colours of the rainbow in the correct order and can be remembered by the phrase

Richard **Of** **Y**ork **G**ave **B**attle **I**n **V**ain.

Many **V**ery **E**nergetic **M**en **J**og **S**lowly **U**pto **N**ewport **P**agnell.

This saying helps us to remember the names of the planets in order from the Sun – **M**ercury, **V**enus, **E**arth, **M**ars, **J**upiter, **S**aturn, **U**ranus, **N**eptune, **P**luto.

You could try making up some of your own, but keep them simple.

The day of the exam

1. Get up at your usual time and go through your normal routine. Your revision is complete, so be confident that you are well prepared. Concentrate on the idea that there are lots of topics you now understand really well. Don't become overworried if there are one or two topics you still haven't really grasped.
2. Make sure you have everything you need for the exam – pens, pencils, a rubber, a ruler, a protractor, a calculator and a spare ink cartridge.
3. Listen to the instructions you are given before the exam starts.
4. Fill in the front of your question sheet as instructed.
5. Try to answer all the questions. If you find one too hard, go on to the next question.
6. Keep an eye on the clock. Try to be about half way through the paper after half the exam time.
7. Follow the instructions given in the question. For example, if the question says 'Use a ruler to ...' then you must use a ruler to gain full marks.
8. If the answer to a question is worth 3 marks, you will probably need to provide **three** pieces of information.

   For example, if the question reads 'Which of the following planets are further from the Sun than the Earth: (A) Saturn, (B) Mercury, (C) Jupiter, (D) Venus, (E) Mars (3 marks)', you can assume that you will have to name three planets (correct answer, A, C and E).
9. Read your answers through when you have finished a question to make sure that you have written what you wanted to write and that you have answered the question asked.
10. Finally, try to write neatly. Everyone wants you to do well in your exams but if the examiner can't read your answer he can't give you the marks.

GOOD LUCK

Scientific investigations

In your tests there will be questions which assess your appreciation of how scientists of the past, using the limited apparatus available at that time, were able to carry out experiments and devise new ideas and theories based upon the evidence they gathered. You will also need to demonstrate that you have gained those skills which will allow you to a) devise a fair experiment, b) carry it out safely, c) manipulate the results so that relevant conclusions can be drawn and lastly d) look back on your experiment and make valid judgements about its value and the direction that future investigations might take.

There are four skill areas that will be tested.

1 Planning an experiment

In this area you will need to demonstrate that you can

★ plan a simple experiment which is safe

★ plan an experiment which is fair, in which you identify dependent and independent variables and control everything that may affect your readings/observations

★ select the right equipment for your experiment

★ make a sensible prediction for your experiment, i.e. say what you think will happen and why.

2 Obtaining evidence

In this area you will need to demonstrate that you can

★ use the apparatus safely and with skill

★ take accurate readings and/or make accurate observations

★ record your readings and observations clearly, accurately and in a suitable manner, for example in a table.

Figure 2 Obtaining reliable evidence safely

Figure 1 Planning a safe and fair test

3 Analysing the evidence and drawing conclusions

In this area you will need to demonstrate that you can

★ construct diagrams, charts or graphs to help in the analysis of your evidence
★ identify trends and patterns in your results
★ draw conclusions from your results
★ look back at your prediction and say whether or not it was correct
★ link your conclusions with some scientific knowledge.

4 Evaluate your experiment and the evidence you have obtained

In this area you will need to demonstrate that you can

★ recognise whether the evidence you have collected can be used to draw any firm conclusions
★ recognise observations and measurements that do not fit a pattern and that may be faulty
★ explain how and why faulty readings have occurred and suggest ways in which the experiment might be changed so that the evidence obtained is more accurate and/or more reliable.

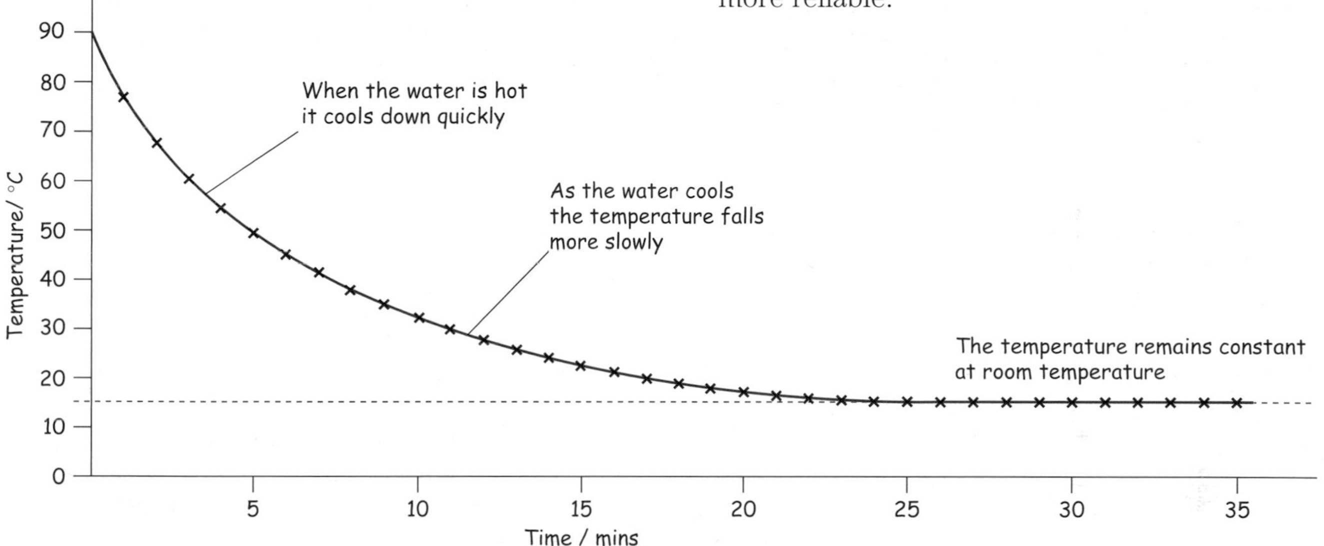

Figure 3 Analysing your results

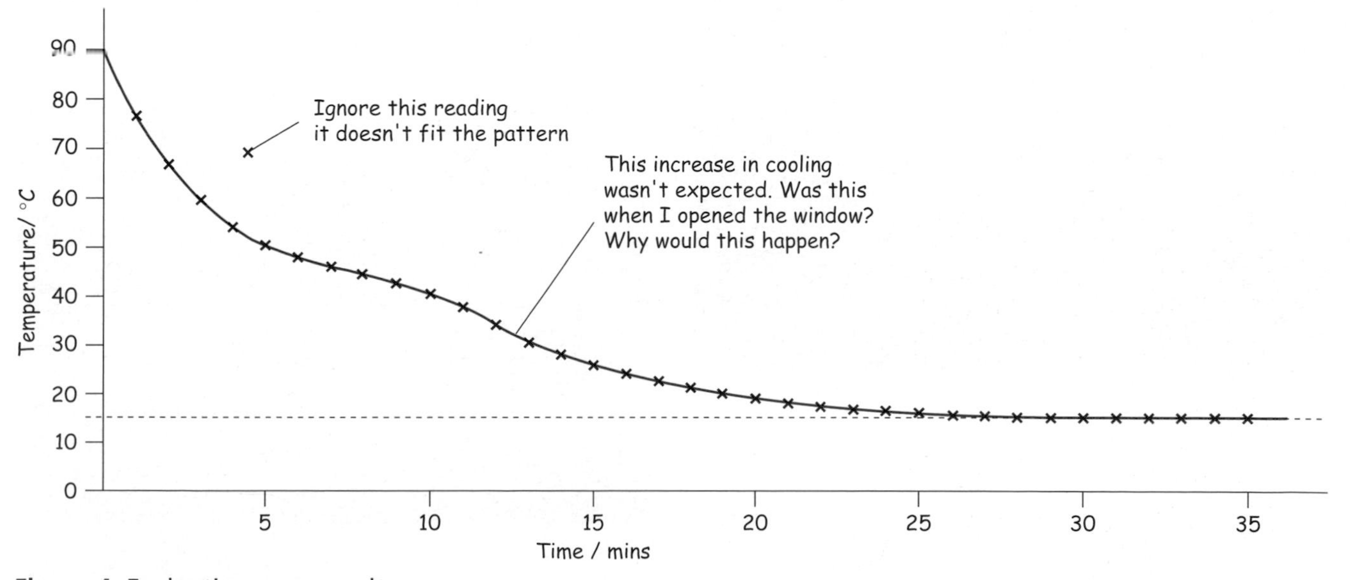

Figure 4 Evaluating your results

1.1 Cells

A living thing is called an **organism**. All organisms are made from simple building blocks called **cells**. The chemical reactions which are needed for living and growing take place inside these cells. It is possible to see the cells that make up plants and animals using a microscope.

(a) Pollen cells

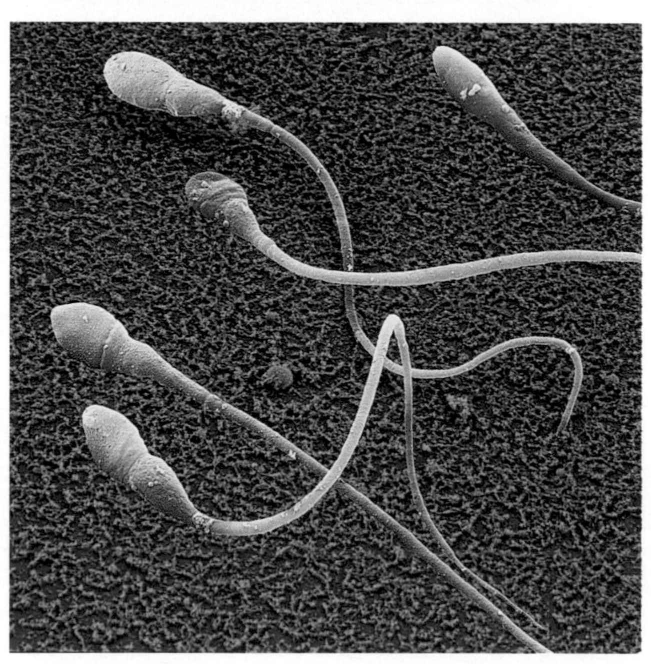

(b) Sperm cells

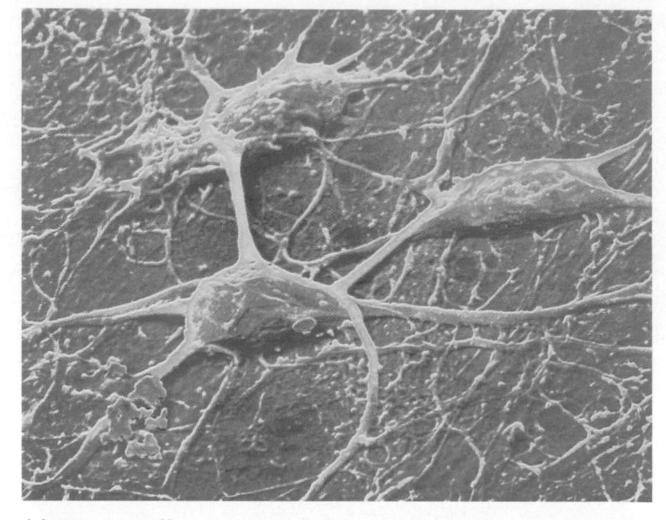

(c) Nerve cells

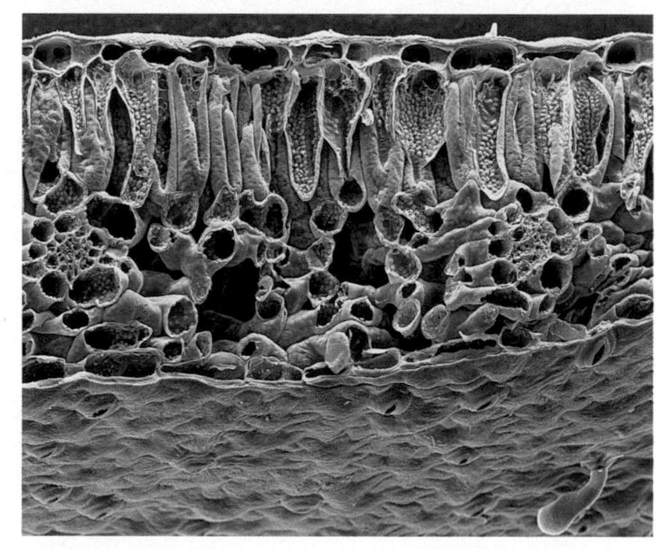

(d) Palisade cells in a leaf

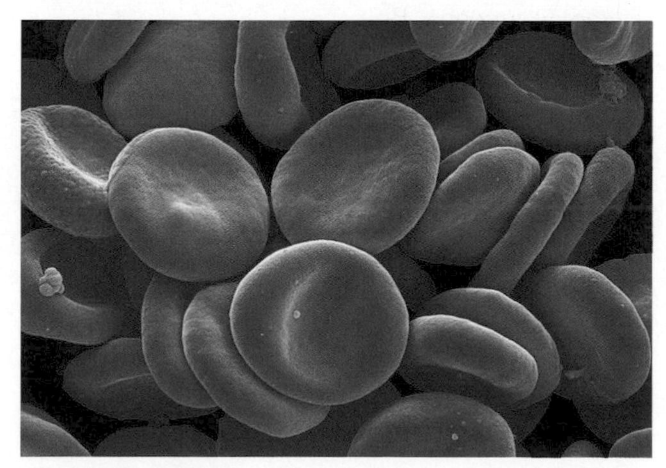

(e) Red blood cells

Figure 1 These photographs show cells in different organisms

There are similarities and differences between animal cells and plant cells.

Animal cells

A typical animal cell has:

★ a **nucleus** which controls the activities of the cell

★ **cytoplasm** – a semi liquid where most of the chemical reactions take place

★ a **cell membrane** – a delicate skin which
  – holds the cell together
  – allows food and oxygen to enter the cell
  – allows waste products to leave the cell.

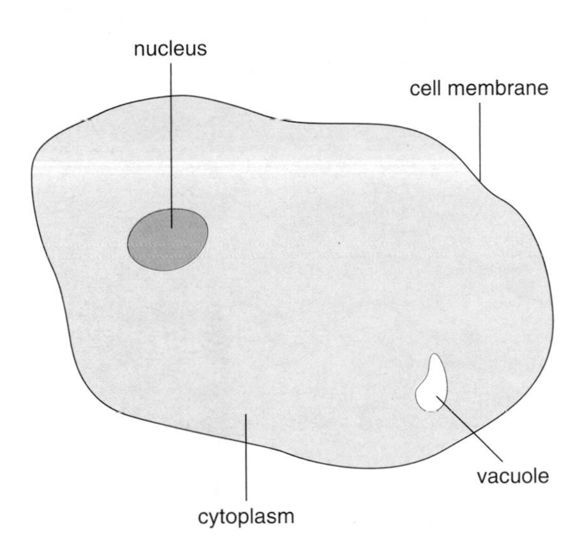

Figure 2 A typical animal cell

Figure 2 shows a simple single-celled organism called an amoeba. Because it has only one cell, this cell has to do all the jobs . . . reproduce, move and feed.

Plant cells

A typical plant cell has:

★ a nucleus

★ cytoplasm

★ a cell membrane

★ a **cell wall** which gives the plant cell shape, strength and support. Plant cell walls are made of **cellulose**.

★ **chloroplasts** which contain **chlorophyll** to absorb light energy so the plant can make food.

★ a **vacuole** – a large central area containing **cell sap**, a watery liquid. The sap contains substances that the plant cell needs to survive.

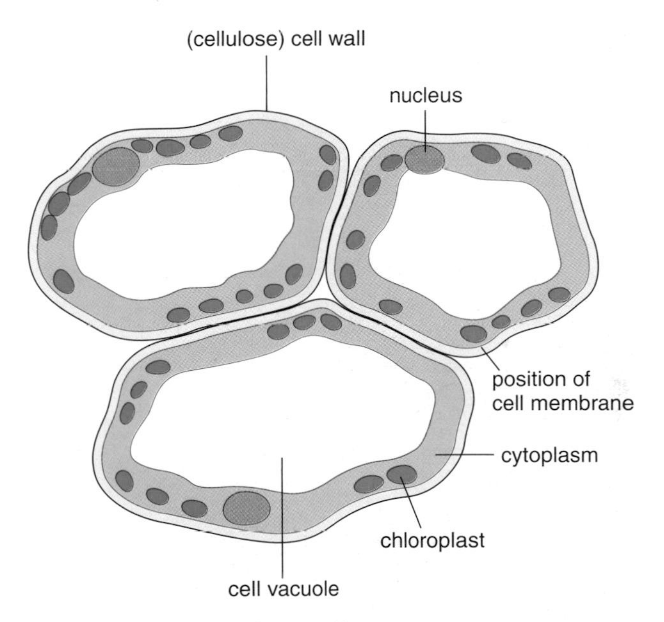

Figure 3 A typical plant cell

1.1 Cells (*continued*)

Bigger and more complicated organisms like you and me are made from a variety of cells.

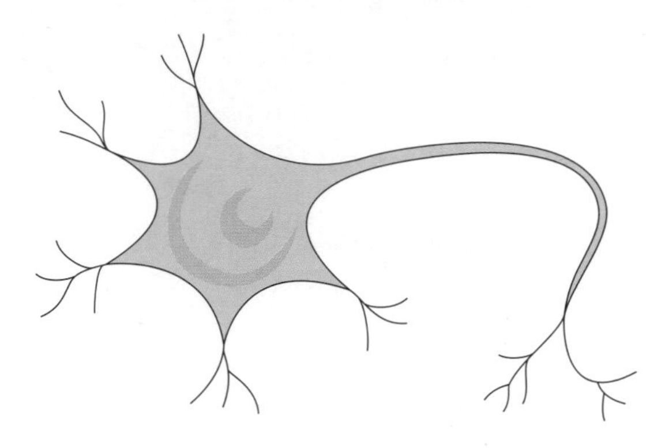

Nerve cells – these cells carry messages around your body

Red blood cells – these cells carry oxygen around your body

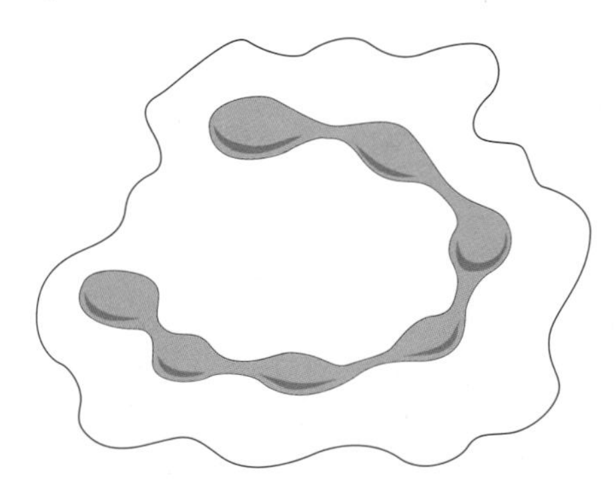

White blood cells – these cells kill germs that get into your body

The different cells carry out all the different jobs. They are **specialised cells**.

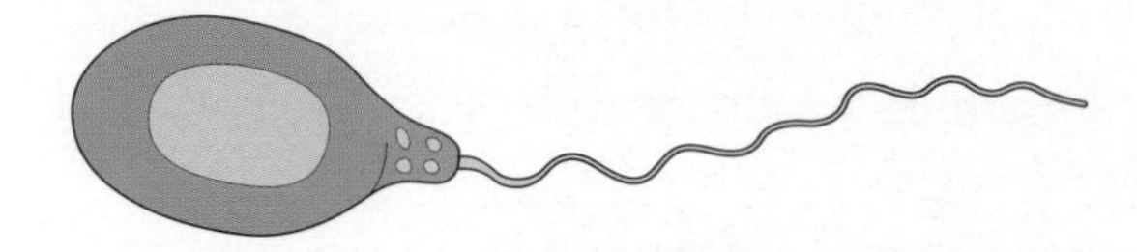

Sperm cells – these cells are produced by the male. They are shaped like tadpoles so that they are able to swim

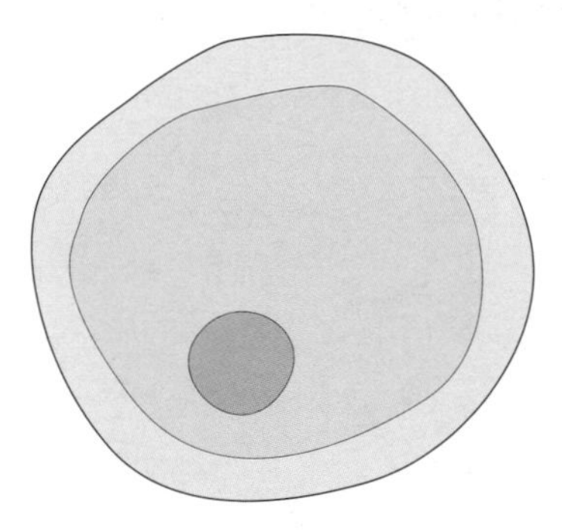

Egg cells or ova – these cells are produced by the female. They are much larger than the sperm and are round like a ball

Figure 4 Examples of specialised animal cells

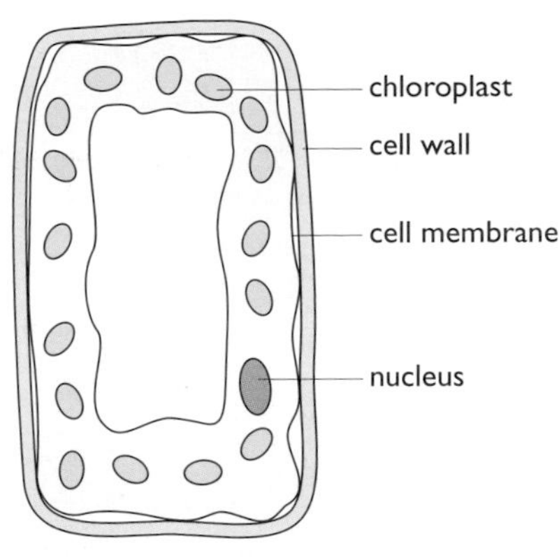

Palisade cells – these cells are found near the upper surface of a leaf. They contain lots of chloroplasts which absorb the sunlight needed for the plant to make food

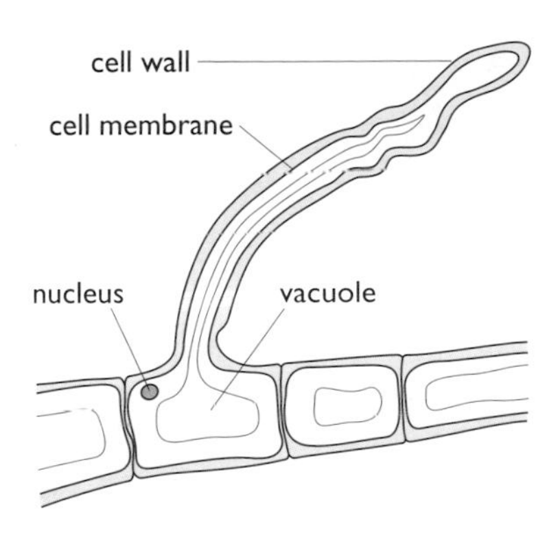

Root hair cells – these cells have a long thin shape which helps them to absorb water and minerals from the soil

Figure 5 Examples of specialised plant cells

Groups of cells

Most plants and animals contain very large numbers of cells. Often cells will form groups in order to do a particular job. These groups are called **tissues**. Examples of tissues include skin tissue, blood tissue and muscle tissue. Tissues may group together to form **organs** such as eyes, lungs and kidneys. Where several organs work together to perform an overall function they form an **organ system** such as the digestive system or the circulatory system.

Cells → Tissues → Organs →
Organ systems → Organism(s)

Where do new cells come from?

Plants and animals grow by making new cells in a process called **mitosis**. A fully grown cell divides into two and the two new cells then increase in size.

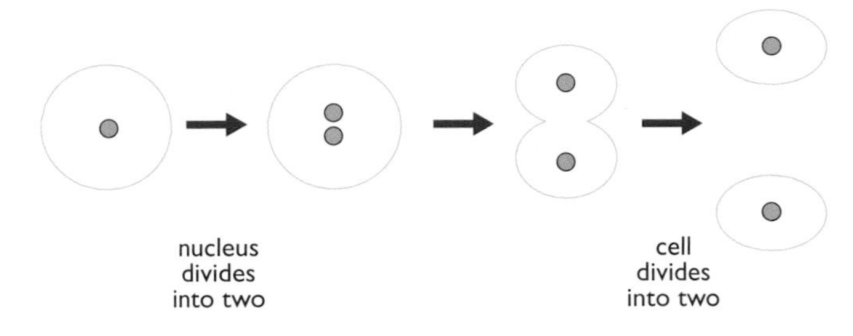

Figure 6 Cell division

Cell division begins with the division of the nucleus. The nucleus contains the information which tells the cell what to grow into and what it is to do. This information is transferred from one generation to the next.

Key terms

Check that you understand and can explain the following terms:

- ★ organism
- ★ cell
- ★ nucleus
- ★ cytoplasm
- ★ cell membrane
- ★ cell wall
- ★ cellulose
- ★ chloroplasts
- ★ chlorophyll
- ★ vacuole
- ★ cell sap
- ★ specialised cells
- ★ tissue
- ★ organ
- ★ organ system
- ★ mitosis

Questions

1. What is an organism? How could you check to see if an organism is made up of cells?
2. Draw a diagram of **a)** a plant cell and **b)** an animal cell.
3. Write down three differences between an animal cell and a plant cell.
4. Give three examples of specialised cells and explain what they do.
5. Explain the difference between a tissue and an organ. Give one example of each.

1.2 Reproduction

For many animals a new life begins when an egg is fertilised by a sperm. This is called **sexual reproduction**. Some animals, such as invertebrates, fish and amphibians, rely on **external fertilisation** in which the female releases eggs into the water and the male then releases his sperm onto the eggs. For this to be effective thousands of eggs and sperm need to be released at the same time. External fertilisation is only possible in water.

Figure 1 As this female frog releases her eggs, the male frog releases his sperm. This is an example of external fertilisation

In other animals, such as **mammals**, birds and reptiles, the eggs are fertilised inside the female. This is known as **internal fertilisation**. Human beings are mammals and use internal fertilisation to create new life.

When an egg has been fertilised, the baby animal begins to develop.

★ Animals that develop outside of the mother's body are at greater risk from predators, adverse conditions such as cold, fast-flowing water and illness. Those that develop inside the mother's body have a greater chance of surviving and becoming independent young.

★ Mammals which have a large number of offspring, like rabbits, cats and mice, keep the developing young inside their bodies for a relatively short amount of time (for example, 8 weeks in the case of guinea pigs) and invest only small amounts of time in aftercare.

★ Others like humans, elephants and horses that produce fewer offspring, keep their developing young in their bodies for much longer and invest more time in rearing them.

Figure 2 It takes just 8 weeks from the fertilisation of her eggs for a cat to give birth to a litter of kittens. Within a very short time they will become independent of her

Figure 3 It takes two years from the fertilisation of the egg for an elephant to give birth to her calf and it will be many years before the calf is independent of its mother

Human reproduction

The male and female reproductive systems are very different.

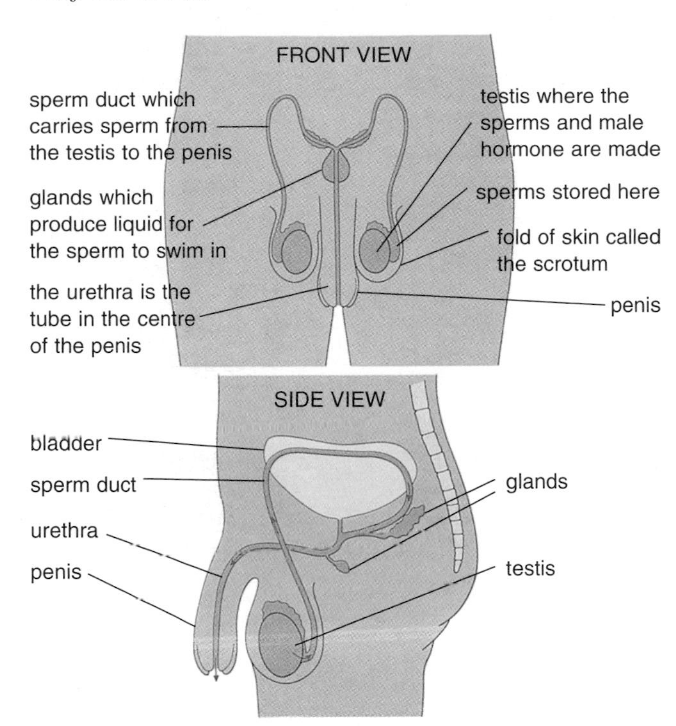

Figure 4 The male reproductive system

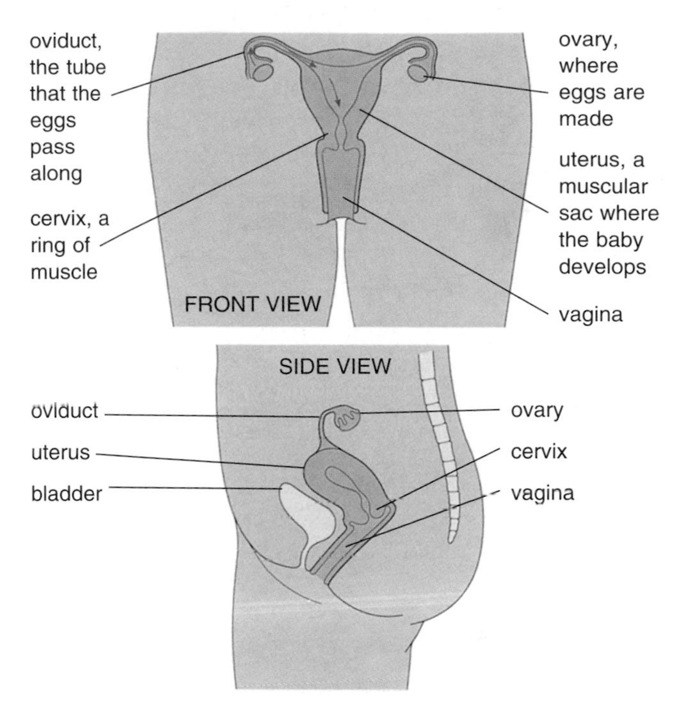

Figure 5 The female reproductive system

Sexual intercourse and fertilisation

When sexual intercourse takes place, the **penis** of the man is inserted into the **vagina** of the woman. **Sperm** from the man's **testes** travel out of the penis, into the vagina, through the **womb** (**uterus**) and into the **oviducts** or **fallopian tubes**.

- ★ Millions of sperm are released by the man to maximise the chance of fertilisation. They all swim towards the egg (**ovum**).
- ★ Just one sperm enters the egg and loses its tail.
- ★ Fertilisation usually takes place in the oviduct.
- ★ The nucleus of the sperm and the nucleus of the egg join together to make the first cell of a new human being. This is called a **zygote**.
- ★ The zygote will contain characteristics of both parents.
- ★ As the **fertilised egg** passes down the oviduct, it divides again and again forming a ball of cells called the **embryo**.
- ★ The embryo embeds itself into the uterus lining and continues to develop here.

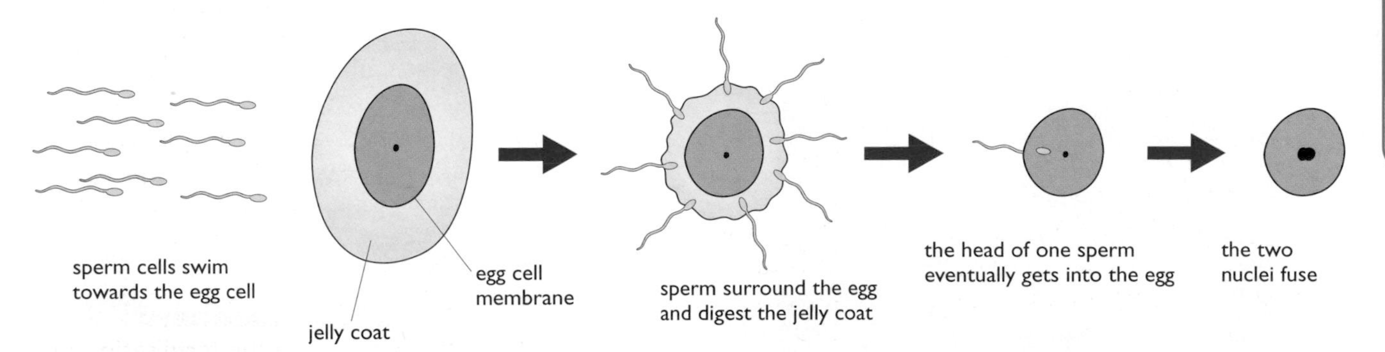

Figure 6 Fertilisation of an egg by a sperm

1.2 Reproduction (*continued*)

Menstrual cycle or period

About every 28 days, a woman's body starts to prepare itself for fertilisation by releasing an egg from one of her **ovaries**. From the ovary, the egg travels down the oviduct. If the egg is fertilised, it attaches itself to the wall of the womb where a thick, blood-filled lining has formed in readiness for the fertilised egg. If the egg is not fertilised, it and the lining pass out of the body through the vagina. This is commonly called having a **period**. The 28-day cycle of releasing an egg, growing the womb lining and discharging it out of the body is called the **menstrual cycle**.

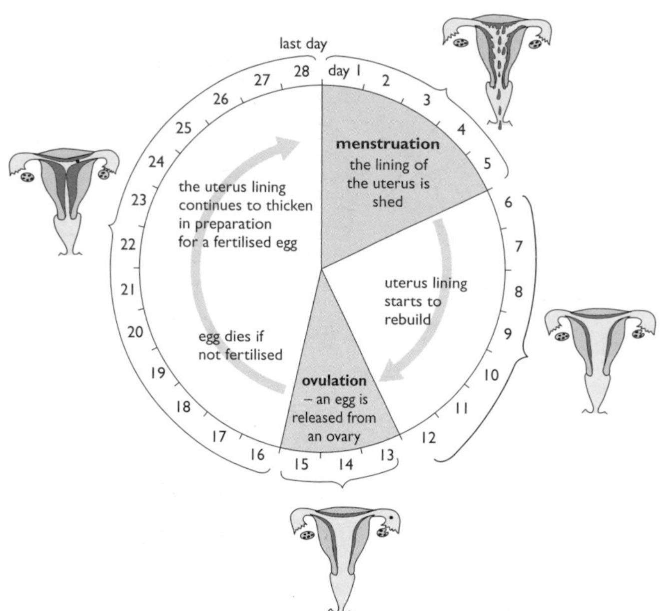

Figure 7 The menstrual cycle

Pregnancy

Around 48 days after the fertilised egg has implanted itself in the womb lining, the developing embryo has many features and begins to look like a human being. It is now called a **fetus**.

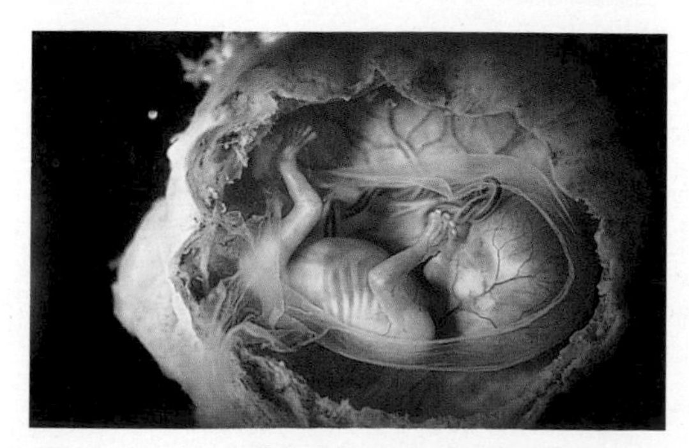

Figure 8 A growing fetus inside its mother's uterus

During **pregnancy** the growing baby receives everything it needs from its mother. It is connected by a tube called the **umbilical cord** to the mother's **placenta**. In the placenta, food and oxygen from the mother's blood can pass into the bloodstream of the embryo. Waste products can pass in the other direction from the baby's blood to the mother's. The placenta acts as a barrier preventing harmful substances reaching the fetus. However, some harmful substances, such as alcohol and certain other drugs, can pass through the placenta and into the fetus. These may affect the development of the fetus, so it is important that pregnant women stop smoking and drinking alcohol.

The fetus is enclosed in a sac of fluid called amniotic fluid. This supports the fetus and protects it from bumps and knocks.

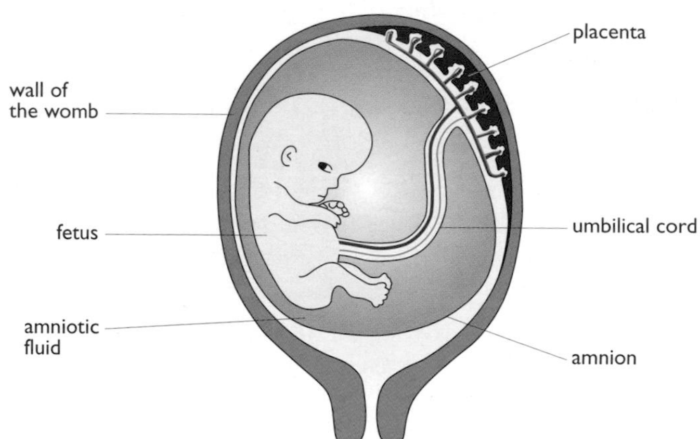

Figure 9 The growing fetus receives everything it needs from its mother's blood

Birth

On average pregnancy lasts 280 days or nine months. Towards the end of this period the baby usually turns so that its head is downwards. The entrance to the vagina, called the **cervix**, widens and strong muscles in the walls of the mother's vagina begin to contract, pushing the baby out, followed by the placenta (afterbirth).

At this point the umbilical cord is cut and the baby begins to breathe for itself. Nourishment is obtained from its mother's breasts (mammary glands) in the form of milk. The milk also contains chemicals called **antibodies** which help the baby to fight infection.

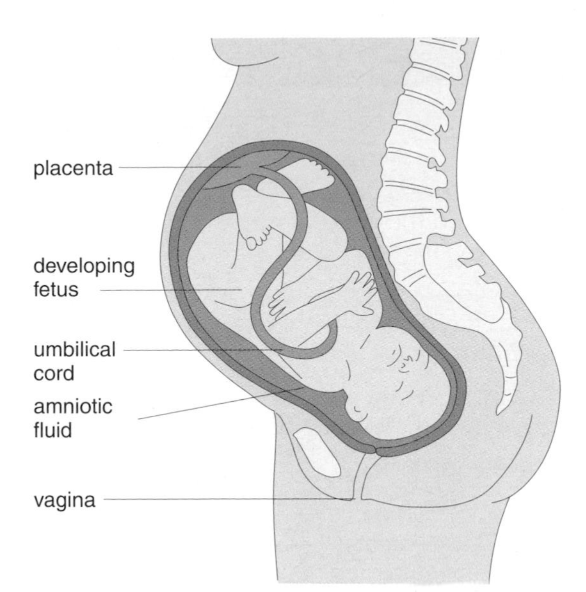

Figure 10 Towards the end of the pregnancy, the baby moves so its head is facing downwards

Growth

As humans pass through the different stages of their life cycle, i.e. infancy, childhood, puberty, **adolescence** and adulthood, their growth rate varies greatly. Figure 11 shows the growth spurts we all experience as very young children and at adolescence.

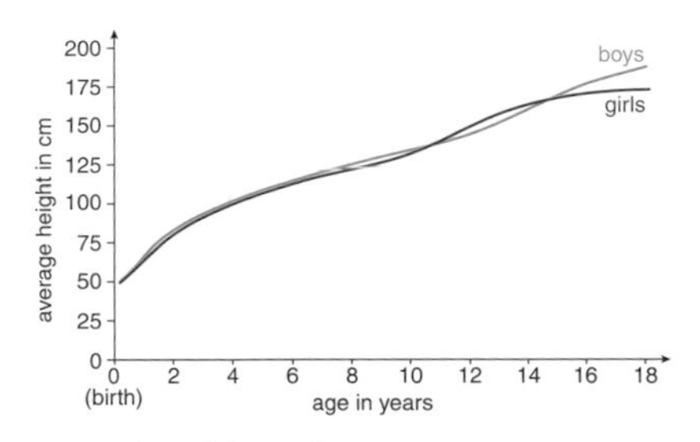

Figure 11 Graph of height against age

Puberty

As we move through our life cycle, our bodies don't just grow, they change too. For example, very young people cannot reproduce, but by the time they are between about 10 and 14 years old, their reproductive organs will have begun to produce eggs or sperm. Once this starts to happen other changes may also take place – these are called the **secondary sexual characteristics**.

Boys (11–14 years old)

- ★ Their voices break and become deeper.
- ★ They begin to grow hair on their face, chest and pubic region.
- ★ They become more muscular.
- ★ Their skin may become greasy and spotty

Girls (10–14 years old)

- ★ They start to grow larger breasts.
- ★ They grow hair around their pubic regions.
- ★ They begin to have periods.

As well as all these physical changes, it is perfectly normal that boys and girls experience some emotional changes such as becoming more self conscious and more aware of the opposite sex. This period of change from boy to man and from girl to woman is called **puberty**. These changes are controlled by chemicals produced by the body called **hormones**.

Key terms

Check that you understand and can explain the following terms:

- ★ sexual reproduction
- ★ external fertilisation
- ★ mammals
- ★ internal fertilisation
- ★ penis
- ★ vagina
- ★ sperm
- ★ testes
- ★ womb
- ★ uterus
- ★ oviduct
- ★ fallopian tube
- ★ ovum (ova)
- ★ zygote
- ★ fertilised egg
- ★ embryo
- ★ ovaries
- ★ period
- ★ menstrual cycle
- ★ fetus
- ★ pregnancy
- ★ umbilical cord
- ★ placenta
- ★ cervix
- ★ antibodies
- ★ adolescence
- ★ secondary sexual characteristics
- ★ puberty
- ★ hormones

Questions

1. Draw a labelled diagram of a) the female reproductive organs and b) the male reproductive organs.
2. Why do sperm have tails?
3. What is the maximum number of sperm that can fertilise one egg?
4. On average, how often is an ovum released from a woman's ovaries?
5. What happens to a zygote as it travels down the oviduct to the uterus?
6. What is the placenta and what does it do?
7. What three physical changes may take place in a) boys and b) girls during puberty?

Chapter 1 Cells and reproduction:

What you need to know

1. Plants and animals are made up of cells.
2. The importance of cell membranes, cytoplasm and nuclei in plant and animal cells.
3. The importance of chloroplasts and cell walls in plant cells.
4. Growth occurs when cells divide and increase in size.
5. Some cells have adapted and evolved so that they can carry out a particular task e.g. ova, sperm, nerve cells and root hair cells.
6. The main parts of the male and female reproductive system and their functions.
7. A human life begins when an egg is fertilised by a sperm.
8. How a fertilised egg develops in the mother's womb and the important role the placenta plays in the embryo's growth.
9. The longer a fetus develops inside its mother and is given aftercare, the greater its chance of survival.
10. The physical and emotional changes that take place during adolescence and puberty.
11. The menstrual cycle.

How much do you know?

1 This diagram shows the structure of a simple single-celled animal.

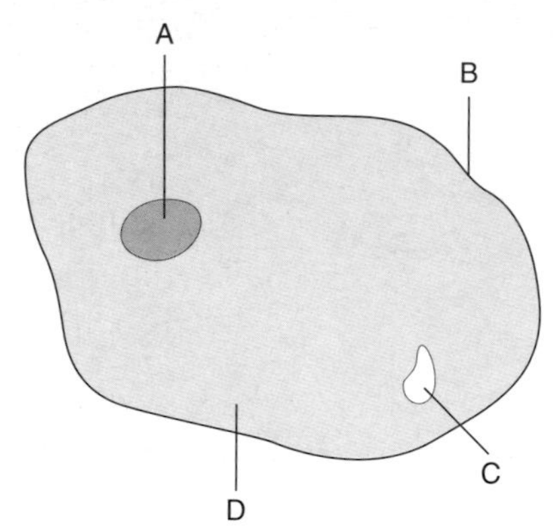

a) Name the parts labelled A, B, C and D.

A ______________________

B ______________________

C ______________________

D ______________________

4 marks

b) Explain what these parts of the animal cell do.

A ______________________

B ______________________

C ______________________

D ______________________

4 marks

2 This diagram shows the structure of some plant cells.

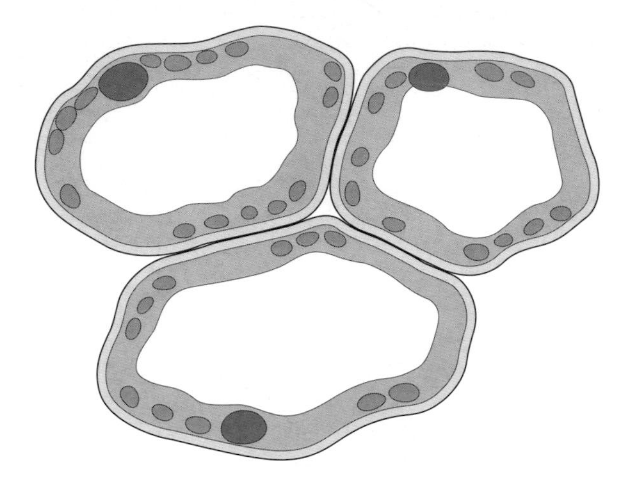

a) On the diagram above label the cell wall and a chloroplast.

2 marks

b) Explain what these parts of the plant cell do.

(i) chloroplasts ______________________

(ii) cell wall ______________________

2 marks

How much do you know? *continued*

3 a) Explain the difference between internal fertilisation and external fertilisation.

____

____

2 marks

b) Name one animal that reproduces using external fertilisation.

____

1 mark

c) Name one animal that reproduces using internal fertilisation.

____

1 mark

d) Explain the difference in the number of eggs released by the female when external fertilisation is taking place compared with when internal fertilisation is taking place.

____

____

2 marks

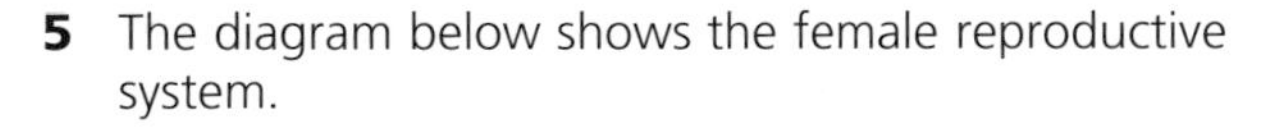

5 The diagram below shows the female reproductive system.

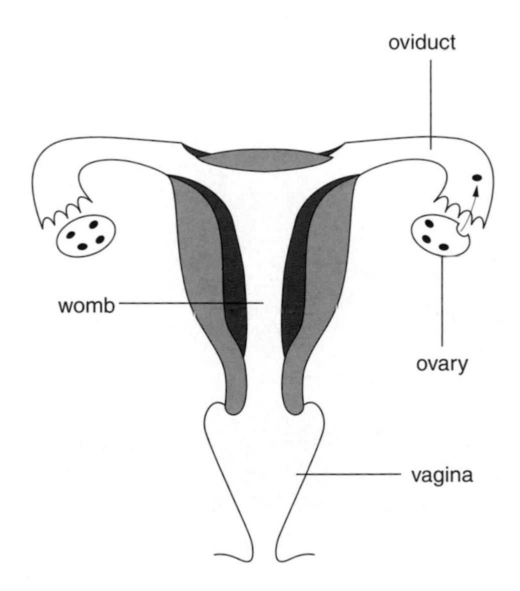

a) Approximately how often is an ovum released from the ovaries?

☐ One every day

☐ One every 7 days

☐ One every 28 days

☐ One every year

1 mark

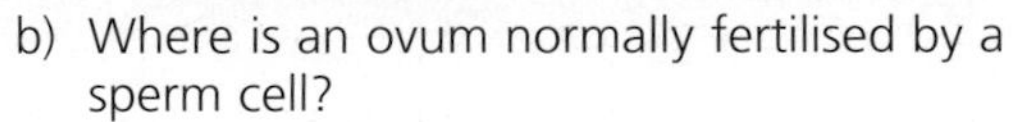

b) Where is an ovum normally fertilised by a sperm cell?

____

1 mark

c) What happens to an ovum if it is not fertilised?

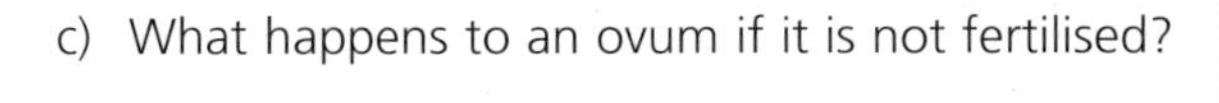

____

1 mark

6 The diagram below shows a developing embryo attached by the umbilical cord to the placenta.

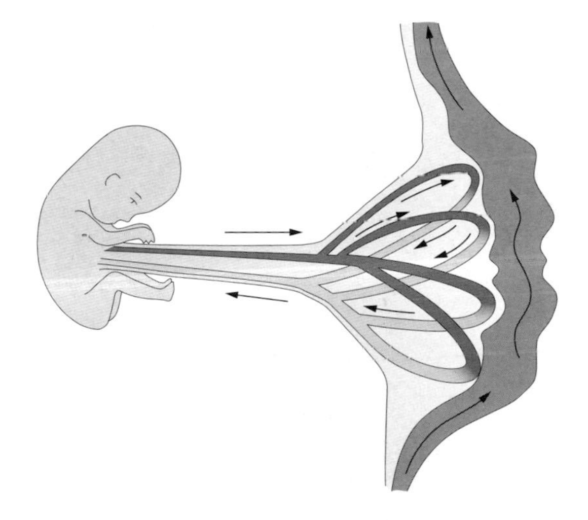

a) What two important substances pass from the mother through the placenta to the embryo?

____

2 marks

b) Why should someone who is pregnant not smoke?

____

2 marks

c) What is amniotic fluid and what does it do?

____

____

2 marks

2.1 Food

Food

Food is vital for life. The kind of food you eat and the amount of food you eat affect your health and well-being.

Healthy eating

There are five main types of food:

★ carbohydrates
★ fats
★ proteins
★ minerals
★ vitamins

It is important that you eat these in the correct amounts. To be healthy you must have a **balanced diet**.

Figure 1 Without the right kinds of food your body will suffer from **malnutrition** like this family

Carbohydrates are energy-giving foods. They can release energy quickly. Bread, rice, pasta, flour, potatoes, chocolate and sugar all contain large amounts of carbohydrate. If you eat more carbohydrates than your body requires, they are stored by your body as fat.

Figure 2 Foods rich in carbohydrates

Figure 3 Carbohydrates can provide energy when it is needed quickly

Fats are also energy-giving foods, but they cannot release their energy quickly. It takes a long time for your body to use energy which is stored as fats. Butter, margarine, meat and fried foods contain lots of fats. Fats are very good insulators. Your skin has a layer of fat underneath it which stops you losing too much heat from your body.

Figure 4 Foods rich in fats

Figure 5 Inuits have a diet which is very rich in fats. When the fats are stored in their bodies it helps to keep them warm

Proteins are body-building foods. Your body uses them to grow and to repair any damaged tissue. Meat, cheese, eggs and nuts all contain lots of protein.

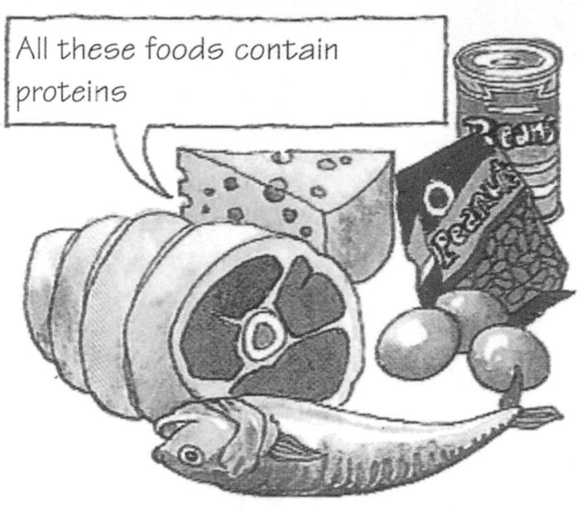

Figure 6 Foods rich in protein

Minerals are simple chemicals found in most foods. Your body needs them in small amounts to work properly. For example, iron is needed to help your red blood cells carry oxygen to all parts of your body efficiently. Calcium and phosphorus are needed for strong, healthy bones and teeth.

Figure 7 All these foods are rich in minerals. Fish contains phosphorus and calcium, milk contains calcium, spinach contains iron, meat contains phosphorus and iron

2.1 Food (*continued*)

Vitamins are complicated chemicals, but they must be present in the food that you eat if the cells of your body are to work properly. Like minerals, they are only needed in very small amounts.

Figure 8 A mixed diet of fresh foods will contain all the vitamins you need

In addition to the five main food types, it is important that your diet also contains **water** and **fibre**.

★ Water is a vital part of your diet. It is needed to transport many different materials around your body and to help flush out toxins. Your body is almost three-quarters water!

★ Fibre helps to keep the digestive system clean and healthy. Brown bread, brown rice, fruit and vegetables contain lots of fibre. A lack of fibre can cause the lower part of the digestive system to become blocked. This is called constipation and can be very painful.

Food tests

Testing for starch

★ Add two or three drops of iodine solution to the food.

★ If starch is present, the iodine will turn blue/black.

Figure 9 Testing for starch using iodine solution

Testing for sugar

★ Add two or three drops of Benedict's solution to a small amount of food in solution.

★ Heat this gently in a water bath until it boils.

★ If sugar (glucose) is present, an orange/red precipitate will form.

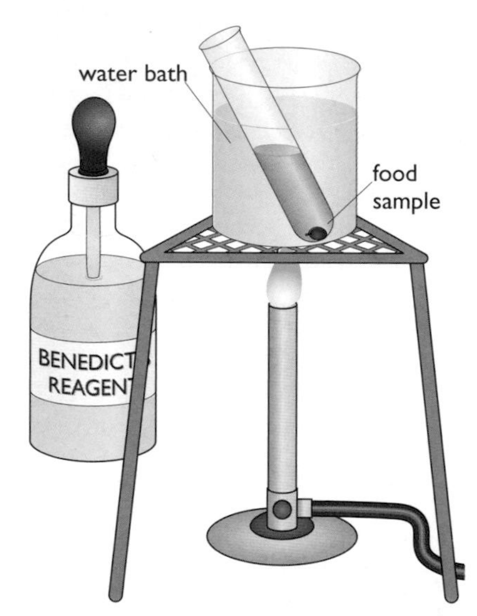

Figure 10 Testing for sugar using Benedict's solution

Testing for fats

★ Add 2 cm^3 of ethanol to a small amount of food in a test tube and shake.

★ Add 2 cm^3 of water to this and shake again.

★ If fat is present, the solution will turn cloudy white.

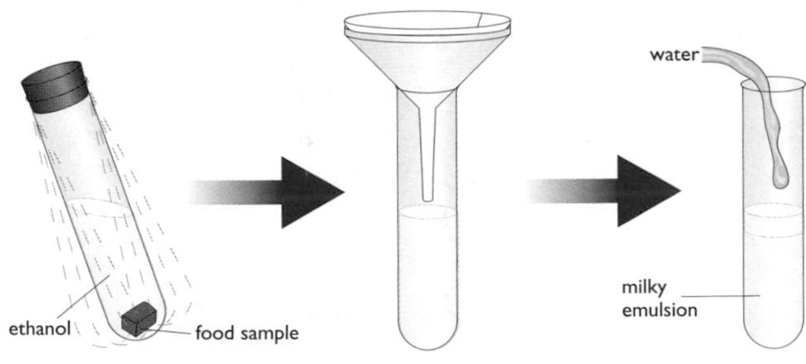

Figure 11 Testing for fats using ethanol

Testing for proteins

★ Add 4 cm^3 of copper sulfate to a small amount of food in solution.

★ Add two or three drops of sodium hydroxide.

★ If protein is present, the solution will turn purple.

Figure 12 Testing for proteins

Key terms

Check that you understand and can explain the following terms:

★ balanced diet
★ malnutrition
★ carbohydrates
★ fats
★ proteins
★ minerals
★ vitamins
★ water
★ fibre

Questions

1a) Name five important types of food which you must have in your diet.

b) Give two examples of foods in which each of these may be found.

2 Explain the phrase 'balanced diet'.

3 What will happen to you if you eat too many energy-giving foods?

4 Why is it important to eat food which contains fibre?

5 Find out the meanings of the words *malnutrition* and *obese*.

6 Describe how you would test some food to see if it contained protein.

7 Name two minerals needed by the body. Explain why these minerals are needed.

8 Write down a list of all the foods you ate yesterday. Try to put all the foods into the five main categories: carbohydrates, fats, proteins, minerals and vitamins. Some foods will be in more than one category.

2.2 Digestion

You eat food to provide your body with substances it needs to live and grow.

To be of any use this food has to be broken down into simpler substances. It can then dissolve in your bloodstream and be carried to those parts of your body where it is needed. This breaking down of large food molecules into smaller molecules is called **digestion**.

Breaking down your food

Large food molecules such as protein and starch are broken down into smaller ones by chemical reactions. Special chemicals called **enzymes** help to speed up these reactions.

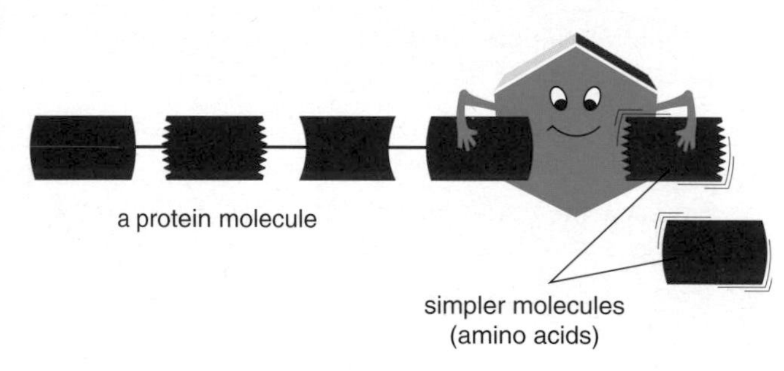

Figure 1 The enzyme pepsin helps to break protein molecules down into simpler molecules called amino acids which can dissolve in blood

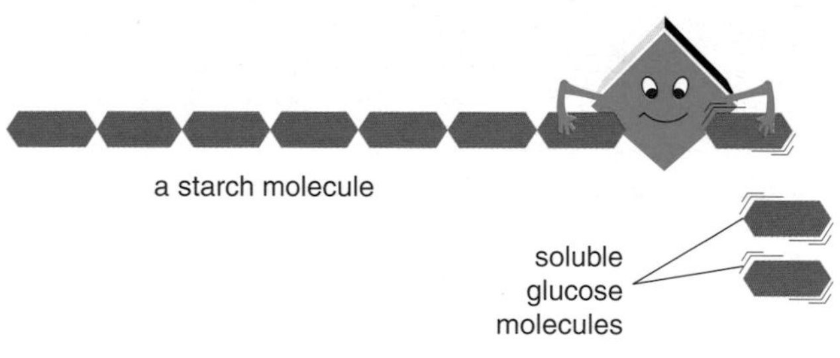

Figure 2 The enzyme amylase helps to break starch molecules down into simpler sugar molecules which are soluble in blood

Enzymes are very sensitive to the temperature and pH of their surroundings. If these are not right for an enzyme, it will not work properly.

The enzymes in your body have adapted so that they work best at body temperature i.e. 37 °C. Below this temperature they work much more slowly and above this temperature they can be destroyed.

Most enzymes work best in neutral surroundings, i.e. pH 7, but some have adapted to work at different pH levels. For example the pH in our stomachs is about pH 2 – it is very acidic. The enzyme called pepsin has adapted to work in these conditions.

Passage of food through your gut

When you swallow your food, it travels along a tube which begins at your mouth and finishes at your anus. This tube is called the **alimentary canal** or **gut**. At various places along the gut **digestive juices** containing enzymes are produced. These react with the food, breaking it down as it passes through. This digested food is then able to pass through the wall of the gut into the blood. Food which has not been digested passes out of the anus.

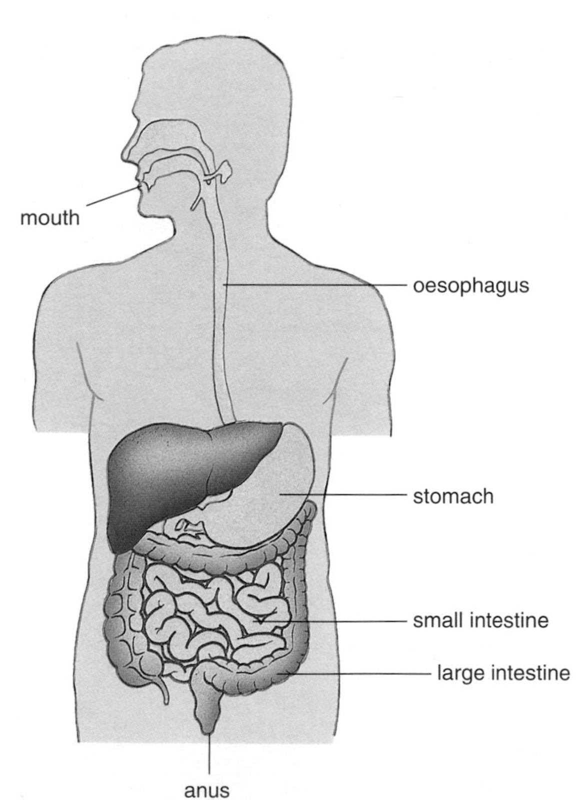

Figure 3 The digestive system

The stages of digestion

1. Food is broken down mechanically in the mouth by the chewing action of teeth.
2. **Saliva** is produced by the salivary glands as you chew. This wets the food making it easier to swallow. Saliva contains an enzyme called amylase which begins the digestion of **starch** molecules.

3 Food passes down the **oesophagus** to the stomach. No mechanical or chemical breakdown takes place in the oesophagus.

4 When food enters the stomach, more chemicals (gastric juices) are added to it. The gastric juices contain enzymes. **Proteins** are digested by an enzyme called pepsin. Food stays in your stomach for about four or five hours. During this time muscles in your stomach churn and mix the food with these gastric juices to help digestion.

5 In the upper part of the small intestine, juices from the pancreas (pancreatic juices), which contain several enzymes, are added. These juices continue to digest starches, proteins and fats. As the digested food travels along the small intestine (two to three hours), it passes through the thin walls and is absorbed by the blood.

6 Undigested food like fibre passes through the large intestine and out of the body through the anus.

Key terms

Check that you understand and can explain the following terms:

- ★ digestion
- ★ enzymes
- ★ gut/alimentary canal
- ★ digestive juices
- ★ saliva
- ★ starch
- ★ oesophagus
- ★ protein

Questions

1 What does your digestive system do for you?

2a) What is an enzyme?
b) Name two enzymes that are produced in your body.

3 What might happen to an enzyme if the temperature or pH of its surroundings are not right?

4 Where does your gut begin and end?

5 Where in your body is food mechanically broken down?

6 Where in your body does digested food enter your bloodstream?

7 What happens to food which is undigested?

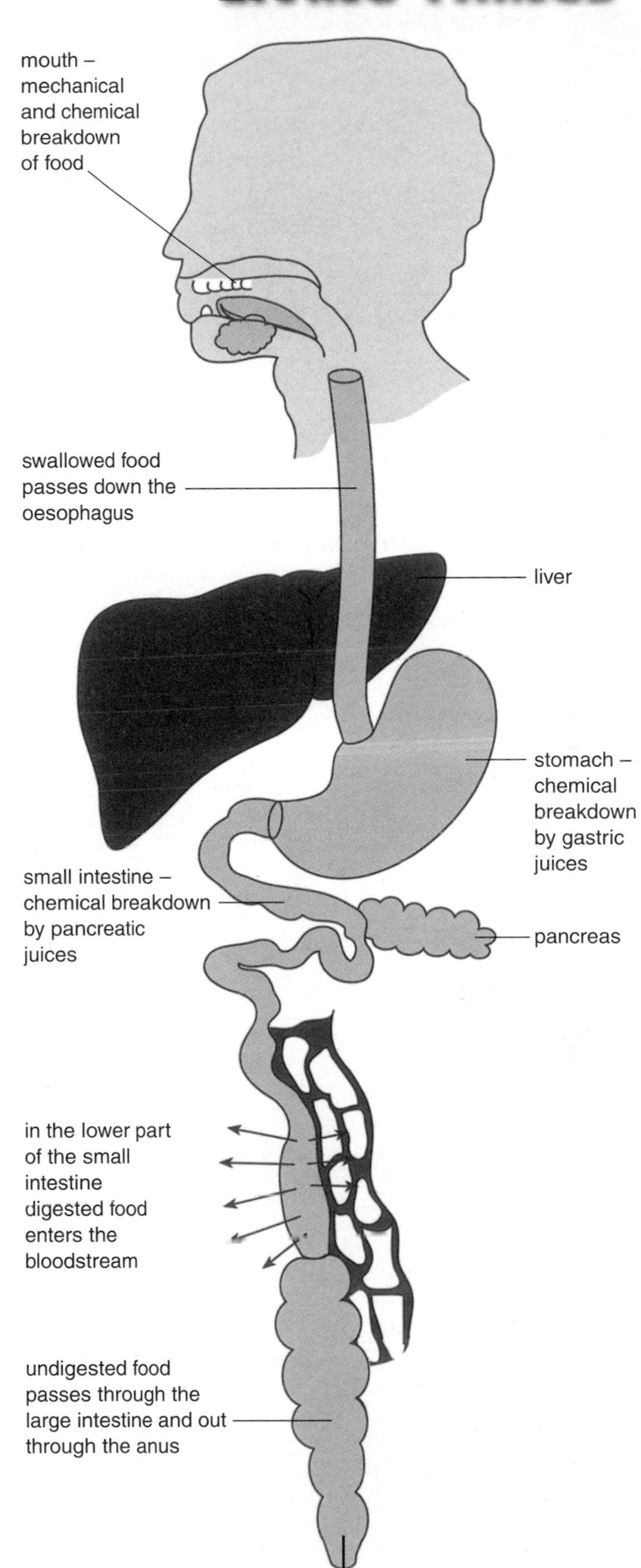

Figure 4 The different stages of digestion

2.3 Respiration

All the living cells in your body need energy to survive. They obtain this energy from a chemical reaction between the food (**glucose**) your body has digested (see page 25) and oxygen which comes from the air which you breathe. This process of obtaining energy is called **respiration**. We can summarise this process by a word equation:

glucose + oxygen → carbon dioxide + water + ENERGY

In everyday life, the rate at which your body needs oxygen for this reaction and the rate at which the blood supplies it are the same. An activity such as digging which keeps this balance is called an aerobic activity.

Figure 1 Your body receives the energy it needs by **aerobic respiration**

In events of short duration like a 100 m sprint, your body will allow you to make greater demands on the supply of oxygen than it can cope with. But at the end of the event you must repay this **oxygen debt**, usually by breathing rapidly, i.e. panting.

Figure 2 This athlete is being given oxygen to help him recover quickly from his oxygen debt

In longer events, such as marathon running or playing football, you may find that there is no opportunity to rest, so your oxygen debt grows. Your body helps you get over this problem by releasing the energy in your food using a different chemical reaction. This reaction does not need oxygen and is called **anaerobic respiration**. This reaction is shown by the following word equation:

glucose → lactic acid + ENERGY

Unfortunately one of the waste products of this reaction is a chemical called **lactic acid**. When lactic acid builds up in your muscles it makes them ache and eventually it may cause cramp.

Figure 3 Sustained anaerobic exercise can cause cramp

The ingredients necessary to produce energy, glucose and oxygen, are carried to where they are needed in the body by your blood.

Circulation of blood

Blood is moved around your body by the pumping action of your heart. The heart pumps the blood to the lungs where it picks up oxygen from inhaled air. This oxygen-rich blood then returns to the heart from where it is pumped to all parts of your body. Your cells use this oxygen to respire (produce energy) using the reaction we saw earlier. Carbon dioxide, which is a **waste product** of respiration, is carried away from the cells by the blood and returned to the lungs. It is then lost from the body when we breathe out.

Blood vessels

Blood is carried around the body in a network of tubes called **blood vessels**. There are three different kinds of blood vessels:

1 **Arteries**. Oxygen-rich blood leaves your heart through tubes called arteries. (This is easy to remember if you think that both the words Arteries and Away start with the letter 'A'.) These tubes are quite wide and have thickish walls as the blood is flowing under high pressure, because it has to travel all around the body. When you feel your pulse, it is the pumping action of the blood through the arteries which you can feel.

2 **Veins**. Blood which has been around the body and so is rich in carbon dioxide returns to the heart through tubes called veins. These tubes are often wider than arteries, but have a thinner wall as the blood is under much less pressure. No pulse can be felt here as the blood flows through veins smoothly.

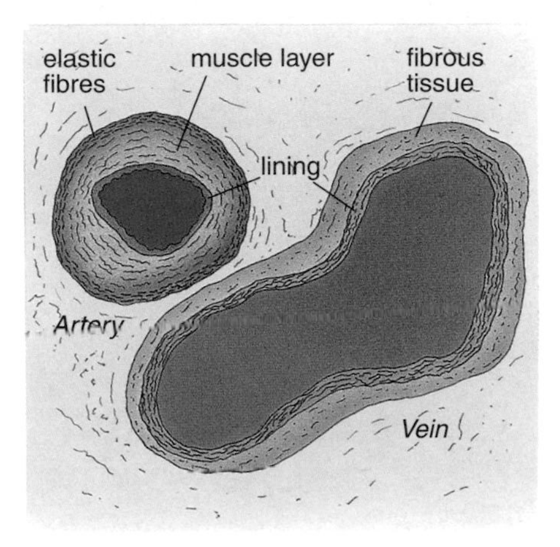

Figure 5 An artery and a vein

3 **Capillaries**. As blood moves away from the heart, it passes through narrower and narrower tubes until it eventually passes through a very fine network of tubes called capillaries. Capillaries have extremely thin walls so substances can easily pass through them. Every living cell in your body has a capillary close by, so glucose and oxygen can **diffuse** into the cell and waste products such as carbon dioxide can diffuse out.

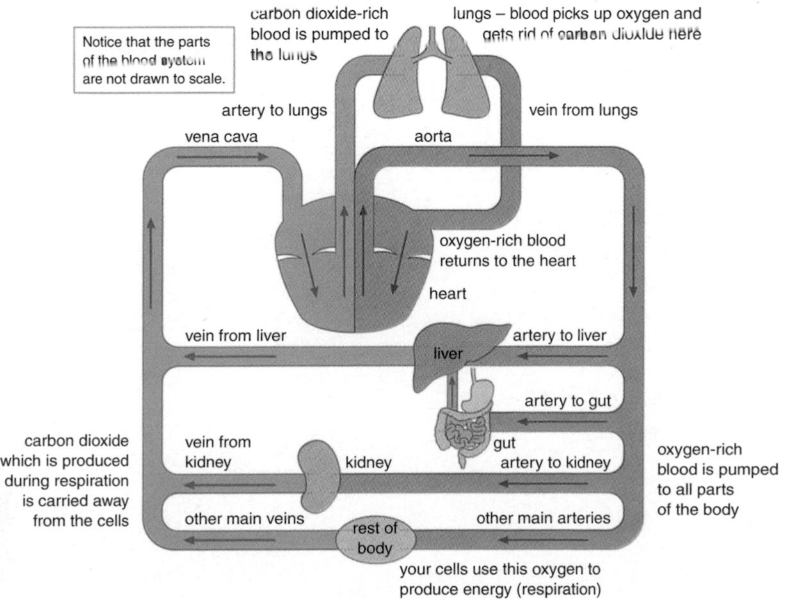

Figure 4 The circulatory system

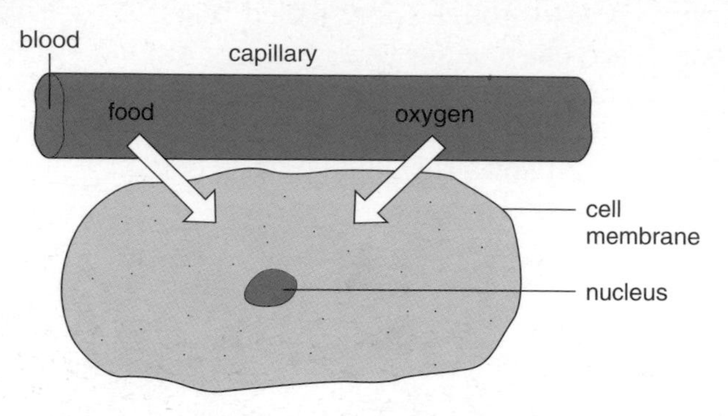

Figure 6 Blood delivers the glucose and oxygen that cells need for respiration.

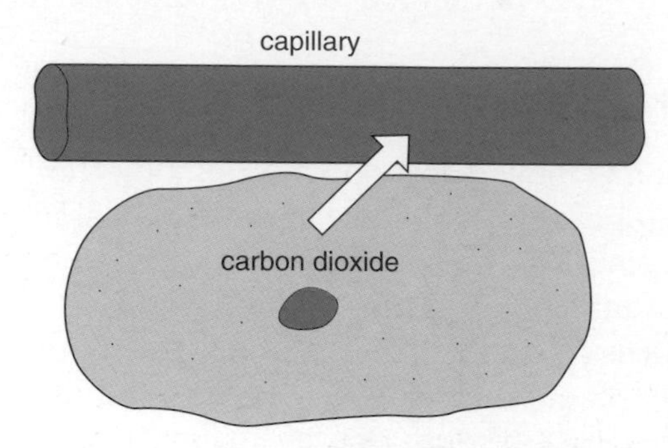

The blood then carries away the carbon dioxide produced by the chemical reaction

How do the ingredients needed for respiration get into our blood?

Glucose

As we saw on pages 24–25, food which is broken down (digested) in the digestive system passes through the thin walls of the small intestine and is absorbed by the blood, which carries it around the body to wherever it is needed.

Oxygen

When we breathe in, air enters our lungs and then travels through a network of fine tubes called **bronchioles** which lead to millions of tiny air sacs called **alveoli** (singular alveolus). The alveoli are surrounded by a dense network of blood vessels (capillaries). **Gas exchange** happens in the alveoli. Oxygen can diffuse through the very thin walls of the alveoli into the blood. Carbon dioxide can diffuse from the blood into the alveoli.

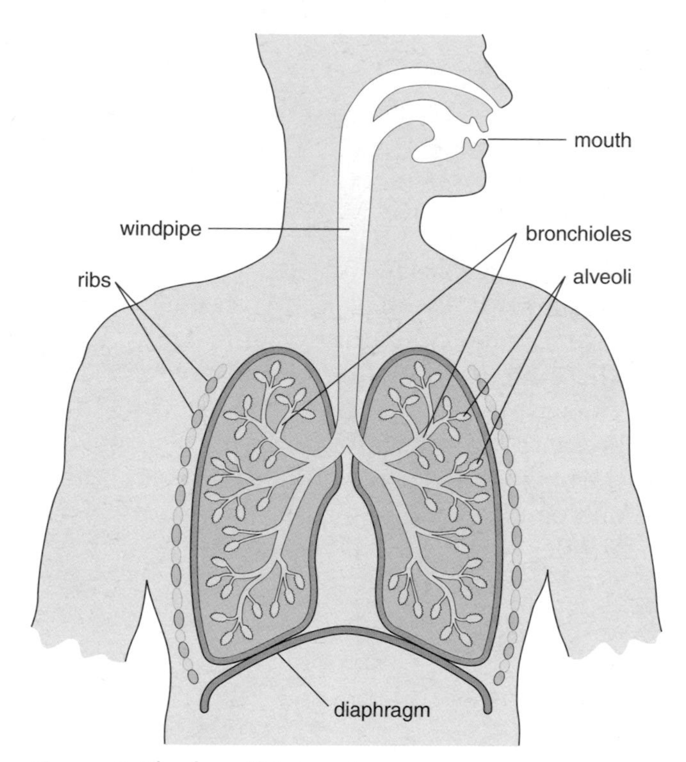

Figure 7 The breathing system

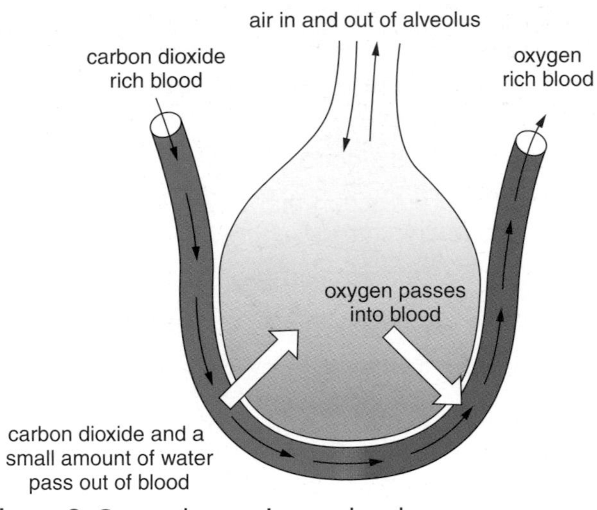

Figure 8 Gas exchange in an alveolus

Composition of inhaled and exhaled air

The air which we exhale contains more carbon dioxide and less oxygen than the air we inhale.

Gases present	Inhaled air	Exhaled air
nitrogen	78 %	78 %
oxygen	21 %	18 %
carbon dioxide	almost 0 %	3 %
other gases	1 %	1 %

The damage caused by smoking

The alveoli in your lungs are very delicate. If you smoke you are likely to damage them permanently.

Chemicals within the tobacco smoke:

★ may cause cancer of the lung cells
★ will aggravate the air passages making them inflamed
★ will weaken the walls of the alveoli. Less oxygen will then be able to diffuse into your bloodstream and you will feel breathless whenever you exert yourself. This can lead to a serious medical condition called emphysema.

Key terms

Check that you understand and can explain the following terms:

★ glucose
★ respiration
★ aerobic respiration
★ oxygen debt
★ anaerobic respiration
★ lactic acid
★ waste product
★ blood vessels
★ artery
★ vein
★ capillary
★ diffuse
★ bronchioles
★ alveoli
★ gas exchange

Questions

1. What is respiration?
2. Write down a word equation which explains what happens when aerobic respiration takes place.
3. Draw a labelled diagram to show how gas exchange takes place between an alveolus and a nearby capillary.
4. Why is smoking likely to cause you to feel out of breath when you run?
5. Name three different types of blood vessel. Write a sentence about each type to explain the differences between them.
6. Draw a diagram and explain what substances are exchanged between the capillaries and the cells in your body.
7. Explain the difference between respiration, gas exchange and breathing.

Chapter 2 Food, digestion and respiration:

What you need to know

1. A balanced diet contains carbohydrates, proteins, fats, minerals, vitamins, fibre and water.
2. Some sources of the different types of food needed in a balanced diet.
3. The role of the different food types in the body.
4. How digestion takes place in your body and how enzymes help this process take place.
5. That digested food is absorbed into your blood.
6. That waste products are passed out of your body.
7. How blood transfers important substances around your body.
8. Why cells need a good supply of blood.
9. The chemical reaction that takes place during aerobic respiration.
10. That gas exchange takes place at the alveoli.
11. How smoking can damage the alveoli.
12. The differences between inhaled and exhaled air.

How much do you know?

1 The diagram below shows a typical meal.

a) Which two parts of this meal are good sources of carbohydrates (energy providers)?

____

2 marks

b) Which part of this meal is rich in protein which is needed for growth and repairing damaged cells?

____

1 mark

c) Brown bread contains lots of fibre. Why is it important that we eat food that is rich in fibre?

____

1 mark

2 Many people have unbalanced diets. This may cause illness or poor health.

Draw a line from each of the unbalanced diets on the left to the illness or health problem it may cause on the right.

not enough carbohydrate	poor growth
too much fat	lacking energy
not enough fibre	weak teeth and bones
not enough protein	unable to get rid of undigested food (constipation)
not enough calcium	heart disease

5 marks

How much do you know? *continued*

3 The diagram below is of a human digestive system.

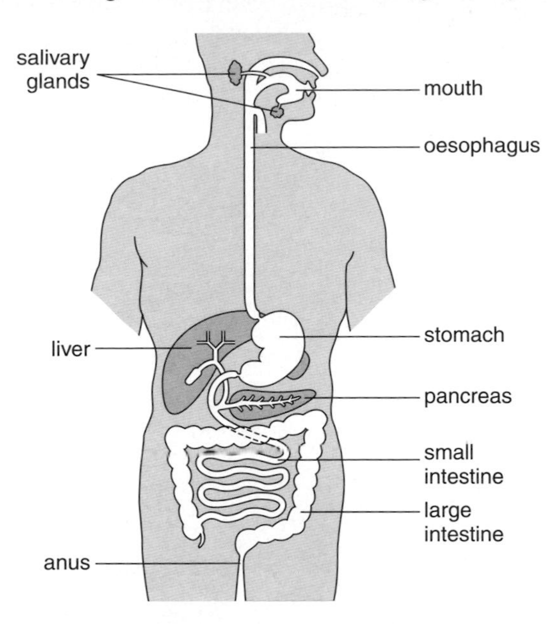

a) Explain why starches and proteins have to be digested before they can be absorbed into the bloodstream.

___

1 mark

b) Where in your digestive system is food broken down mechanically?

___

1 mark

c) Explain the function of enzymes in your digestive system.

___

1 mark

d) In which part of your digestive system does digested food pass into your bloodstream?

___

1 mark

4 The diagram below shows a hiker walking along a flat track.

a) Name the process which is providing her with the energy she needs to walk.

___

1 mark

b) Write down a word equation showing how she obtains this energy.

___

1 mark

c) Explain why her heart pumps faster if the track takes her up a steep hill.

___

___

___

3 marks

5 The diagram below shows an alveolus and a capillary.

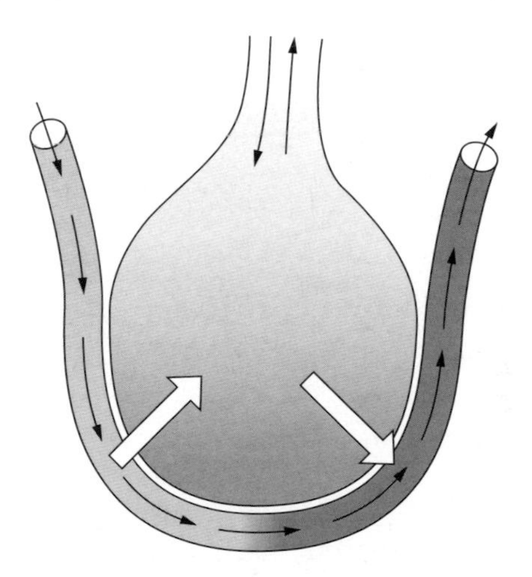

a) Which gas diffuses from the alveolus into the blood?

___

1 mark

b) Which gas diffuses from the blood into the alveolus?

___

1 mark

c) Explain why someone who smokes may feel out of breath when they exercise.

___

___

2 marks

3.1 Living things in their environment

Habitats and adaptation

A **habitat** is a place where plants and animals live. The conditions of a habitat are called its **environment**. The environment, for example temperature, amount of light, oxygen and moisture, determines what lives in a particular habitat.

Figure 1 A woodland habitat

The habitat in Figure 1 has the right conditions for trees, birds and woodland animals to live and grow.

Figure 2 A cold 'arctic' habitat

The habitat of the polar bear in Figure 2 is very different from the woodland habitat. In order to survive here the polar bear has had to **adapt** to the extreme cold.

To survive here polar bears must:

★ have a very thick fur coat to help keep them warm

★ have layers of fat beneath their fur. Fat is an excellent insulator and so prevents too much heat loss

★ **hibernate** during the winter when the conditions become too severe.

Figure 3 A very hot and dry 'desert' habitat

The habitat for camels is an extremely hot and dry one.

Camels are able to survive in the desert because they:

★ are able to store fat in their humps

★ have large feet so that they can walk over the sand without sinking into it

★ have long eye lashes which help to stop sand being blown into their eyes.

For plants to survive in a desert habitat it is important that they do not lose too much water by evaporation from their leaves. Cacti have almost no leaves. Their water is stored in the centre of the plant from where there is little evaporation.

All the different organisms that live in a habitat are called a **community**. In Figure 1 the deer, the badger, the squirrel and the bat are all part of the woodland community. The number of one kind of organism in a habitat is the **population** of that organism. For example, if there are ten badgers in the wood, the population of badgers in the wood is ten.

Even within one kind of habitat there may be differences in conditions which have caused adaptations.

In a land habitat these may include differences in:

★ temperature
★ moisture
★ pH of soil
★ organic content of soil
★ light intensity
★ wind strength.

In an aquatic habitat these may include differences in:

★ temperature at different depths
★ flow rate of the water
★ availability of oxygen
★ light intensity.

Figure 4 Different organisms have adapted to live in different parts of this pond

3.1 Living things in their environment *(continued)*

Figure 5 Deer live above the ground. They have keen eyesight and are able to blend in with their surroundings. Because they live below the ground, it is not important for moles to have good eyesight. Instead they have adapted to have strong shovel-like paws that enable them to burrow well and a very keen sense of smell and touch to help them detect their prey.

Changing conditions

The conditions in a habitat never remain the same. They change each day with the rising and setting of the Sun, and they change with the seasons. These changes affect the plants and animals that live there.

Day and night

The word daisy comes from Latin and means 'eye of the day'. Like many other flowers, the petals of a daisy open out during the day when there is lots of light, but close up again at night.

Figure 6 A daisy opens its petals in the day, but closes them at night

Most animals will hunt, eat and move around their habitat in the daytime. At night they hide away and sleep. **Nocturnal** animals such as owls, bats and badgers react differently to these changes. They are more active at night and sleep during the day.

Figure 7 This owl has adapted to hunting in the dark. It has very keen hearing and eyesight and flies almost silently

Seasonal changes

As the seasons change, the temperature and length of day alter. Plants and animals adapt their behaviour to suit these changing conditions. For example in winter:

★ birds like swallows **migrate** – they fly to warmer countries
★ many trees lose their leaves and stop growing to save energy
★ some animals grow thicker coats and then get rid of them when summer approaches
★ some animals like hedgehogs hibernate
★ some plants like daffodils lie **dormant** below the ground as bulbs or seeds until the conditions become warmer.

Figure 8 Trees lose their leaves in winter to save energy

Key terms

Check that you understand and can explain the following terms:

★ habitat
★ environment
★ adapt
★ hibernate
★ community
★ population
★ nocturnal
★ migrate
★ dormant

Questions

1. What is a habitat? Give one example of a habitat
2. What is the environment of a habitat?
3. What conditions might affect the plants and animals that live in an aquatic habitat?
4. Give one example of an animal which has adapted to **a)** a cold habitat and **b)** a habitat which is underground. Give examples of how each has adapted.
5. Name two animals that are nocturnal. How have these animals adapted to these conditions?
6. Name four ways in which plants or animals alter their behaviour to suit the changes in their environment when winter approaches.

3.2 Feeding relationships

Plants make their own food using a process called **photosynthesis** (see page 60). Plants are called **producers**.

Animals feed by eating other organisms. They are **consumers**. Animals such as cows, sheep and rabbits eat only plants. They are called **herbivores**. Animals such as lions, tigers and crocodiles eat only other animals. They are called **carnivores**. Animals such as human beings, which eat both plants and animals, are called **omnivores**.

Food chains

All organisms feed in order to obtain the energy they need to live. We can draw **food chains** to show how energy is transferred from one organism to the next.

Figure 1 Energy from the Sun is used by the grass to make food. The rabbit obtains its energy by eating the grass. It is the **primary consumer** in this chain. The fox obtains its energy by eating the rabbit. The fox is the **secondary consumer** in this chain

Here are some more examples of food chains:

weeds → tadpole → fish

grass → insects → hedgehog

wheat → mouse → owl

leaves → caterpillars → blue tits → kestrel

weeds → minnow → trout → man

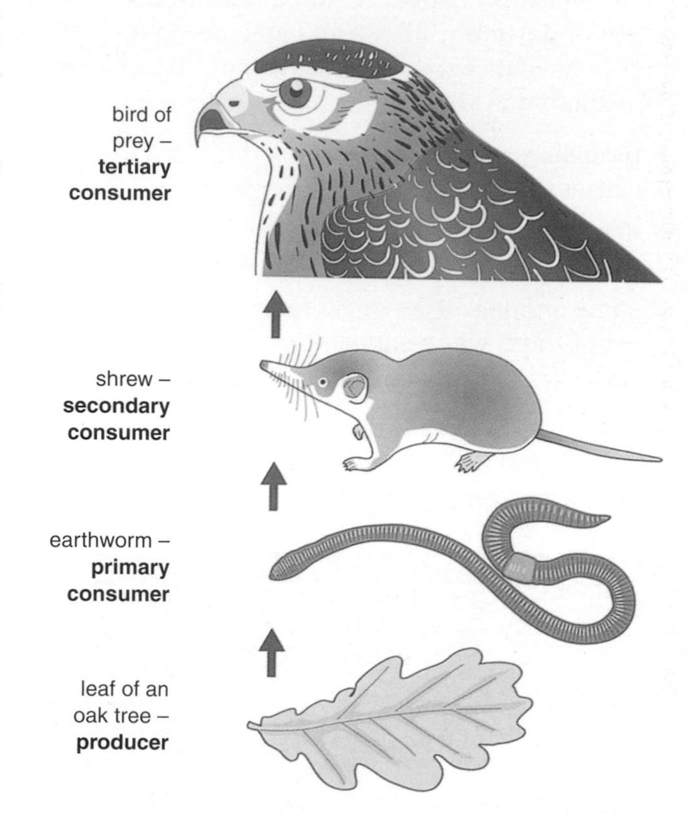

Figure 2 Energy moves along a food chain in the direction shown by the arrows

In all food chains some energy is lost along the way. The energy is used by the individual organism to stay alive, to move, to keep warm and to carry out all of life's processes.

Pyramids of numbers

A **pyramid of numbers** tells us how many organisms are involved at the different stages of a food chain. Pyramids of numbers can take different shapes (see Figure 3).

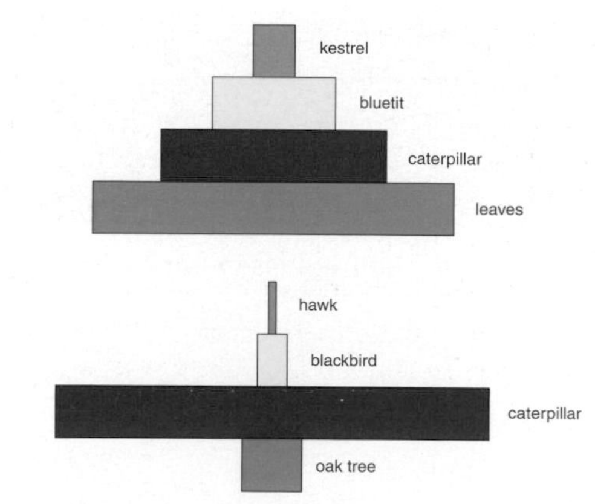

Figure 3 Pyramids of numbers

Food webs

Plants and animals rarely belong to just one food chain. Several connected food chains are called a **food web**. Food webs show how the feeding relationships between organisms in a habitat are interconnected.

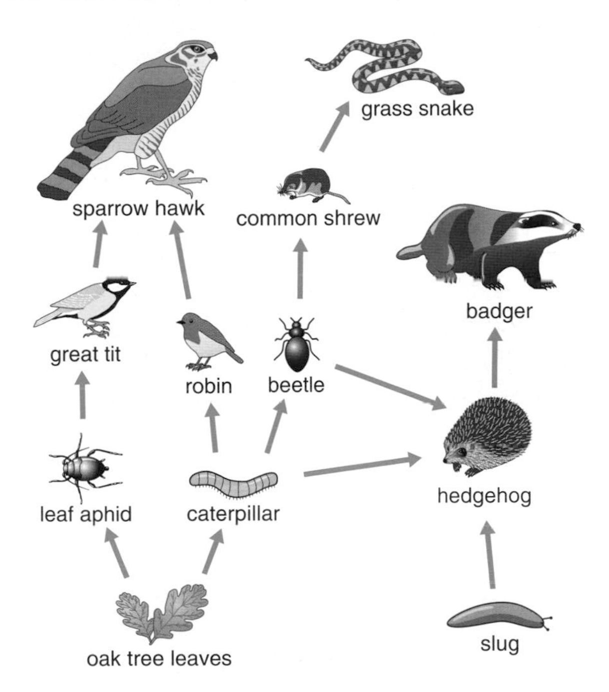

Figure 4 A simple food web

If the number of one of the organisms in a food web changes, it can have a big effect on other organisms in the web. For example, if a very cold winter has caused lots of great tits to die, this might result in:

★ there being more leaf aphids
★ there being fewer sparrowhawks
★ more robins being eaten by the sparrowhawks
★ there being more caterpillars.

Poisons in food chains and food webs

Farmers will often spray their crops with chemicals to increase their harvest or protect it from insects and disease. Animals that eat the crops will absorb the chemicals sprayed on to them. At the beginning of a food chain, in the animals that eat the crops directly, the concentration of these chemicals is quite low. But higher up the food chain, the concentration of the chemical increases and may be high enough to kill.

In the food chain in Figure 5, it is the heron which is in greatest danger of being poisoned. The large fish eats many small fish. The heron eats lots of large fish, so the heron will consume the most poison.

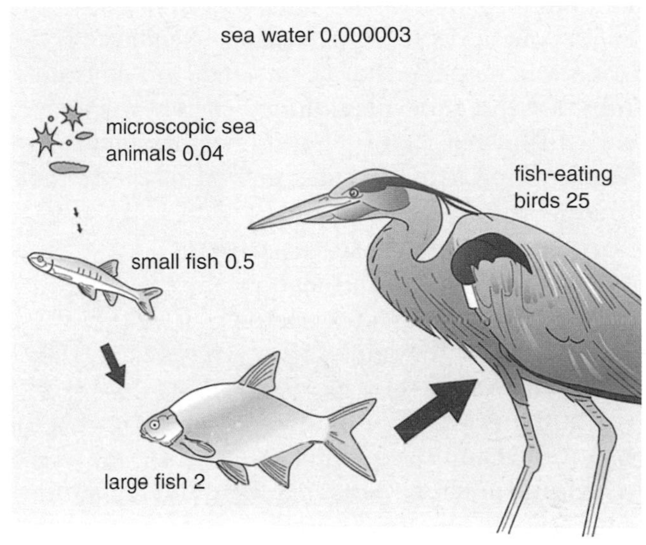

Figure 5 The numbers compare the concentration of a chemical in the different stages of a food chain

Key terms

Check that you understand and can explain the following terms:

★ producer
★ consumer
★ herbivore
★ carnivore
★ omnivore
★ food chain
★ primary consumer
★ secondary consumer
★ pyramid of numbers
★ food web
★ photosynthesis

Questions

1 What is a producer? Give two examples of producers.
2 What is a consumer? Give two examples of consumers.
3 Give one example of a food chain which contains at least two consumers.
4 Give one example of a pyramid of numbers. What does this pyramid show?
5 What is a food web? Why are animals at the end of food chains and webs most likely to die from any poison which is present lower down the chain?

3.3 Competition and survival

Limited resources

All habitats have limited amounts of water, food, light and places to live. Plants and animals compete with each other for these **resources**. Some will adapt so that their needs are different from those of their neighbours. Others will adapt so that they can compete more successfully. **Competition** leads to the **survival of the fittest**.

Competition between plants

Trees in the South American rainforests (Figure 1) compete with each other for the sunlight which they need to survive. This is the main reason why they grow so tall. At the very top of the trees where there is bright sunlight there is a thick canopy of leaves. On the floor of the forest, only plants and shrubs which can survive with very little light can grow.

Figure 1 These trees in the South American rainforest are competing for light

Figure 2 Competition between plants

Figure 2 shows how plants have adapted to outcompete other plants in the same habitat.

- ★ The plants are of different heights. Plants that are tall will receive most light. Only small plants that require little light will survive between the larger ones.
- ★ The plants have roots of different lengths. Grasses and dandelions survive next to each other on lawns because their roots take water from different depths of soil.
- ★ Plants that need insects to help in pollination may have a few large, colourful, strongly-scented flowers. Other plants that rely on the wind for pollination may have lots of small, less colourful flowers.

Competition between animals

Animals may compete for food, water, space or territory, or mating partners.

Figure 3 These two birds are competing for the same bit of food

Living in the same habitat as the tiger in Figure 4 could be hazardous. It is a **predator**. It hunts **prey**. Tigers have had to adapt in order to become successful predators.

- ★ They are well **camouflaged**. Their striped coat helps them to stalk their prey whilst hidden in forests and long grasses.

★ They have developed very strong, powerful muscles so that they can run quickly and pounce on their prey.
★ They have long claws and sharp teeth to grip and eat their prey.

Figure 4 A magnificent predator

Numbers of predators and prey

Figure 5 shows the relationship between numbers of lynx (a type of cat) and hares. The main food source for a lynx is the hare population.

From A to B: The number of hares increases.
From C to D: There is more food for the lynx population so their numbers increase.
From B to E: Because the number of lynx increases, more hares are eaten so the hare population declines.
From D to F: Because the hare numbers decrease, there is less food for the lynx, so their numbers decrease.
From E to G: As the number of lynx declines, the hare population again increases in size and the cycle begins again.

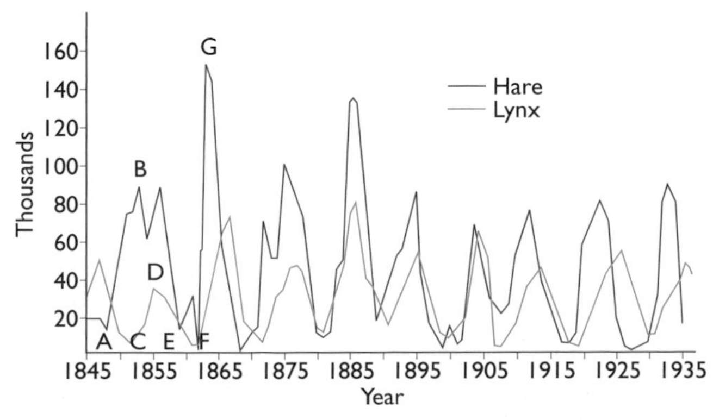

Figure 5 The relationship between numbers of lynx and hare

Passing on the trick

Plants and animals that have a 'trick' or **adaptation** which helps them survive will pass this on to many of their offspring. Those without this adaptation are less likely to survive. Eventually this adaptation becomes a natural feature for that organism. This is called **natural selection**.

Figure 6 Adaptations to avoid predators include strong muscles to run fast, a sting, spines and living in big groups

Organisms within a habitat don't just depend upon each other for their food, there are many other kinds of relationships:

★ birds depend on trees to provide nesting places
★ fish depend upon plants to provide the water with oxygen
★ some plants depend on insects to pollinate them
★ some fruit trees and bushes depend on animals to spread their seeds.

Key terms

Check that you understand and can explain the following terms:

★ resources
★ competition
★ survival of the fittest
★ predator
★ prey
★ camouflage
★ adaptation
★ natural selection

Questions

1 What resources might **a)** plants and **b)** animals compete for?
2 Explain how grasses and dandelions have adapted so that they can avoid competing for water.
3 Give three examples of animals that are predators. For each predator give one example of its prey.
4 Explain the phrase 'natural selection' in your own words.

Chapter 3 Living and feeding:

What you need to know

1. Different habitats support different plants and animals.
2. Plants and animals are adapted to live in a particular habitat.
3. Plants and animals have adapted to suit daily and seasonal changes that take place in their habitat.
4. Feeding relationships – food chains, food webs and food pyramids.
5. Toxic materials can build up in food chains.
6. What factors affect the size of populations e.g. predators, competition for food, water and light?
7. Plants and animals that compete successfully for resources pass their qualities on to the next generation.

How much do you know?

1 a) Give three examples of different habitats.

____

____

____

3 marks

b) Animals that live in the polar regions have to be able to survive in very cold conditions.

Give two ways in which this polar bear has adapted in order to live in these conditions.

____

____

2 marks

2 A group of pupils studied the living organisms in a wood for several months. They made the following observations.

★ Weasels eat wood mice

★ Blue tits and great tits eat caterpillars

★ Sparrowhawks eat blue tits and great tits

★ Caterpillars eat oak leaves

★ Wood mice eat acorns from oak trees

★ Aphids eat oak leaves

★ Blue tits and great tits eat aphids

Use this information to complete these food chains:

a) Oak leaves → __________ → __________

1 mark

b) Acorns → __________ → __________

1 mark

c) What is the producer for both of these chains?

____

1 mark

d) From the observations name one consumer.

____

1 mark

e) Draw a pyramid of numbers for each of the following food chains.

(i) oak tree → caterpillars → blue tits → kestrel

1 mark

(ii) weeds → minnow → trout → man

1 mark

3 The food web below shows the feeding relationships between several organisms.

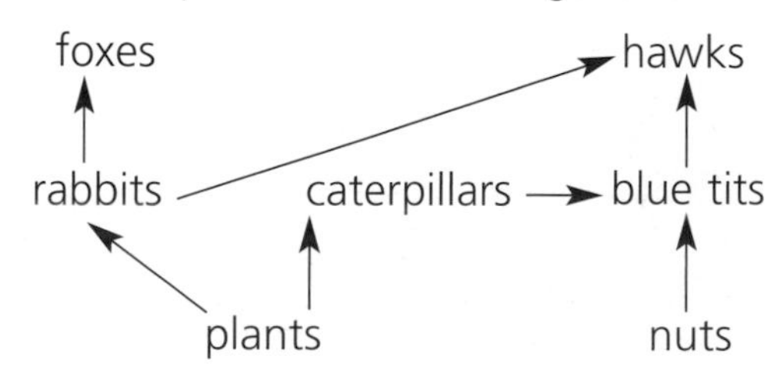

a) Name one herbivore.

____

1 mark

b) Name one omnivore.

____

1 mark

How much do you know? *continued*

c) If the number of rabbits decreases:

(i) How will this affect the number of foxes.

1 mark

(ii) How will this in turn affect the number of plants?

1 mark

4 This is a food chain for organisms that live in a pond.

weeds → beetles → frogs

The graph below shows how the numbers of these organisms change over a period of time.

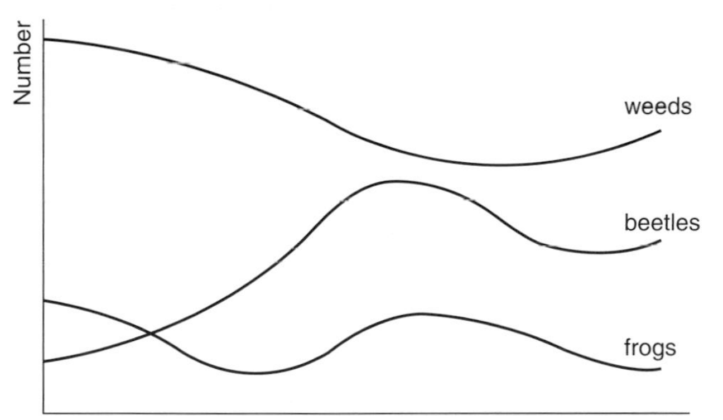

a) What happens to the beetle population as the number of frogs increases?

1 mark

b) What is happening to the weed population during this time?

1 mark

c) Explain why the weed population is changing in this way.

1 mark

d) Pike are predators, they eat frogs. Explain what would happen to the beetle population if a pike was put into this small pond.

2 marks

5 The diagram below shows four different birds that have successfully adapted to their habitats.

Write down the name of the bird which has adapted so that it can:

a) catch and eat mice

1 mark

b) hunt for food deep in the sand

1 mark

c) hunt for ants and beetles beneath the bark of trees

1 mark

d) fly at high speeds

1 mark

4.1 Microbes and disease

Microbes are very small organisms. We call them **micro-organisms** because they can normally only be seen using a microscope. There are three types of microbes: **viruses**, **bacteria** and **fungi**.

Viruses are the smallest micro-organisms. They are so small that they cannot be seen using a normal light microscope. Viruses are not made of cells.

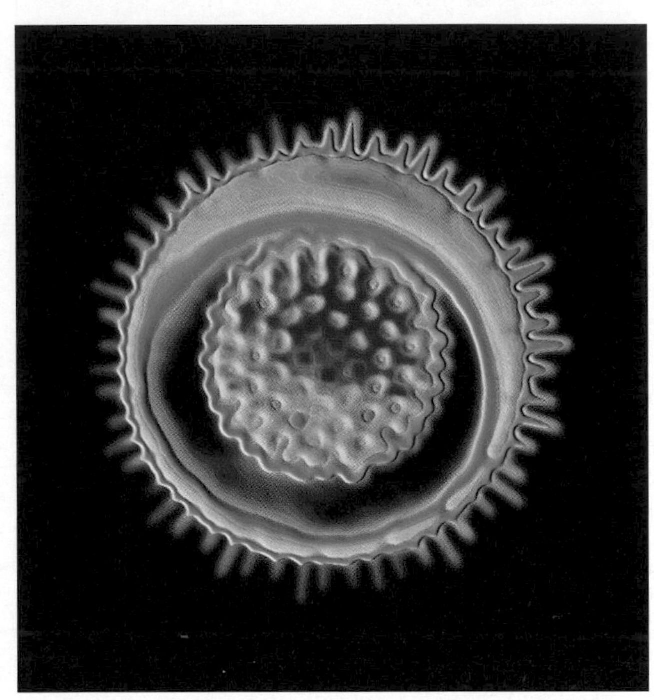

Figure 1 This picture of the herpes virus was taken using a very powerful microscope called an electron microscope

Bacteria are the next largest type of micro-organism. They are about a thousand times larger than viruses, but are still much smaller than a single plant or animal cell. A bacterium consists of one cell which has a cell membrane, a cell wall and cytoplasm, but no nucleus.

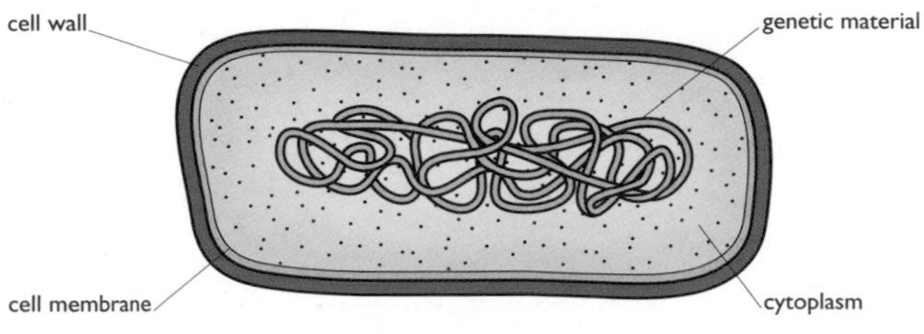

Figure 2 A simple bacterium

Fungi are the largest type of micro-organism. Athlete's foot is an example of a disease caused by a fungus.

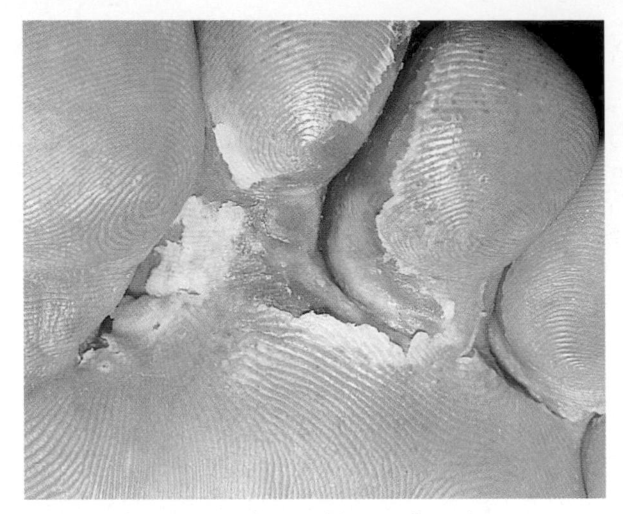

Figure 3 This person has athlete's foot between their toes. This infection is caused by a fungus

Uses of microbes

Yeast is another example of a fungus. It is used in the making of bread. When yeast is mixed with flour, water and sugar it respires, releasing the gas carbon dioxide (see page 26). It is the release of this gas which make bread rise, giving it a light texture.

Figure 4 This man is kneading bread dough before baking it in the oven. The yeast in the dough will make the bread rise

Yeast is also used to make wine. By a process called **fermentation**, the yeast produces carbon dioxide and alcohol.

Some bacteria are used in the making of cheese and yoghurt.

Harmful microbes

Not all microbes are useful. Some of them are harmful and can cause illness. Microbes that cause disease are called **pathogens**. Most pathogens are viruses or bacteria. Some common diseases are shown in the table below.

Diseases caused by viruses	Diseases caused by bacteria
measles	food poisoning
'flu	tetanus
cold	tuberculosis
polio	bacterial meningitis

Pathogens can be passed or **transmitted** from person to person, or from animal to person. There are several ways in which pathogens can enter your body.

★ You may breathe in pathogens which have been released into the air, for example by someone coughing and sneezing. Cold and 'flu viruses are transmitted in this way.

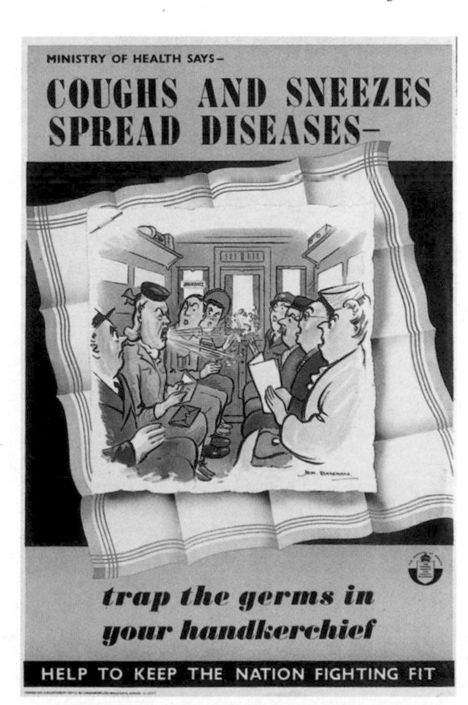

Figure 5 A World War II poster encouraging people to be aware of the dangers of germs (pathogens)

★ You may eat food which is contaminated. Food poisoning is spread in this way.

★ You may drink water which is contaminated. Typhoid, cholera, polio and dysentery are spread in this way.

★ You may have a cut or sore which allows pathogens such as tetanus to enter your body through your skin.

★ You may be bitten by an animal. Mosquitos spread malaria in this way. The rabies virus can be spread from the bite of rabid dogs or foxes.

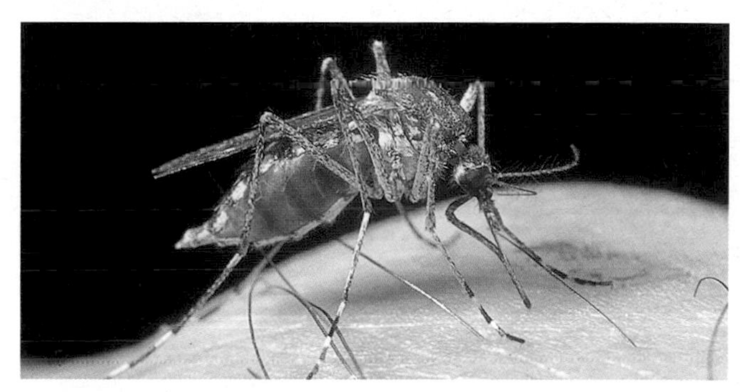

Figure 6 Micro-organisms can enter the bloodstream as a mosquito sucks blood

Why do I feel ill?

Once pathogens have entered your body they reproduce very rapidly. As they do so, they feed on your body's cells, killing them. Some pathogens produce chemicals called **toxins** which are poisonous to your cells. You feel unwell because the pathogens are interfering with the workings of your body.

Fighting against infectious diseases

Your body has several different ways of defending itself against invading pathogens.

★ Your skin is the first line of defence and unless you have a cut or open sore, most pathogens cannot get through it.

★ Your breathing airways are lined with **mucus** which traps any pathogens that you breathe in. Tiny hairs called **cilia** then sweep them back up your throat away from your lungs.

★ Your stomach produces a very strong acid, which kills many of the pathogens that enter the body through your food.

4.1 Microbes and disease (*continued*)

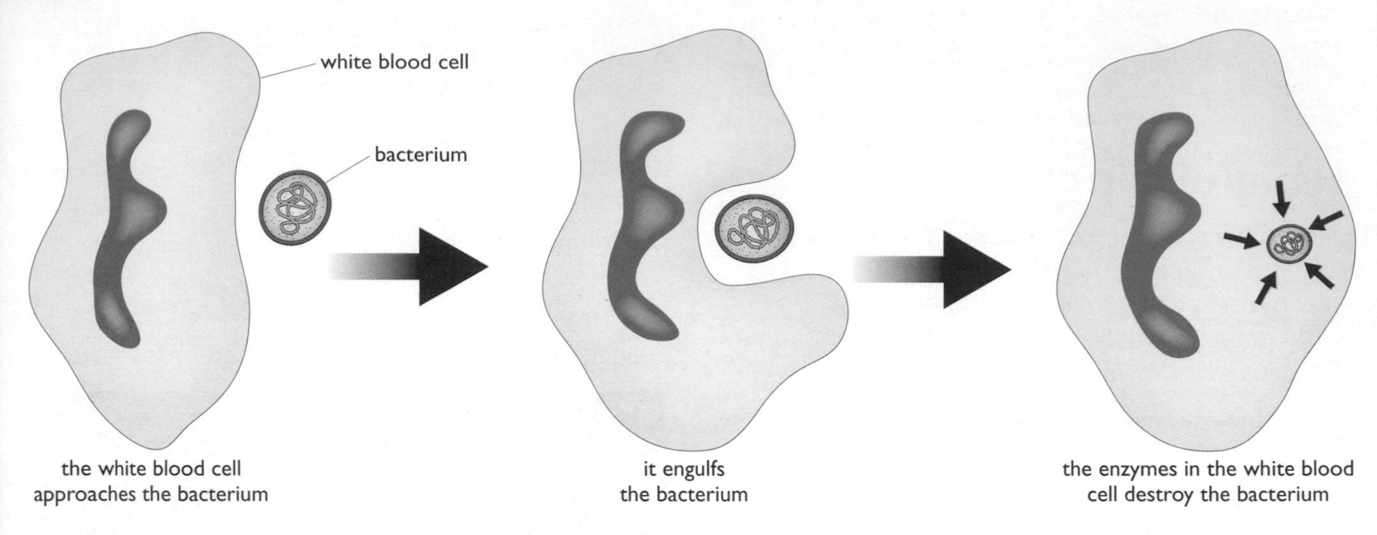

Figure 7 The white blood cell engulfs the bacterium and destroys it

If any pathogens are able to get past your body's first lines of defence, it has an **immune system** to deal with them. This is how it works:

Your blood contains several types of **white blood cells**. These seek out and attack invading microbes.

Some white blood cells (called phagocytes) surround the invading microbes and break them down using enzymes (Figure 7).

Other white blood cells (called lymphocytes) make substances called **antibodies**. These mark infecting microbes, making it easier for them to be destroyed by other white blood cells (Figure 8).

Once your body has been infected by a pathogen and antibodies have been produced to kill it, some of your white blood cells *remember* this pathogen and will be able to make the antibody much more quickly if the same microbe enters your body in the future. We say that you have become **immune** to the disease caused by that pathogen. This happens with diseases such as measles, chickenpox and mumps.

Unfortunately there are thousands of different types of pathogens and each requires a different antibody to kill it. Each white blood cell can only make one kind of antibody, so each pathogen needs a different set of antibody-making white blood cells.

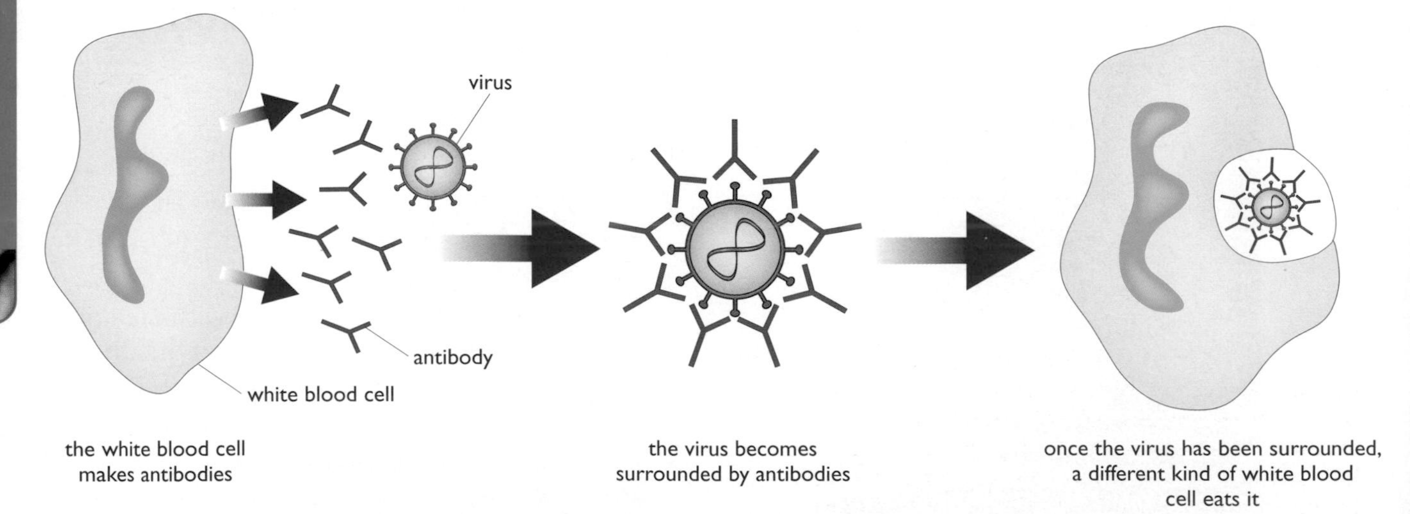

Figure 8 Antibodies help you fight off infection by neutralising microbes like viruses

Antibiotics and vaccination

Although your body is normally very good at dealing with invading micro-organisms, there are times when drugs and medicines are also needed. These help to speed up recovery from the illness and act as reinforcements for your body's natural defences when the illness is life-threatening.

Antibiotics such as penicillin are drugs that will kill bacteria. Unfortunately antibiotics have no effect on viruses.

Vaccination prepares your immune system for a particular pathogen. It involves the injection of a small amount of the pathogen which stimulates your immune system to develop antibodies against it. If you then become infected by this pathogen at a later date, your body is already prepared to fight it.

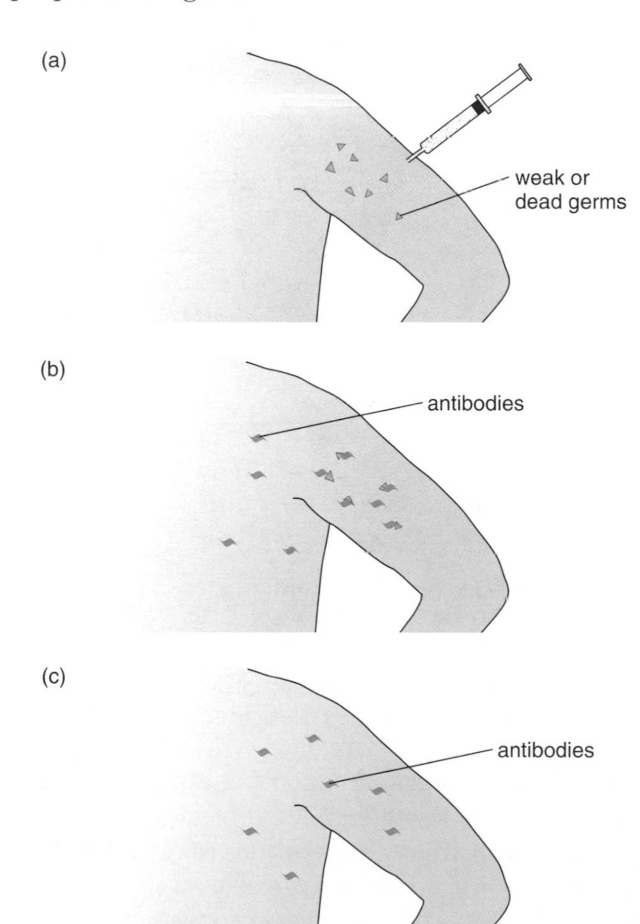

Figure 9 a) Partially-destroyed microbes, such as the measles virus, are injected into the bloodstream b) The body produces antibodies to fight the virus c) Even when all the infecting viruses have been killed, some white blood cells remember which antibody to produce for this particular disease

In the UK we vaccinate children against a variety of diseases:

- ★ diphtheria, tetanus, whooping cough and polio at 6, 8 and 12 months
- ★ measles at 2 years
- ★ diathermy, tetanus and polio at 5 years
- ★ German measles at 11–13 years
- ★ tuberculosis (TB) at 12–13 years

Key terms

Check that you understand and can explain the following terms:

- ★ microbe
- ★ micro-organism
- ★ virus
- ★ bacteria
- ★ fungi
- ★ fermentation
- ★ pathogen
- ★ transmitted
- ★ toxins
- ★ mucus
- ★ infectious disease
- ★ cilia
- ★ immune system
- ★ white blood cells
- ★ antibodies
- ★ immune
- ★ antibiotics
- ★ vaccination

Questions

1. Name three different types of micro-organism.
2. Give two uses of microbes.
3. Name three diseases caused by **a)** viruses and **b)** bacteria.
4. Name three ways in which a pathogen may enter your body.
5. What is a toxin and how is it produced?
6. How does penicillin help your body to overcome certain illnesses?
7. How does your immune system react if a pathogen invades your body?

4.2 Fit and healthy

What does being fit actually mean?

It is difficult to explain exactly what being fit means. It depends upon your age, sex, medical history and the kinds of activities you normally take part in. Being able to run a fast 100 m race does not mean you are fit enough to run a marathon! Being fit means being able to cope with the demands of your life and activities without your body becoming distressed.

Figure 1 Fit for playing darts but what else?

In order to function properly our muscles must have a good blood supply. Dissolved in the blood are oxygen from the lungs and digested food (glucose). These two chemicals react together to produce the energy muscles need to work. This process is called respiration (see page 26)

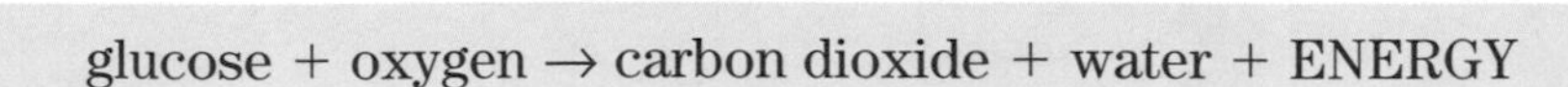

glucose + oxygen → carbon dioxide + water + ENERGY

If you exercise a group of muscles, they need a greater supply of blood. This is achieved by the heart beating faster. A fit person who takes **regular exercise** will have a heart which copes well with the increased demand. However, the heart of someone who is unfit will not cope so well and they may feel breathless and possibly distressed.

Figure 2 A fit athlete is able to cope with the increased energy demands of his/her muscles

All those systems inside your body which play a part in respiration will affect your fitness, for example your **digestive system**, **respiratory system** and **circulatory system** as well as the condition of your skeleton and your joints.

Respiratory system

How we breathe

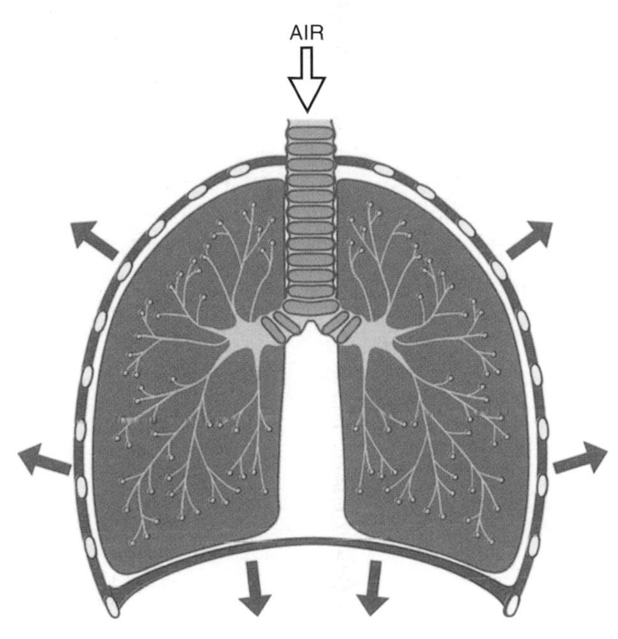

Figure 3 When you breathe in, your rib cage moves upwards and outwards. A layer of muscle called the **diaphragm** moves downwards. These movements draw fresh air into your lungs

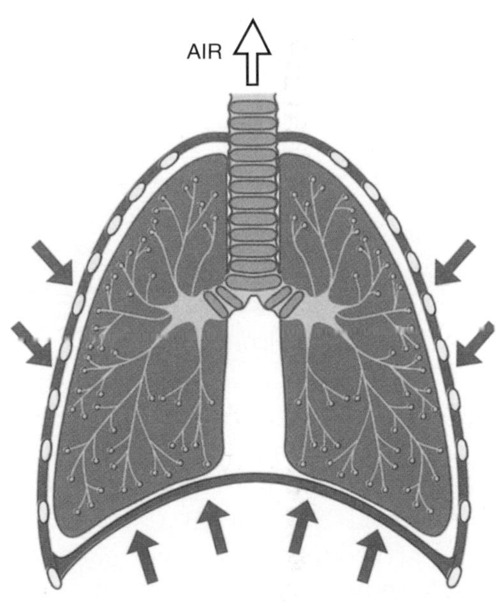

Figure 4 When you breathe out, your rib cage moves downwards and inwards and your diaphragm moves up. These movements push the stale air out of your lungs

Anything which increases our lung capacity, such as playing a brass instrument or swimming, will improve our fitness. In contrast, someone who suffers from asthma is unlikely to be able to draw large quantities of air into their lungs and so will not be as fit.

Gaseous exchange and smoking

Having got the air into your lungs, it's important that the oxygen can pass through the walls of the **alveoli** into the bloodstream and that carbon dioxide can pass in the opposite direction (see pages 28–29). Smoking can damage the walls of the alveoli, greatly reducing the surface area through which **gaseous exchange** can take place. This condition is called **emphysema**.

a)

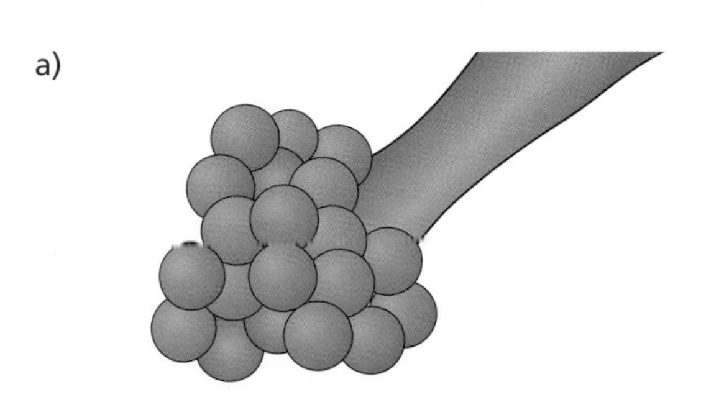

b)

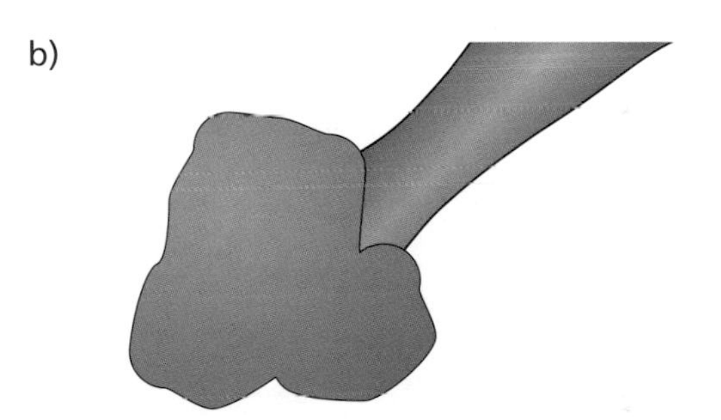

Figure 5 a) The healthy alveoli of a non-smoker b) The damaged alveoli of a heavy smoker

People who have this condition are unable to get enough oxygen into their bloodstream. The slightest exertion, causing an increase in the body's demand for oxygen, leaves them breathless and distressed. Where the damage is extreme, sufferers may have to have a supply of oxygen available to them 24 hours a day.

Figure 6 This person has severe emphysema from smoking, so needs to breathe pure oxygen

Fit and healthy (*continued*)

Other ways in which smoking affects your health:

★ cigarette smoke contains an **addictive** drug called nicotine. Nicotine can thicken your blood and narrow your blood vessels, both of which make it harder for your heart to pump the blood around your body

★ tar in cigarette smoke can cause cancer

★ carbon monoxide in cigarette smoke prevents red blood cells from carrying as much oxygen around the body

★ some conditions such as bronchitis are aggravated by smoking.

Your body is an incredible 'human machine'. It has a highly advanced computer (your brain) controlling everything it does – breathing, digesting, moving. To keep your machine in good condition you must look after it. As well as exercising, it is important that you have a healthy, **balanced diet** as described on page 20.

A poor diet, for example eating too much of just one kind of food, can lead to poor health because parts of your body are not receiving all the substances they need to work properly. Imagine what would happen to a machine if no oil or grease was applied to its moving parts!

Too much fatty food can cause **obesity** and heart disease.

Effects of alcohol on your body

Alcohol is regarded by most people as a sociable drink, but in excess it can damage your body permanently.

★ It is a **depressant**.

★ It affects reaction times and alters behaviour.

★ It damages brain cells.

★ It can damage the liver.

★ It can affect the development of a fetus.

★ It can be very addictive.

Figure 7 What effect is alcohol having on these people?

Drugs

Under proper medical supervision drugs can help people who are ill. When drugs are taken without medical supervision, however, they can seriously harm you, both mentally and physically. Ultimately, misuse of drugs can kill.

There are four main types of drugs.

1 **Painkillers** – e.g. aspirin, paracetamol, heroin and morphine. These work by blocking out the messages of pain that go to the brain.

2 **Stimulants** – e.g. amphetamines, cocaine and caffeine. These also work by affecting messages that go to the brain, causing it to work harder. This has the effect of making a person feel more alert.
3 **Tranquillisers** – e.g. sleeping pills and barbiturates. These work by slowing down the activity of the brain. They can cause drowsiness, and affect a person's **reaction time** and **co-ordination**.
4 **Hallucinogens** – e.g. ecstasy, LSD and cannabis. These cause people to feel, to see and to hear things that don't really exist.

The cost of drug abuse

Figure 8 A drug addict damaging a machine that can never be replaced

When the first effects of a drug wear off, people often feel depressed and weary. To overcome these feelings they feel the need to take the drug again. They may then become **addicted** to the drug – unable to live without it. If too much of a drug is taken – an **overdose** – this can cause permanent damage to the heart, liver and kidneys and may cause death.

Figure 9 After the light-headed feelings pass, this boy will feel sick, suffer from acne around the face, possibly become unconscious and may even die

Fit and healthy (*continued*)

Movement

The Skeleton

An important aspect of being fit is having a body which allows you to move easily. This means having a strong **skeleton** and efficient joints.

Your skeleton has several important functions:

★ it supports your body, giving it shape.
★ it protects your **vital organs**. For example, your heart and lungs are protected by the rib cage, your brain is protected by the cranium.
★ it is jointed in such a way as to allow movement.

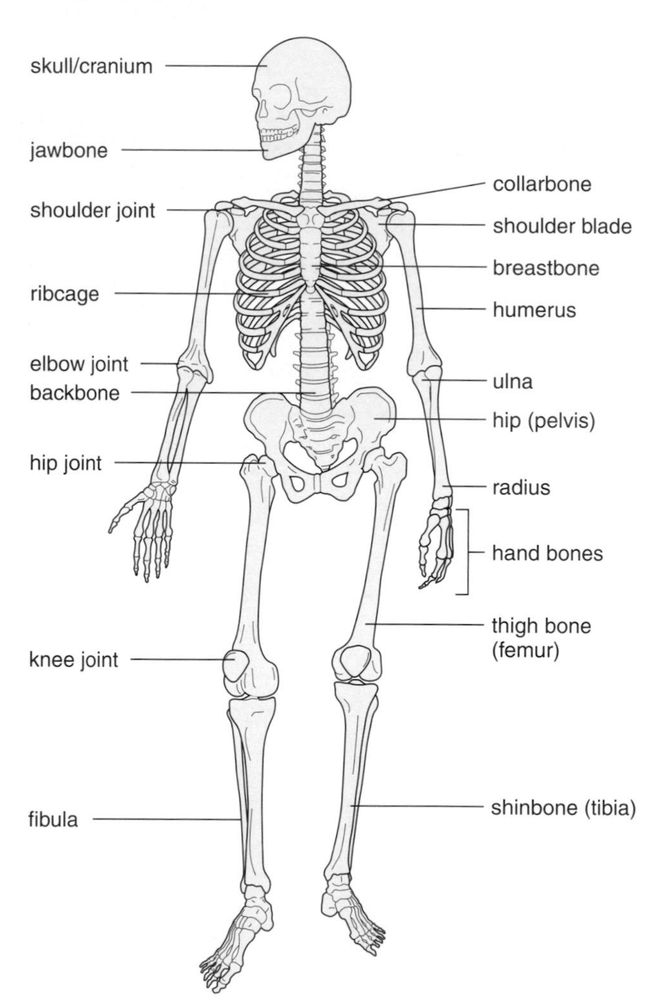

Figure 10 The human skeleton

In order to carry out all these functions, your bones have to be strong enough to withstand knocks and not break or bend. Yet they must also be quite light so that movement is easy.

Joints

You are able to move your arms and legs because they are attached to the rest of your skeleton at **joints**. There are two types of joint in your body which allow movement.

Hinge joints

Hinge joints allow a pivoting movement in just one plane. Your elbows and knees are hinge joints.

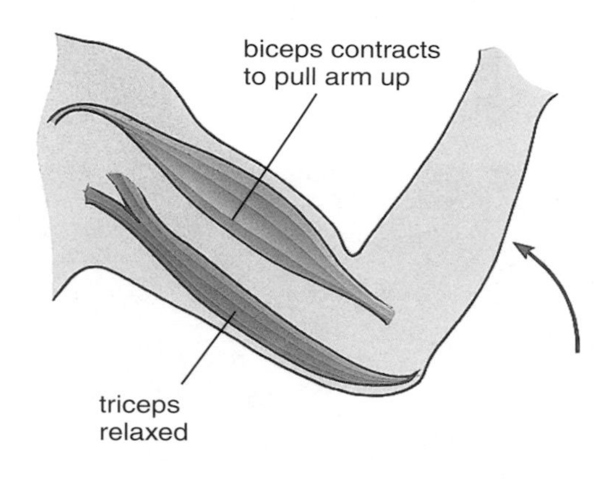

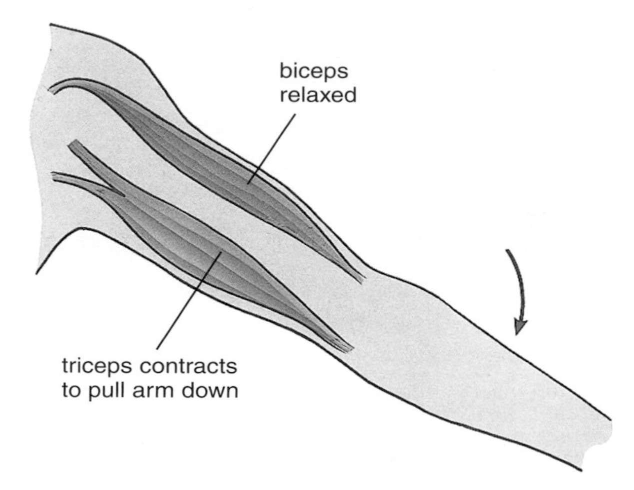

Figure 11 The elbow, a hinge joint

The two **muscles** which control the movement of your lower arm are called the **biceps** and the **triceps**. When you bend your arm upwards, the biceps contracts (shortens) and the triceps relaxes. When you straighten your arm, the biceps relaxes and the triceps contracts. This co-operative behaviour of muscles is the basis of movement. Pairs of muscles which behave like this are called **antagonistic pairs**.

Ball and socket joints

Ball and socket joints allow pivoting and rotational movement. Your shoulders and hips are ball and socket joints.

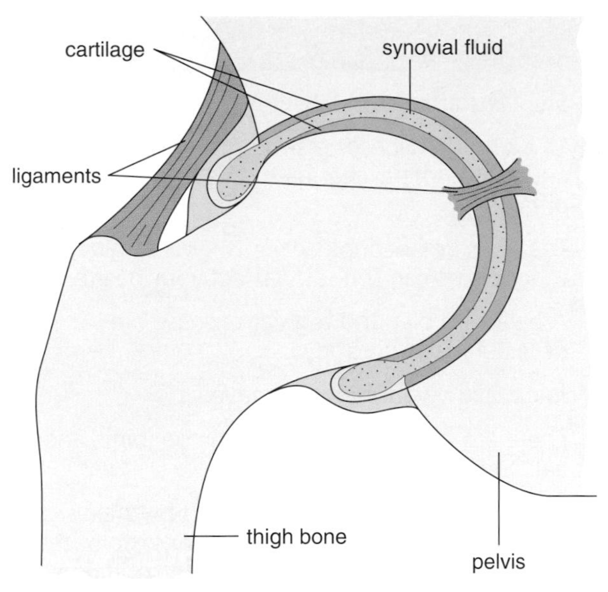

Figure 12 The hip, a ball and socket joint

Although these joints allow you different degrees of movement, they do have certain features in common.

a) The two bones forming the joint are held in place by fibrous tissue called **ligaments**. These limit your movements so that the bones don't move out of position, for example dislocating your shoulder.

b) The bones are connected to the two muscles by a tissue called a **tendon**.

c) In order to ensure that the bones can hinge and rotate smoothly, there is a lubricating liquid called **synovial fluid** within each joint.

d) Within the joint the bones are covered with a tough, gristly material called **cartilage**. Like the synovial fluid, it encourages smooth movement between the bones of a joint.

Common injuries to joints

★ **Sprains** – caused when a person tries to move the bones of a joint beyond their normal range of movement, resulting in the ligaments holding the bones in place being overstretched.

★ **Dislocation** – when one of the bones forming the joint becomes completely dislodged from its correct position.

★ **Pulls, strains and tears** – often caused by a muscle contracting violently. The muscle and/or the tendon may be damaged.

★ **Cartilage damage** – because of wear and tear or a sudden knock, some of the cartilage covering one of the bones within the joint is damaged and now inhibits the movement of the joint.

Key terms

Check that you understand and can explain the following terms:

★ regular exercise
★ digestive system
★ respiratory system
★ circulatory system
★ diaphragm
★ alveoli
★ gaseous exchange
★ emphysema
★ addictive
★ balanced diet
★ obesity
★ depressant
★ painkillers
★ stimulants
★ tranquillisers
★ reaction time
★ co-ordination
★ hallucinogens
★ addicted
★ overdose
★ skeleton
★ vital organs
★ joints
★ hinge joints
★ muscles
★ biceps
★ triceps
★ antagonistic pairs
★ ball and socket joints
★ ligaments
★ tendons
★ synovial fluid
★ cartilage
★ sprain
★ dislocation

Questions

1 Suggest three ways in which you can keep your body healthy.

2 Name the four main groups of drugs. Explain the effect each of these drugs would have on your body.

3 Write down the equation which describes how your body obtains the energy it needs.

4 Describe three ways in which alcohol can affect your body.

5 Name two different types of joint. Name one way in which these joints differ. Name three ways in which these joints are similar.

6 Name three ways in which a joint may be damaged.

Chapter 4 Illness and health:

What you need to know

1. There are three types of micro-organisms: viruses, bacteria and fungi.
2. Some micro-organisms, such as yeast, are useful.
3. Some micro-organisms (pathogens) cause disease.
4. How pathogens can be transmitted.
5. How the body fights against disease.
6. How immunisation and antibiotics help the body fight against disease.
7. How the body obtains the energy it needs by respiration.
8. How oxygen is drawn into the lungs.
9. Oxygen enters the bloodstream at the alveoli.
10. Carbon dioxide leaves the bloodstream and enters the lungs through the alveoli.
11. Smoking affects your health.
12. Smoking damages the structure of the lungs and makes the exchange of the gases more difficult.
13. The abuse of alcohol, solvents such as glue, and other drugs can seriously affect your health.
14. A balanced diet and regular exercise are necessary for a fit body.
15. Joints allow your body to move.
16. There are two types of joints, hinge joints and ball and socket joints.
17. The fitness of a person is affected by their respiratory system, their circulatory system, their digestive system, their diet and the condition of their bones and joints.

How much do you know?

1 The diagram below shows the rib cage, diaphragm and lungs of a human.

a) Explain how air is drawn into the lungs.

____

____

____

3 marks

b) Explain how stale air is expelled from the lungs.

____

____

____

3 marks

c) Describe two differences between the air which is inhaled and the air which is exhaled.

____

____

2 marks

2 a) What is a pathogen?

____

1 mark

b) Describe three ways in which a pathogen can be transmitted.

____

____

____

3 marks

c) Describe how cells in your bloodstream seek out and destroy pathogens.

____

____

2 marks

d) Explain how vaccination prepares your immune system for when a particular pathogen such as the measles virus invades your body.

____

____

____

____

4 marks

How much do you know? *continued*

3 a) Name three ways in which smoking can damage your health.

____

____

____

3 marks

b) Name three ways in which alcohol can damage your health.

____

____

____

3 marks

c) Explain why exercising makes a heavy smoker feel out of breath?

____

____

2 marks

4 For each of the four types of drug listed below

(i) give one example of the drug and

(ii) one effect it has on the person taking it.

a) Painkiller

____

____

2 marks

b) Stimulant

____

____

2 marks

c) Tranquilliser

____

____

2 marks

d) Hallucinogen

____

____

2 marks

5 The diagram below shows a fit athlete.

a) Name three systems within his body which will affect his fitness.

____

____

3 marks

b) Explain why it is important to have a balanced diet.

____

____

____

3 marks

c) Give two examples of illnesses caused by someone not having a balanced diet.

____

____

2 marks

d) Explain why someone who exercises regularly is likely to be fit.

____

____

2 marks

Classification and keys

Figure 1 Finding items in a large supermarket

Trying to find a tin of soup or a piece of meat in this supermarket is really easy. Similar items such as drinks, meats or vegetables have all been gathered together into groups. These groups are then sorted into even smaller groups containing the different types of drinks, meats or vegetables.

Let's suppose you want to find a tin of soup. You will probably use the signs and labels to do the following:

★ enter the food hall
★ find the shelves with the tinned food
★ find the shelf with the soups
★ find the part of the shelf which has the flavour of soup you want.

It's easy. But it would be a nightmare if all the items in the store were not arranged in groups! This sorting into groups is called **classifying**.

Classifying organisms

There are over one million things that live on the Earth. In order to study them more easily scientists have classified them.

The five largest groups of living things are called **kingdoms**. These kingdoms are then sorted into smaller and smaller groups just like the items in the supermarket.

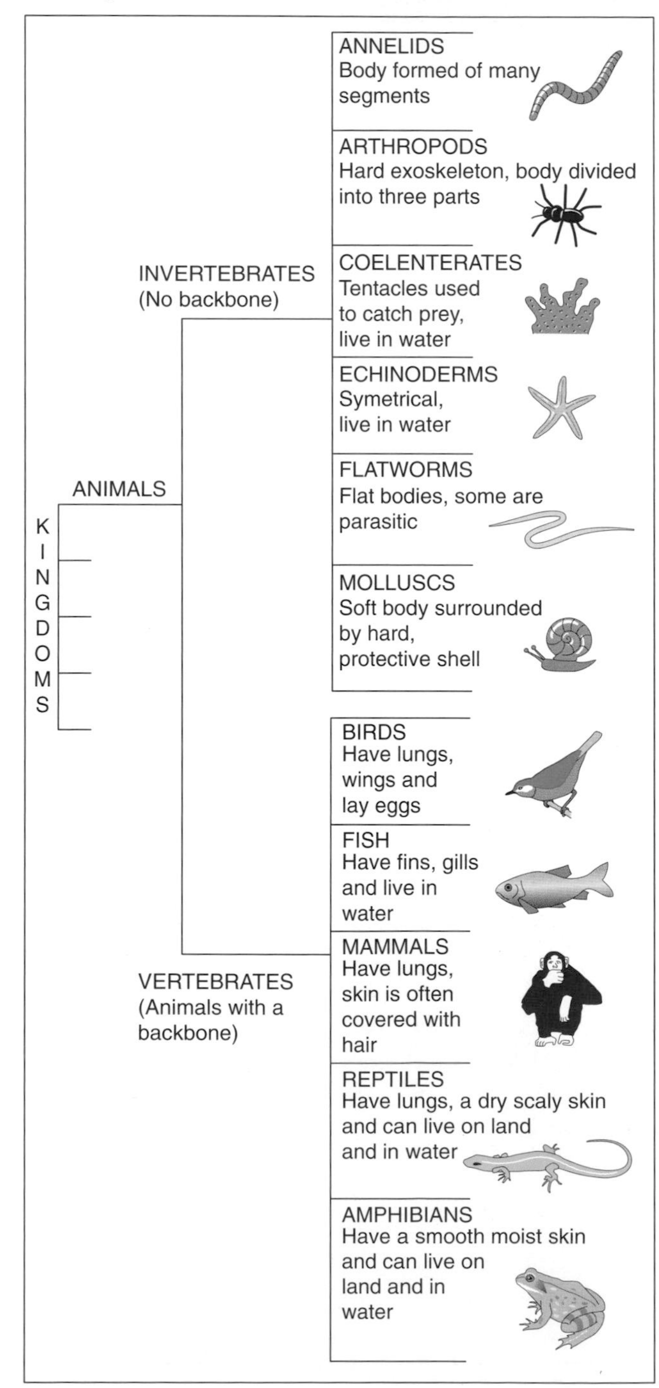

Figure 2 The different parts of the animal kingdom

Keys

If you want to identify a plant or an animal you can do so using a **key**.

See if you can identify the tree from which each of these leaves came from using the key given below.

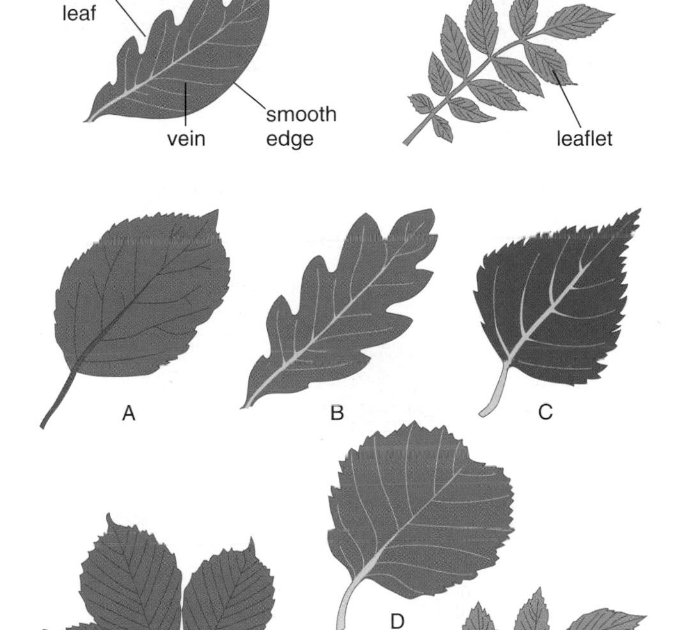

1 Is the leaf simple (no leaflets)? Y.........GO TO 2
N.........GO TO 5

2 Is the leaf lobed? Y.........GO TO 3
N.........GO TO 4

3 Do the veins spread out from a single point? Y.........MAPLE
N.........OAK

4 Is the leaf triangular? Y.........BIRCH
N.........ALDER

5 Are the leaflets in pairs? Y.........ASH
N.........HORSE CHESTNUT

Figure 3 A leaf key

Use the key below to identify the beetles.

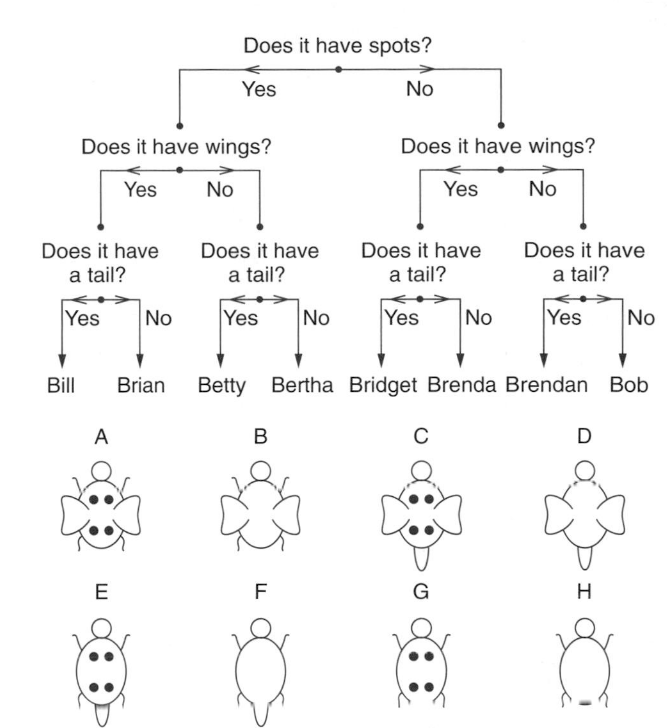

Figure 4 A beetle key

Key terms

Check that you understand and can explain the following terms:

- ★ classification
- ★ kingdom
- ★ invertebrate
- ★ vertebrate
- ★ bird
- ★ fish
- ★ mammal
- ★ reptile
- ★ amphibian
- ★ key

Questions

1. Explain why the books in a library are classified.
2. Explain why scientists classify organisms.
3. What is a vertebrate? Give one example of a vertebrate.
4. What is an invertebrate? Give one example of an invertebrate.
5. What are the main differences between a reptile and an amphibian? Give one example of each.
6. What are the main differences between insects and arachnids?
7. Choose any eight books then make up a key to identify each of them.

5.2 Variation and inheritance

Variation

Living things can look very different. There is a lot of **variation** between them.

Figure 1 The wide variety of life

Even within a single **species** there are often differences. But these variations are usually much smaller than those between species.

To understand how these variations occur we must first recognise that there are two distinct types of variation. These are **continuous variation** and **discontinuous variation**.

Continuous variations

Your height, weight, foot size and arm span are all examples of continuous variations. The sizes of each can be anywhere within a range of values. They are influenced by

* the circumstances and conditions of your upbringing, i.e. your environment
* the **characteristics** you have **inherited** from your parents.

Figure 2 The heights of each of these pupils will depend upon their diet and the characteristics they have inherited from their parents

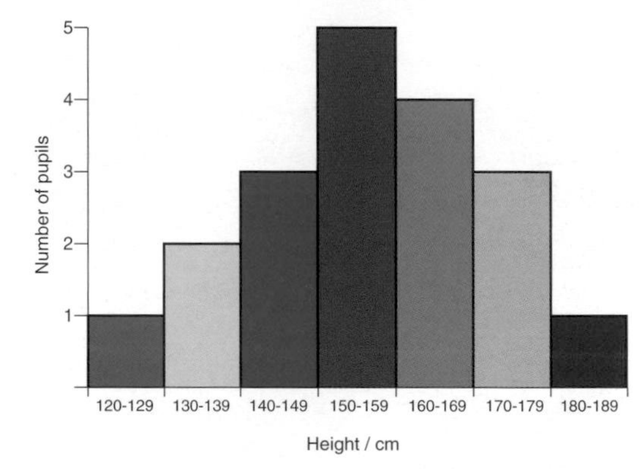

Figure 3 Height is an example of continuous variation because there are a range of values

Discontinuous variations

Your blood group and eye colour are examples of discontinuous variation. They are only influenced by the characteristics you inherited from your parents. Unlike continuous variations, discontinuous variations have no in-between values. You are either one blood group or another – you can't be half of one and half of another.

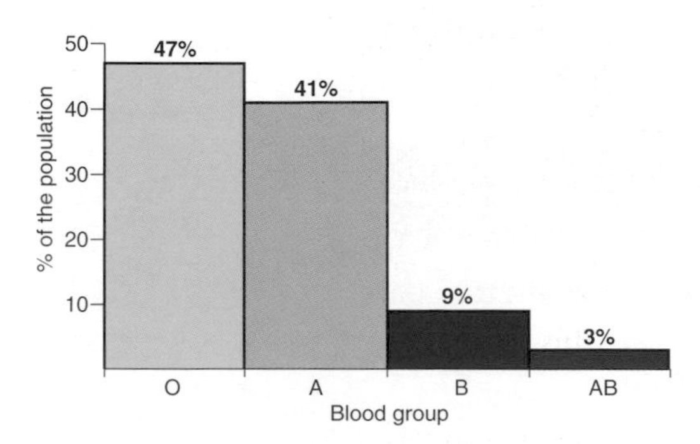

Figure 4 Blood type is an example of discontinuous variation, as you can only be in one of four blood groups

Why are offspring similar but not identical to their parents?

During fertilisation the nuclei of the male cell and the female cell fuse to produce a new individual. Within the two nuclei are chemicals called **genes** which carry the instructions that control the development of characteristics in the new individual. Because it has inherited half of the genes from its mother and half from its father, the new individual will have some of the characteristics from each parent, but it will not be identical to either.

Figure 5 You can see that this boy has inherited features from both parents

Selective breeding

Selective breeding takes advantage of variations between organisms. It is used to try to breed plants and animals with a particular characteristic, for example a cow that produces more milk, a sheep that produces more wool or a plant that is resistant to disease.

Figure 6 This racehorse has been bred for speed and stamina

To produce this magnificent animal, the horse breeder has mated a stallion and a mare that had some of the features which he considered important for a horse to run quickly.

He may, for example, have chosen a mare with long legs and a stallion with strong leg muscles. The first offspring of this mating may or may not have had the desired features. It is a matter of luck. Eventually, however, it is likely that one of the offspring will have inherited both the desired characteristics. The breeder may then have a winner!

Plants can also be selectively bred. It is possible to **cross pollinate** two varieties in order to try to produce a new variety with the desired characteristics, for example resistance to cold, sweetness and long life on shelf.

With some plants it is possible to take all the genetic information from one parent. The new individual is genetically identical to its parent and is called a **clone**.

Key terms

Check that you understand and can explain the following terms:

- ★ variations
- ★ species
- ★ continuous variation
- ★ discontinuous variation
- ★ characteristics
- ★ inherit
- ★ genes
- ★ selective breeding
- ★ cross pollination
- ★ clone

Questions

1. What factors affect the colour of your eyes?
2. What factors affect your height?
3. Explain the difference between continuous variation and discontinuous variation. Give one example of each.
4. What two characteristics would you want to breed into
   a) sheep which are grown for both their wool and their meat
   b) wheat which is grown in a hot dry country
   c) a blackberry plant?
5. Make a note of the colour of everyone's eyes in your class. Now draw a graph of the colour of the eyes against the number of pupils who have that colour.

Chapter 5 Variation, classification and inheritance:

What you need to know

1 There is variation within species and between species.

2 Variation within a species can be caused by both environmental and inherited factors.

3 Selective breeding can be used to produce new, improved varieties.

4 Living things can be classified into groups.

5 Plants and animals have different characteristics which can be used to classify them.

6 Keys can be used to identify plants and animals.

How much do you know?

1 The diagrams below show different types of animals.

The animals can be put into four groups. The table below describes the main characteristics of each group.

Group A	Birds	Wings, feathers, two legs and a beak
Group B	Mammals	Body hair
Group C	Reptiles	Dry scales
Group D	Fish	Fins, gills and wet scales

Using the information in the table decide to which group each belongs.

Animal number 1 Group ________

Animal number 2 Group ________

Animal number 3 Group ________

Animal number 4 Group ________

4 marks

2 a) What are genes and what do they do?

____

____

2 marks

The diagram below shows a child and her two parents.

b) Which characteristics has the child inherited from

(i) her mother?

____

2 marks

(ii) her father?

____

2 marks

c) Name one characteristic of the child which may have been affected by her environment.

____

1 mark

How much do you know? *continued*

3 The graph below shows the percentage of people who have a particular blood type i.e. 41% of people have blood type A.

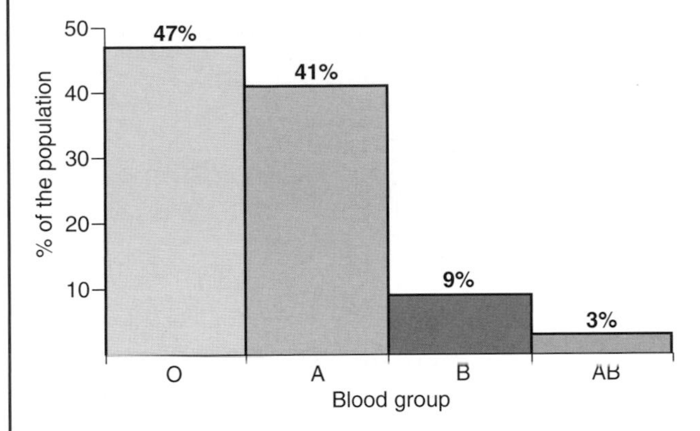

a) What kind of variation is this?

____

1 mark

b) 'This kind of variation is **inherited** from parents.'

Explain what this sentence means.

____

____

1 mark

c) A dog breeder wants to breed a dog which has no spots. Which two dogs shown below should he choose.

__________ and __________

2 marks

4 The diagram below shows one way in which living things can be classified.

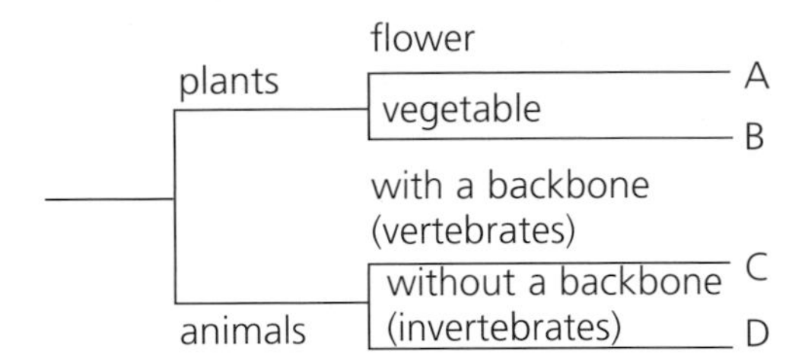

To which group, A, B, C or D, do each of the following living things belong?

dandelion __________

lion __________

jellyfish __________

worm __________

tulip __________

monkey __________

6 marks

6.1 Photosynthesis and respiration

All plants and animals need food in order to live and grow. Animals obtain their energy from the food which they eat. Most plants make their own food. This process is called **photosynthesis**.

Figure 1 When photosynthesis takes place plants grow and make new plant material or **biomass**

Photosynthesis

To make food by photosynthesis plants need water from the soil, carbon dioxide from the air and light energy from the Sun.

The water is drawn up through the roots and the stem into the leaves. The carbon dioxide enters the leaves through tiny holes or pores called **stomata**. The light energy is absorbed by a green substance called **chlorophyll**. The chlorophyll is found in **chloroplasts** in the plant's leaf cells.

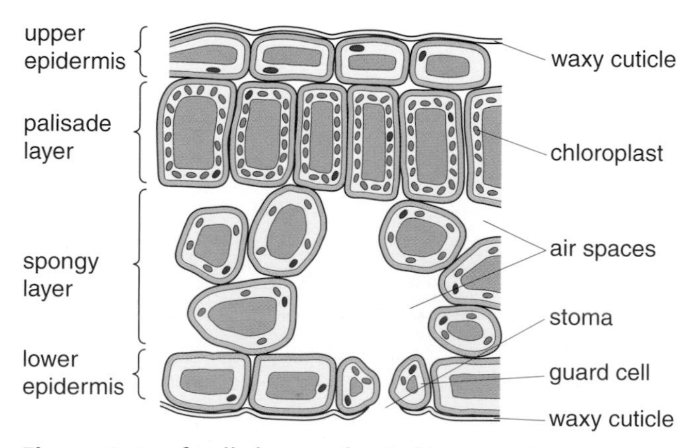

Figure 2 Leaf cells have adapted so that they contain lots of chloroplasts. They are therefore able to capture lots of light energy

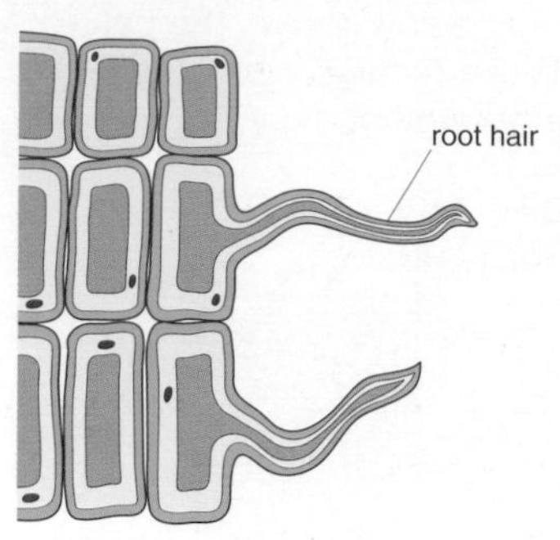

Figure 3 Root cells have also adapted to fulfil their role. They contain no chloroplasts, as they are not used to capture light energy. The cells are long and thin which helps them to absorb water from the soil. The roots also spread out over a large area, increasing their chance of absorbing water and helping to anchor the plant in place

The reaction which produces the plant's food is given below.

$$\text{water} + \text{carbon dioxide} \xrightarrow{\text{sunlight}} \text{glucose (food)} + \text{oxygen}$$

$$6H_2O + 6CO_2 \rightarrow C_6H_{12}O_6 + 6O_2$$

The chlorophyll is unchanged in the reaction. It simply absorbs the sunlight which is needed to make the reaction happen.

The oxygen released by this reaction escapes into the atmosphere through the stomata.

Factors affecting the rate of photosynthesis:

★ light – photosynthesis will take place more rapidly in bright light than in dull conditions
★ carbon dioxide – if the concentration of carbon dioxide is high, the rate of photosynthesis is also high
★ temperature – the optimum temperature for photosynthesis is approximately 30 °C. If temperatures are much higher or lower than this, the rate of photosynthesis decreases.

All three of these factors can be readily controlled in greenhouses like the one shown in Figure 1.

Respiration

The glucose produced by photosynthesis may be used by the plant straight away for **respiration** to produce the energy it needs to live.

The reaction which produces this is the same as that for respiration in animals: (see page 26).

The oxygen needed for this reaction is taken from the atmosphere. The carbon dioxide produced is released through the leaves into the atmosphere.

Plants make more oxygen during the daytime than they use at night. As a result they are continually removing carbon dioxide from the atmosphere and replacing it with oxygen. This oxygen is needed by all animals which live on the Earth.

What happens to the glucose not used for respiration?

Some glucose is used to make more complicated substances, such as proteins required for plant growth and **cellulose** which is used for cell walls. Some glucose is changed into **starch** which is **stored food** the plant can use later e.g. in the winter.

Transpiration

Plants lose a large amount of water from their leaves by evaporation. This loss of water is called **transpiration**.

Transpiration occurs most rapidly in hot, dry, windy conditions. To prevent too much water from being lost in this way

- ★ most plants have a waxy layer on their leaves
- ★ the size of the stomata on the surface of the leaves can be decreased.

The loss of water from the leaves draws new water and important minerals, such as nitrates, potassium and phosphates, up from the **roots** through the **stem**. This upward flow of water through the stem is called the **transpiration stream**.

Many of the minerals needed by plants can be provided by **fertilisers**.

- ★ Nitrates encourage the growth of leaves.
- ★ Potassium (potash) is needed to encourage flowering.
- ★ Phosphates are needed to encourage the growth of roots.

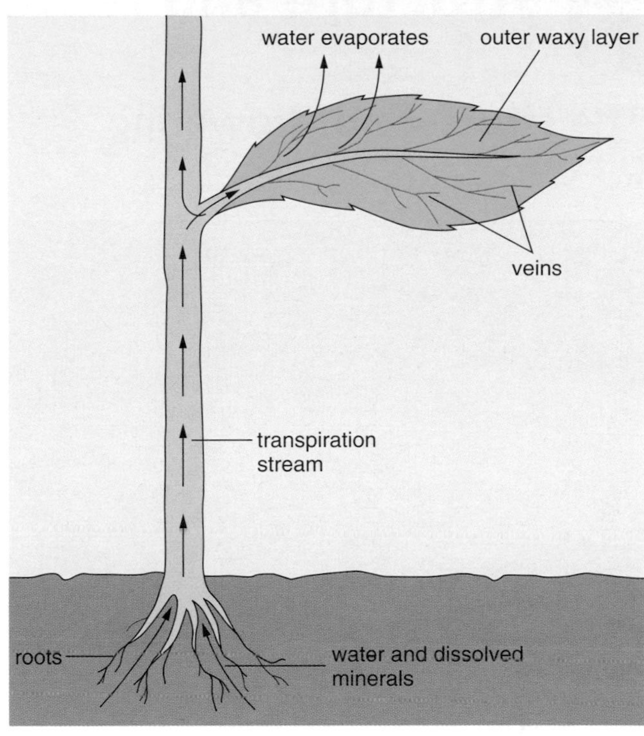

Figure 4 Transpiration

Key terms

Check that you understand and can explain the following terms:

- ★ photosynthesis
- ★ biomass
- ★ stomata
- ★ chlorophyll
- ★ chloroplasts
- ★ respiration
- ★ cellulose
- ★ starch
- ★ stored food
- ★ transpiration
- ★ roots
- ★ stem
- ★ transpiration stream
- ★ fertiliser

Questions

1. By what process do plants make their own food? Write down a word equation and a symbol/chemical equation which explain how this food is made.
2. By what process do plants convert their food into the energy they need? Write down a word equation which explains how this happens.
3. Explain the different functions carried out by leaf cells and root cells. Explain how these cells have adapted for their function.
4. Explain why plants and trees have an important role to play in reducing the levels of carbon dioxide in the atmosphere.

6.2 Structure and reproduction

The structure of flowering plants

Figure 1 A flowering plant

Plants have a structure which allows them to make and store the food they need to live, grow and reproduce. They have roots, a stem and leaves. At certain times of the year they will also have buds and flowers.

Leaves

The leaves of a plant are the chemical factories where photosynthesis takes place, producing food.

Leaves are often coated with a thin waxy layer on their upper surface to prevent them from losing too much water. This layer tends to be thicker for plants living in dry conditions. If a plant is losing water too quickly, it may reduce the size of the stomata on its leaves to help slow down the rate of evaporation.

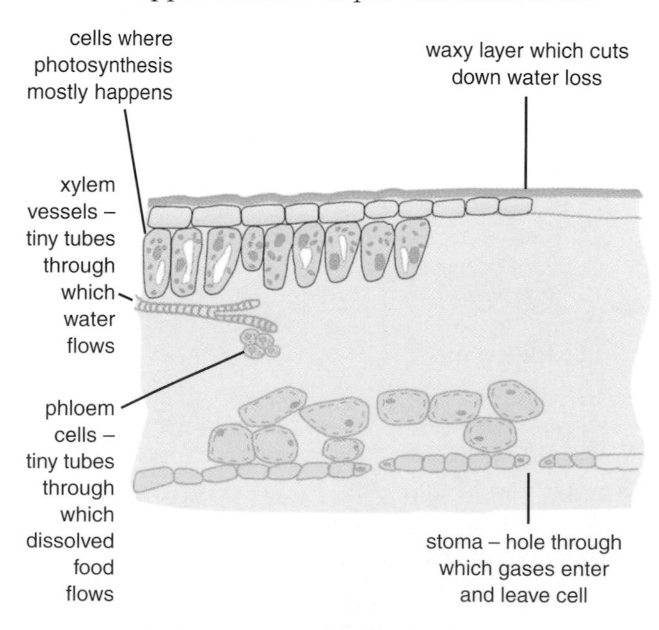

Figure 2 The cellular structure of a leaf

Stem

The stem is the main support for the plant. The cells here contain a lot of water which gives the stem 'stiffness'. If the plant is short of water the stem loses this stiffness and the plant wilts.

Separate tubes carry water and food through the stem. **Xylem** tissue carries water from the roots to the leaves. **Phloem** tissue carries food from the leaves to the rest of the plant.

Roots

The roots of a plant anchor it firmly in the ground. The root hairs take in water and minerals from the soil. In some plants the roots are used to store energy e.g. turnips and carrots.

Figure 3 Carrots are roots

Sexual reproduction in plants

Plants have flowers so that they can reproduce. Most flowers contain both the male and female parts.

The **anthers** produce grains of **pollen** which contain male reproductive cells. The **carpels** contain ovaries which produce plant eggs, the female reproductive cells, in structures called **ovules**.

If a grain of pollen is moved from an anther to the **stigma** of a carpel, this is called **pollination**.

If the pollen has been moved by the wind, the plant has been **wind-pollinated**. If the pollen has been moved by insects, the plant has been **insect-pollinated**.

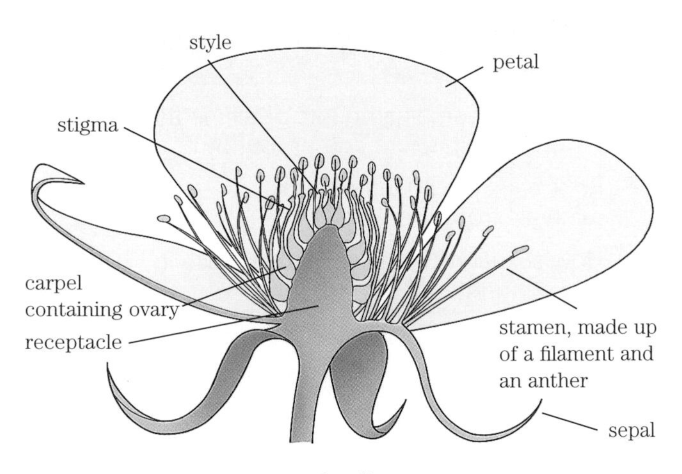

Figure 4 The structure of a flower

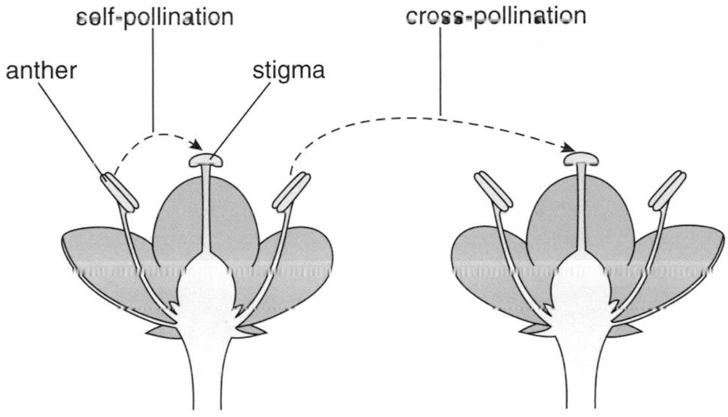

Figure 5 Two types of pollination – self-pollination and cross-pollination

When a grain of pollen sticks to the stigma, a tube grows down through the ovary. The nucleus of the pollen then travels along this tube and fertilises a plant egg. The fertilised egg grows to form a **seed** from which a new plant can be grown. The ovary of a flower often grows into a **fruit** which protects the developing seeds.

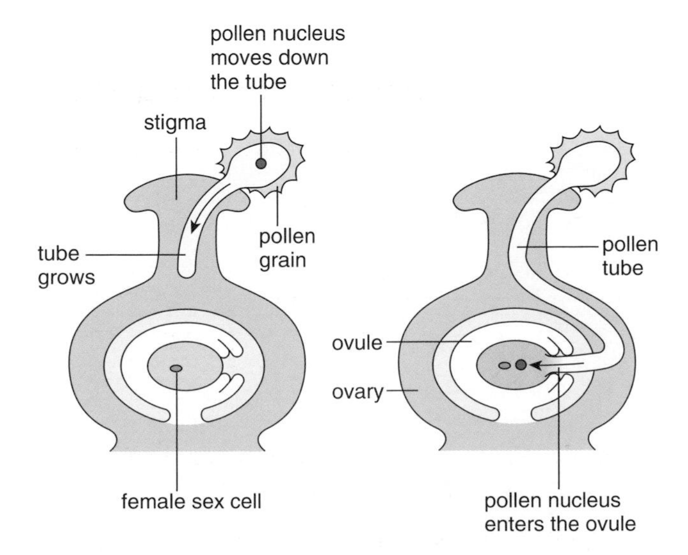

Figure 6 Fertilisation in a flowering plant

Spreading the seeds

To ensure that seeds have a good chance of survival they need to be scattered over as large an area as possible.

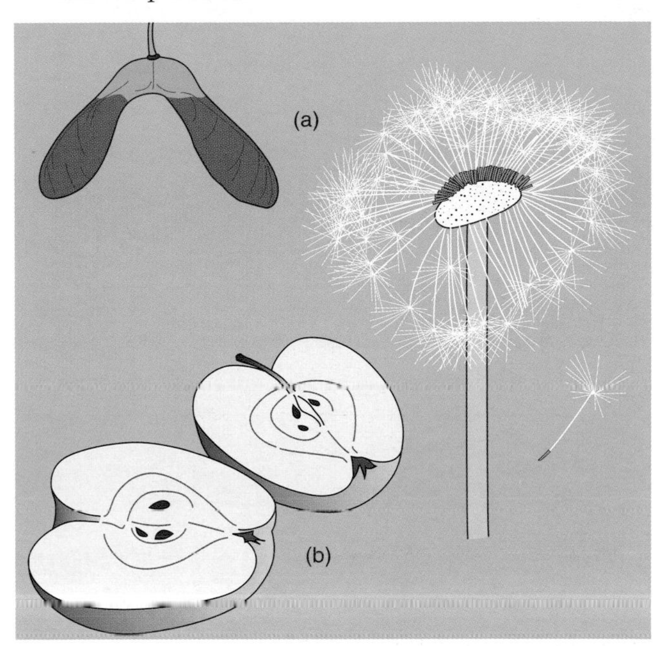

Figure 7 **Seed dispersal** The sycamore and dandelion seeds are shaped so that they are easily dispersed by the wind. The seeds in the apple are spread when an animal eats the fruit

Key terms

Check that you understand and can explain the following terms:

- ★ xylem
- ★ phloem
- ★ anther
- ★ pollen
- ★ carpel
- ★ ovule
- ★ stigma
- ★ pollination
- ★ wind-pollination
- ★ insect-pollination
- ★ seed
- ★ fruit
- ★ seed dispersal

Questions

1. What causes a plant to wilt?
2. Which part of a plant anchors it into the ground?
3. What is the name of the male part of a flower and what does it produce?
4. What is the name of the female part of a flower and what does it produce?
5. Explain the difference between pollination and fertilisation.
6. Explain one way in which the seeds of a plant can be spread.

Chapter 6 Plants:

What you need to know

1. Photosynthesis produces food for plants.
2. Plants need carbon dioxide, water and light for photosynthesis to take place.
3. The equation for photosynthesis:

   water + carbon dioxide $\xrightarrow{\text{sunlight}}$ glucose + oxygen
4. Leaves have chlorophyll for photosynthesis.
5. Plants, like all other organisms, obtain their energy by respiration i.e. glucose + oxygen → water + carbon dioxide + energy.
6. Plants need small amounts of elements such as nitrogen, potassium and phosphorus.
7. Root hairs are adapted to absorb water and minerals from the soil.
8. How sexual reproduction takes place in flowering plants.
9. How fertilisation occurs and seeds develop.

How much do you know?

1 The diagrams below show the cellular structure of a leaf and a root.

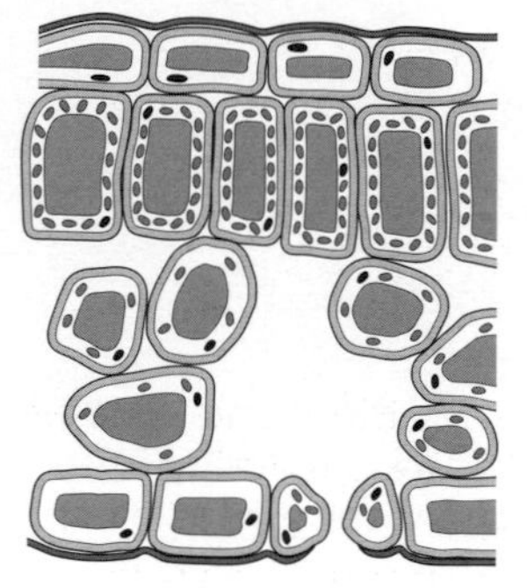

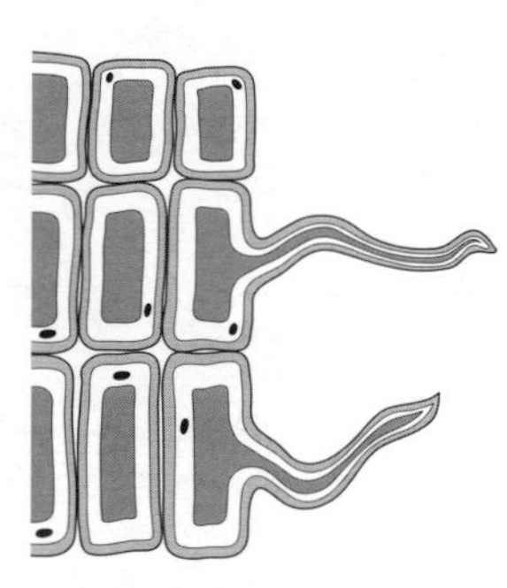

a) Describe how the leaf has adapted for photosynthesis.

____

____

2 marks

b) Describe how the cellular structure of a root has adapted to absorb water.

____

____

2 marks

c) Explain the difference between photosynthesis and respiration in plants.

____

____

2 marks

d) Write down three conditions which would encourage the healthy growth of a plant.

____

____

____

3 marks

e) What are fertilisers and what do they do?

____

____

2 marks

2 The diagram below shows an insect visiting a flower in search of food.

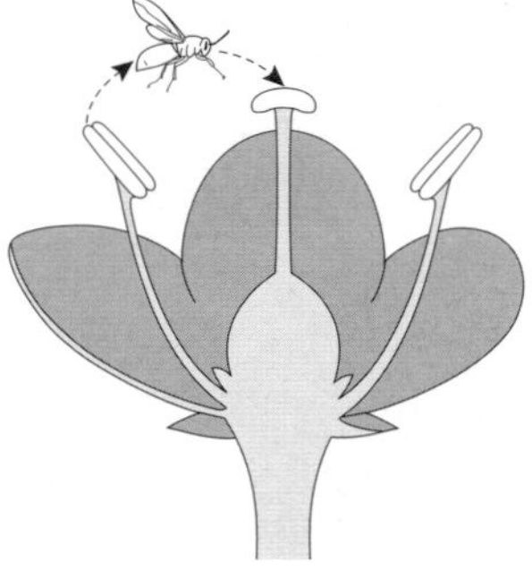

a) Whilst the insect is feeding, some pollen from the ________________ is transferred to the stigma. This process is called ____________.

2 marks

b) Suggest a second way in which the pollen could be moved.

____

1 mark

c) Which part of the flower contains the male cells?

____

1 mark

d) Which part of the flower contains the female cells?

____

1 mark

How much do you know? *continued*

3 The diagram below shows a strawberry plant. A, B, C and D are different parts of the plant.

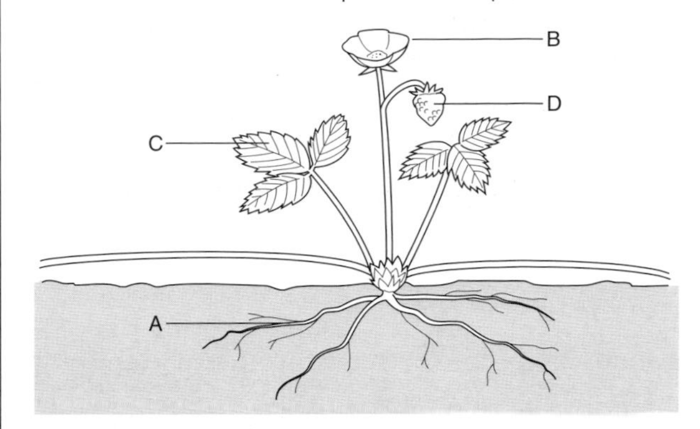

a) Put the correct letter next to the part of the plant in this table

Part of plant	Letter
flower	
root	
leaf	
fruit	

4 marks

b) Which part of the plant

(i) produces the food the plant needs?

____

1 mark

(ii) protects the seeds?

____

1 mark

(iii) draws water and minerals from the soil into the plant?

____

1 mark

(iv) attracts insects to the plant?

____

1 mark

4 a) Write down a word equation which describes how a plant produces food.

____

____

5 marks

b) When campers take their tents down after a fortnights holiday in the summer, they notice that the grass where the tent has been is yellow. Explain why this happens.

____

____

3 marks

c) What happens to the grass after the tent has been removed?

____

____

2 marks

d) What would happen to the grass if the tent was left in the same place for the whole of the summer?

____

1 mark

5 a) Write down the word equation which describes how a plant obtains its energy from food.

____

____

4 marks

b) What is this process called?

____

1 mark

c) What happens to food which is not immediately used to produce energy for a plant?

____

1 mark

6 a) Name two gases that enter or leave the leaf through the stomata.

____

____

2 marks

b) Name one liquid which escapes through the stomata.

____

1 mark

7.1 Solids, liquids and gases

The structure of solids, liquids and gases

Scientists believe that **solids**, **liquids** and **gases** are made up of extremely small **particles** – they are so small it is impossible to see them, even with a strong microscope. These particles are arranged differently in solids, liquids and gases. It is these **arrangements** and the way in which the particles move which give rise to the different properties.

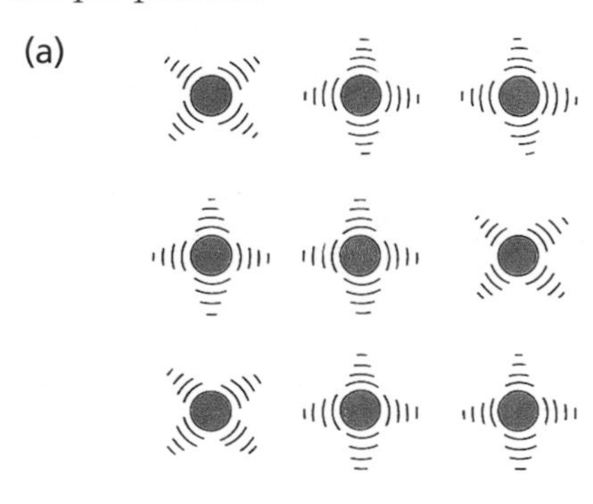

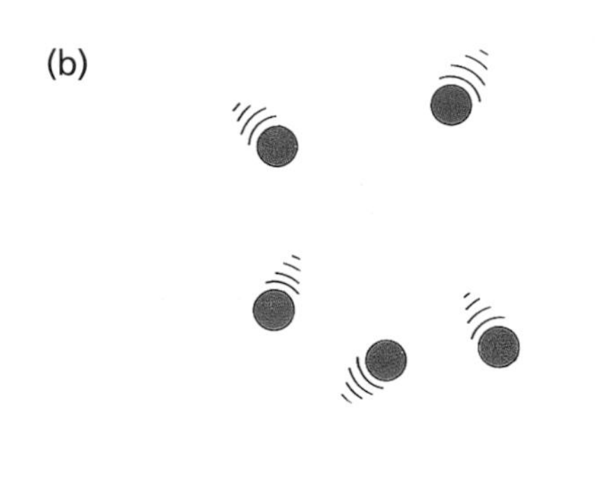

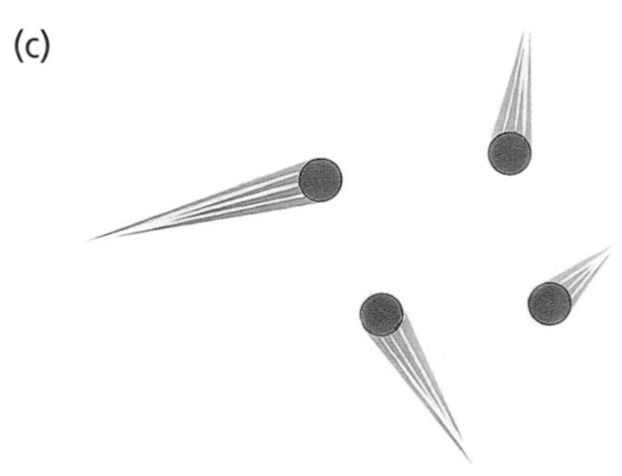

Figure 1 The arrangement of particles in a) a solid, b) a liquid and c) a gas

In solids the particles are:

★ close together
★ in layers
★ in fixed positions, but able to **vibrate** from side to side
★ held together by strong forces.

In liquids the particles are:

★ a little further apart than the particles of a solid
★ not in fixed positions, but able to move around a little
★ held together by weaker forces than the particles in a solid.

In gases the particles are:

★ far apart
★ completely free to move around within their container
★ not held together by any forces.

The properties of solids, liquids and gases

Shape

Solids have a definite shape.

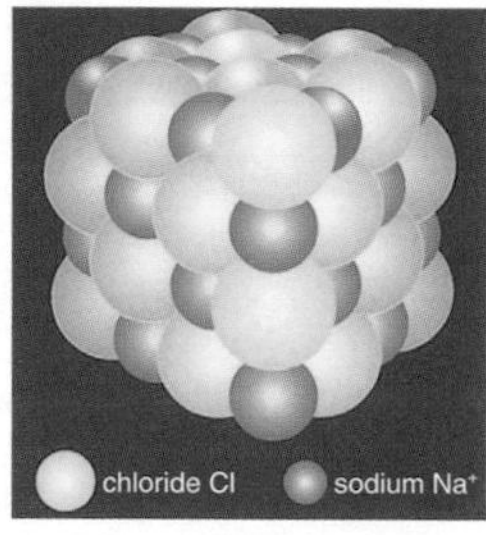

Figure 2 **Crystal structures** like this form because the particles in solids are arranged in a regular pattern

Liquids do not have a definite shape.

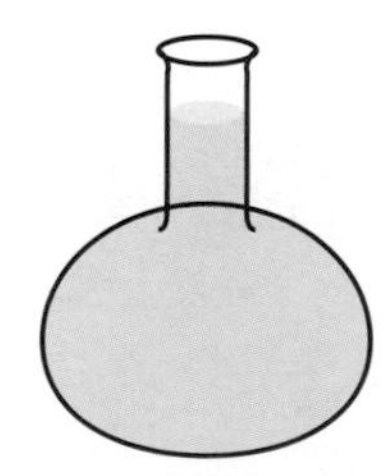

Figure 3 The particles in a liquid are not in fixed positions so the liquid can take the shape of the container

Gases do not have a definite shape.

Figure 4 The particles in a gas are completely free and will fill any container into which they are placed

Strength

Because of the strong forces between the particles, solids are firm and can support things. Liquids and gases have no **strength** or **firmness**.

The ability to flow

Because the forces between the particles in liquids and gases are weak or non-existent, they are able to **flow**. The strong forces between the particles in solids prevent them from flowing.

Squashability

Because their particles are far apart, gases can be squashed. The particles of solids and liquids are close together and they are therefore not easily squashed.

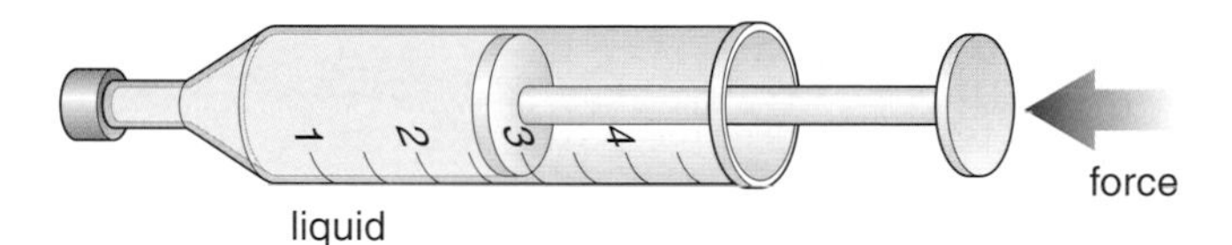

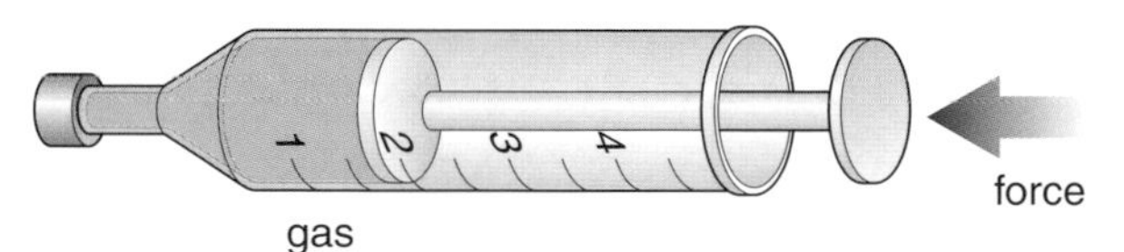

Figure 5 Gases can be **compressed**. Liquids and solids are **incompressible**

Diffusion

The particles of liquids and gases are able to move and mix. This mixing is called **diffusion**.

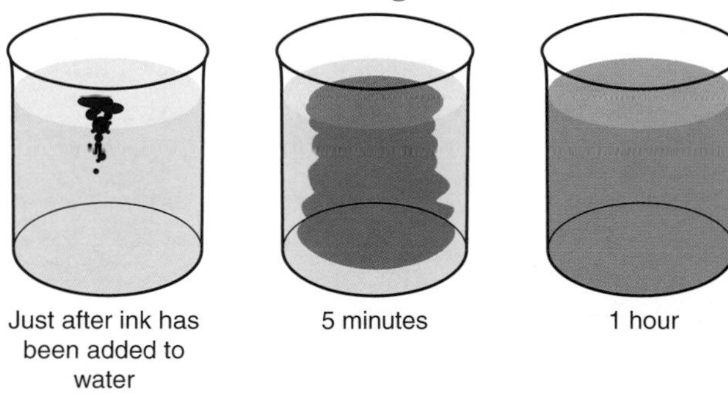

Figure 6 The drop of ink gradually mixes with the water in the beaker. This is an example of diffusion

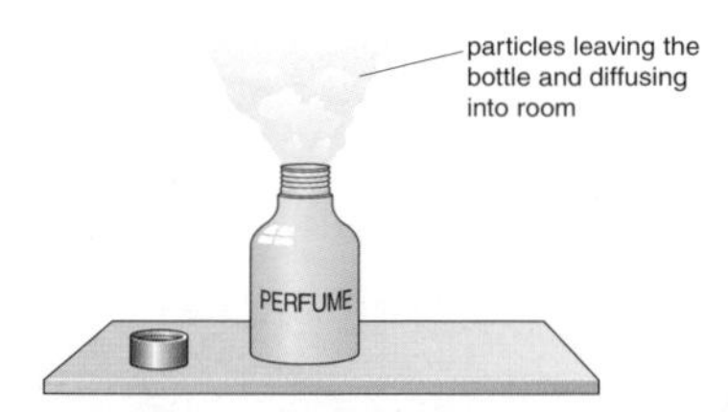

Figure 7 When the top of this bottle is removed, perfume particles spread throughout the room, helped by the motion of the air particles. This is another example of diffusion

Gas pressure

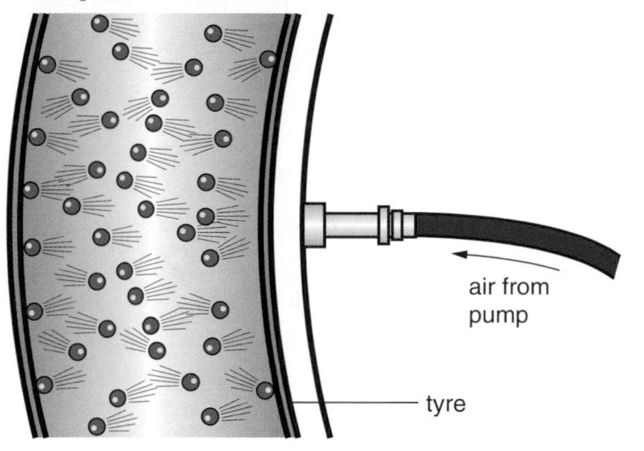

Figure 8 The gas particles inside this tyre are continually bouncing off the sides. It is these collisions which create the pressure inside the tyre

If more air is pumped into the tyre there will be more particles and more collisions, so the **pressure** in the tyre will increase.

Key terms

Check that you understand and can explain the following terms:

- ★ solid
- ★ liquid
- ★ gas
- ★ particle
- ★ arrangement of particles
- ★ vibrate
- ★ crystal structure
- ★ strength
- ★ firmness
- ★ flow
- ★ compress
- ★ incompressible
- ★ diffusion
- ★ pressure of a gas

Questions

1. List the properties of a) solids, b) liquids and c) gases. Use these properties to design a key which can be used to decide if an object is a solid, a liquid or a gas.
2. Explain why it might be difficult to decide whether the following are solids or liquids: sand, jelly, flour, toothpaste and tomato sauce.
3. Draw three labelled diagrams to show how the particles are arranged in **a)** a solid, **b)** a liquid and **c)** a gas.
4. Name three materials or substances that cannot easily be squashed.
5. What is the name of the process by which the smell of food spreads around a kitchen?
6. Draw a diagram to show how the air particles inside a plastic bottle create pressure.
7. What might happen to the plastic bottle in Question 6 if all the air particles inside it were removed? Explain your answer.
8. Explain why dust particles seen in a beam of light seem to be moving about randomly.

7.2 Changing state

Melting

If a solid is heated gently, its particles start to vibrate more vigorously. The more it is heated, the more vigorously the particles will move. Eventually they vibrate so violently that the regular structure of the solid breaks apart and the particles are able to move around freely. The solid has now become a liquid, it has **melted**. The temperature at which this happens is called the **melting point** of the solid.

Figure 1 Iron has a high melting point. It melts at a temperature of 1539 °C.

If a liquid is cooled, the vibrations of its particles become less and less vigorous as its temperature falls. Eventually the forces of attraction between neighbouring particles are strong enough for a regular structure to begin to form. The liquid is **solidifying** or **freezing**. The temperature at which a liquid does this is called its **freezing point**.

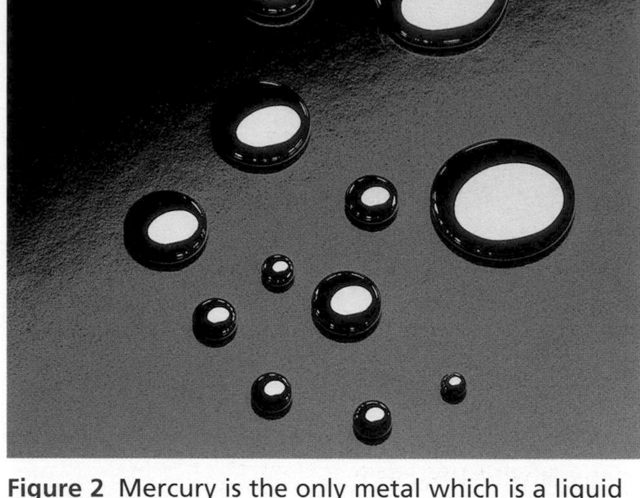

Figure 2 Mercury is the only metal which is a liquid at room temperature. It freezes at a temperature of −38.8 °C

The freezing point of a liquid and the melting point of its solid are the same. For example, pure water freezes at 0 °C and pure ice melts at 0 °C.

Boiling

If a liquid is heated, the vibrations of its particles may become so violent that they have enough energy to become completely free from the surface of the liquid. The liquid is **boiling** and becomes a gas. The temperature at which this happens is called the **boiling point** of the liquid.

If a gas is cooled, its particles lose energy and start to move around less vigorously. At a particular temperature, their energy is so low that attractions can develop between particles, forming groups. At this point, the gas changes into a liquid. We say that the gas has **condensed**. The temperature at which a gas condenses is the same temperature at which the liquid changes into a gas. For example, pure water boils at 100 °C and steam condenses at 100 °C.

Figure 3 summarises how the **particle theory** explains changes of state.

Figure 3 Changes of state

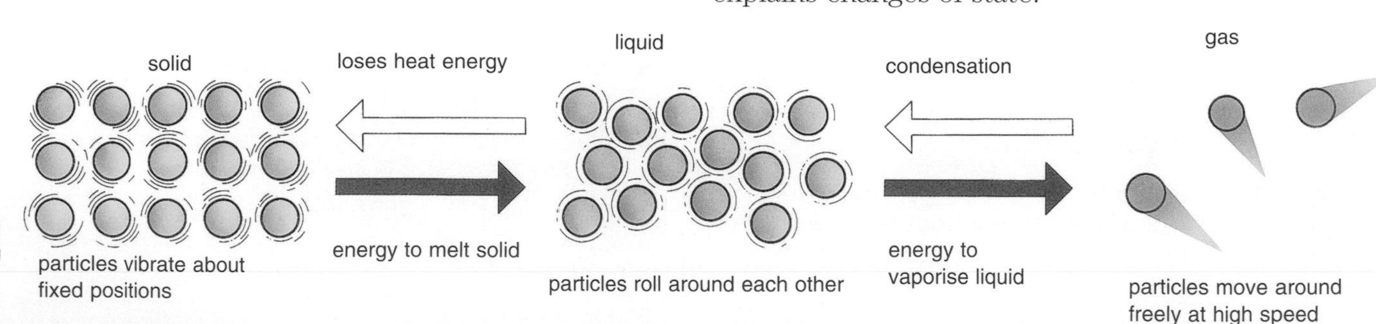

Figure 4 We normally think of nitrogen as being a gas but if it is cooled to below a temperature of −196 °C, it condenses and becomes a liquid.

Figure 5 shows how the temperature of ice/water changes with time as it is being heated.

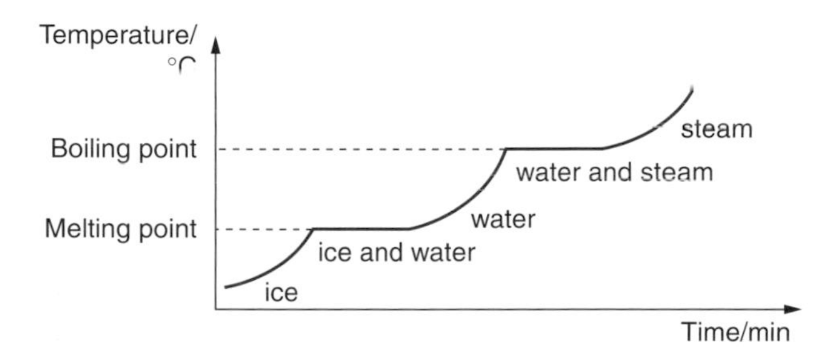

Figure 5 Heating graph for ice/water

The table below gives some examples of the melting points and boiling points of different materials.

Substance	Melting point °C	Boiling point °C
water	U	100
alcohol	2117	78
tungsten	3410	5500
iron	1539	2887
oxygen	2219	2183

Key terms

Check that you understand and can explain the following terms:

- ★ melting
- ★ melting point
- ★ solidifying
- ★ freezing
- ★ freezing point
- ★ boiling
- ★ boiling point
- ★ condensing
- ★ particle theory

Questions

1 Draw diagrams to show what happens to the structure of a solid when it melts.

2 Draw diagrams to show what happens to the particles of a liquid when it boils.

3 Explain the differences between **a)** melting and freezing, and **b)** boiling and condensing.

4 Look carefully at the table below then answer the questions.

Material	Melting point °C	Boiling point °C
Water	0	100
Common salt	801	1420
Copper	1083	2582
Iron	1539	2887

a) What happens to common salt at i) 801 °C and ii) 1420 °C?

b) How are the particles arranged in iron at 1400 °C?

c) What happens to water if its temperature changes from 120 °C to 80 °C?

d) Explain in detail what happens to the particles in a sample of copper if its temperature is increased from 1000 °C to 1200 °C.

7.3 Solutions

If you add salt or sugar to pure water it **dissolves** to form a **solution**. Solutions are **mixtures**.

The substance which is dissolved is called the **solute**. The liquid into which a substance dissolves is called the **solvent**. The solute and the solvent form a solution. In Figure 1 sugar is the solute, water is the solvent and sugar water is the solution.

Not all solids dissolve in water. If you put a piece of chalk or copper in water, a mixture is again created, but it is not a solution. Salt and sugar are **soluble** in water. Chalk and copper are **insoluble** in water.

Figure 1 Making a solution

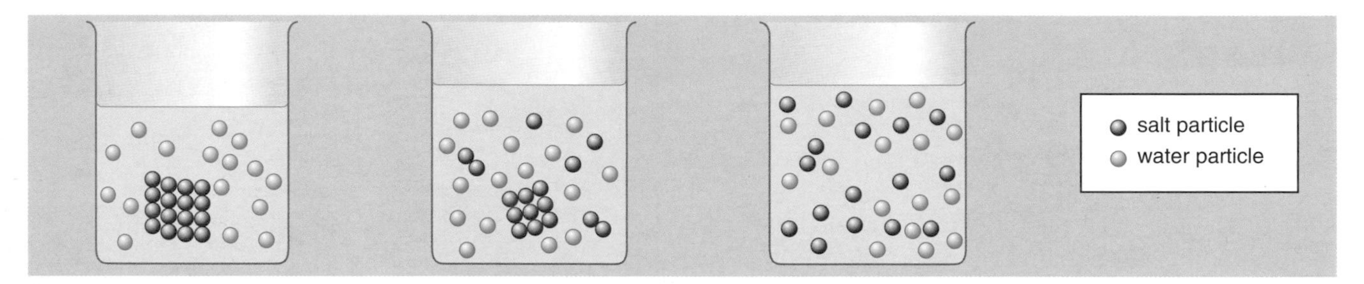

Figure 2 When a substance such as salt dissolves, it breaks up into extremely small particles, too small to see, which spread throughout the liquid

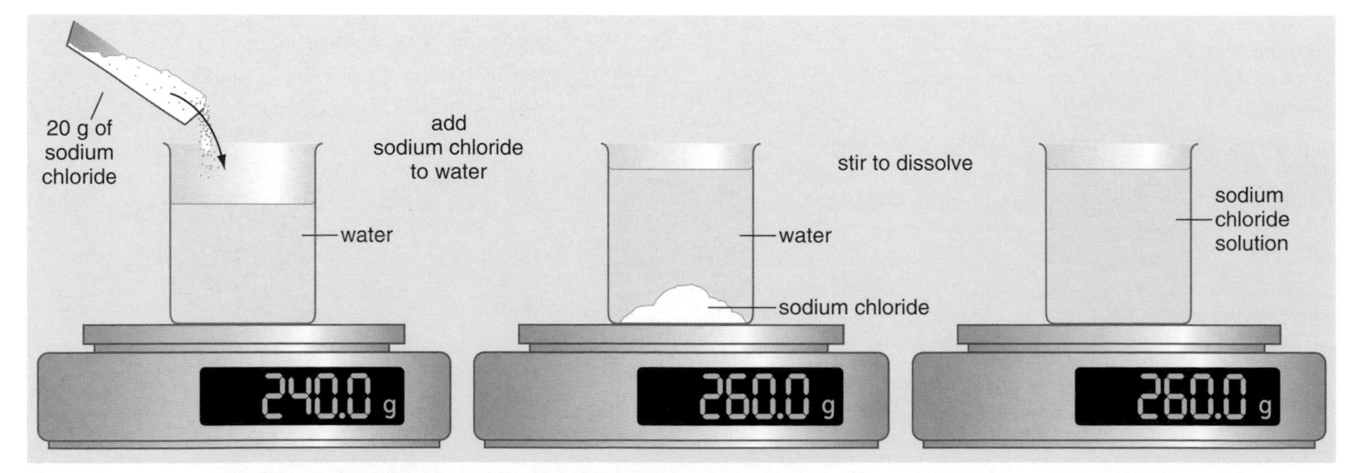

Figure 3 Although the salt seems to have disappeared when it has dissolved in water, it is still there. The mass of the solute + the mass of the solvent = the mass of the solution

Separating mixtures

Filtering

Sand is insoluble in water. Its particles do not dissolve. A mixture like this can be separated by **filtering**. The water passes through tiny holes in the filter paper but the sand particles are too large and so remain on the filter paper.

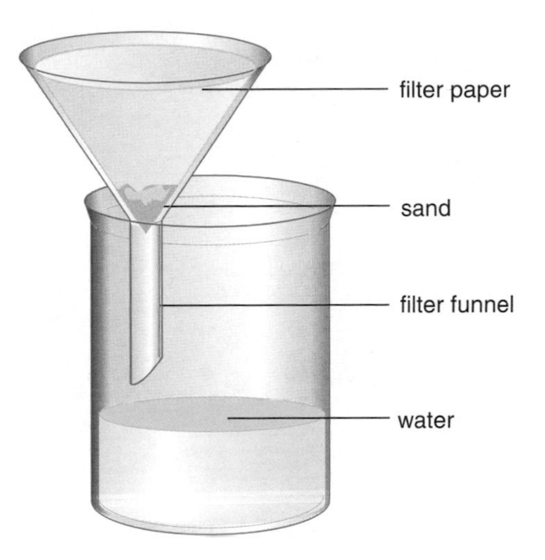

Figure 4 The separation of sand from water by filtration

Evaporating

A solid which is dissolved in a liquid, such as salt dissolved in water, can be recovered by **evaporating** the water off. This can be done by heating the solution or just leaving a small amount in the air.

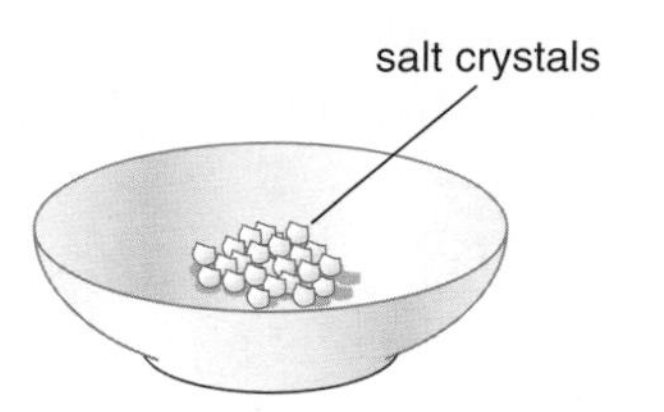

Figure 5 If the water from the solution is left to evaporate, the salt is left behind

Distilling

If salt water is heated in the apparatus shown in Figure 6, salt is left behind in the round bottomed flask and pure water is collected in the conical flask. This process is called **distillation**.

Distillation can also be used to separate liquids with different boiling points, for example water and alcohol.

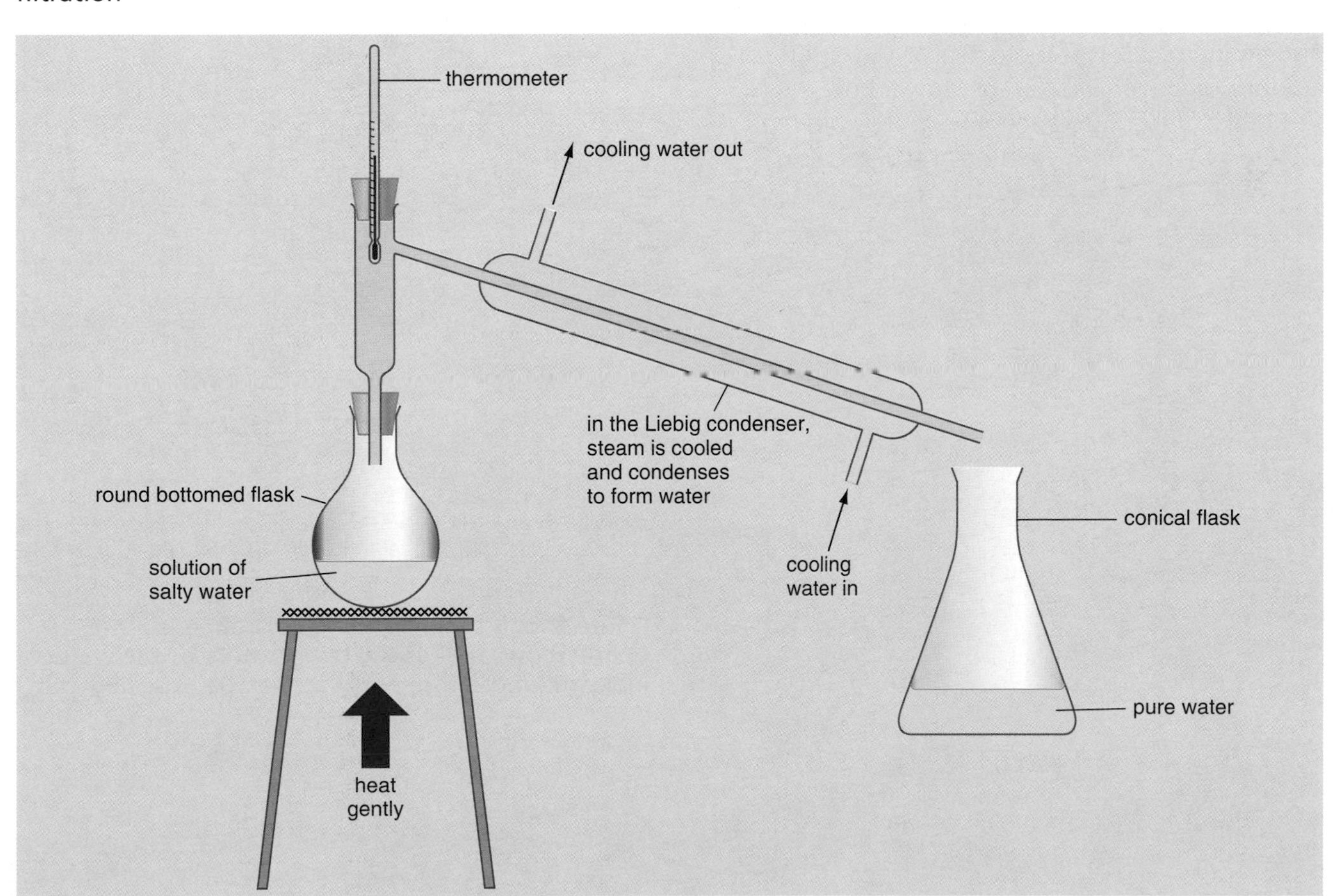

Figure 6 Using distillation to obtain pure water from salty water

Chromatography

If a solution contains more than one solute, they can be separated using a technique called **chromatography**. Inks are often mixtures of several dyes (liquids). We can prove this by separating the different dyes using the technique described below.

A small drop of each of the different inks is placed on the bottom of a piece of filter paper. The paper is then suspended over a beaker of water (the solvent). The solvent is absorbed by the paper and drawn upwards, taking the different dyes with it. The dyes are carried at different speeds, with the most soluble dye travelling the fastest and therefore travelling the furthest over a given time.

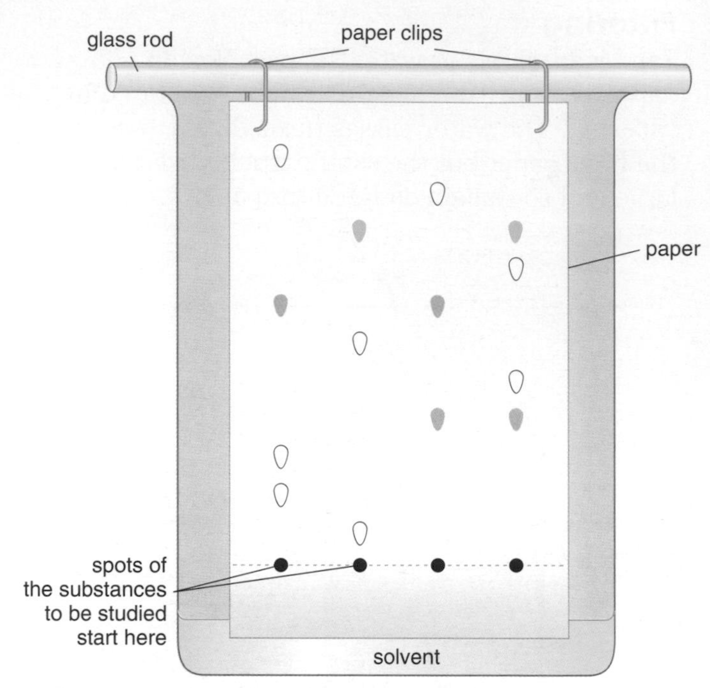

Figure 7 Comparing the dyes in different inks

Solubility

If we add more and more salt to a beaker of water, there will come a point when no more salt will dissolve. The salt solution is then described as being **saturated**. **Solubility** is a measure of how much solute can be dissolved in a certain amount of solvent. Figure 8 shows how the solubility of a solute such as copper sulfate is affected by the temperature of the solvent.

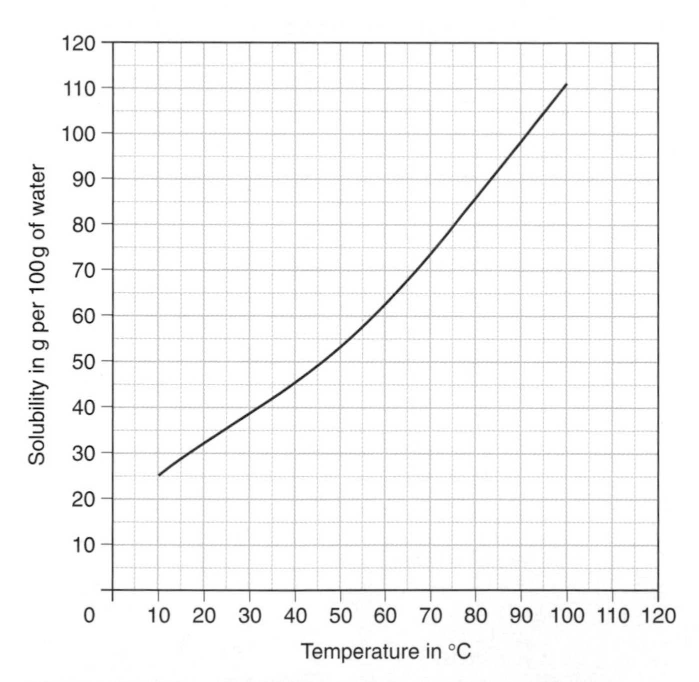

Figure 8 This solubility curve for copper sulfate clearly shows that as the temperature of the water increases, more copper sulfate can be dissolved in it

Crystallisation

When a solution cools, particularly a saturated solution, the forces of attraction between the solute particles may be strong enough to pull them together to form a regular structure. This process is called **crystallisation**.

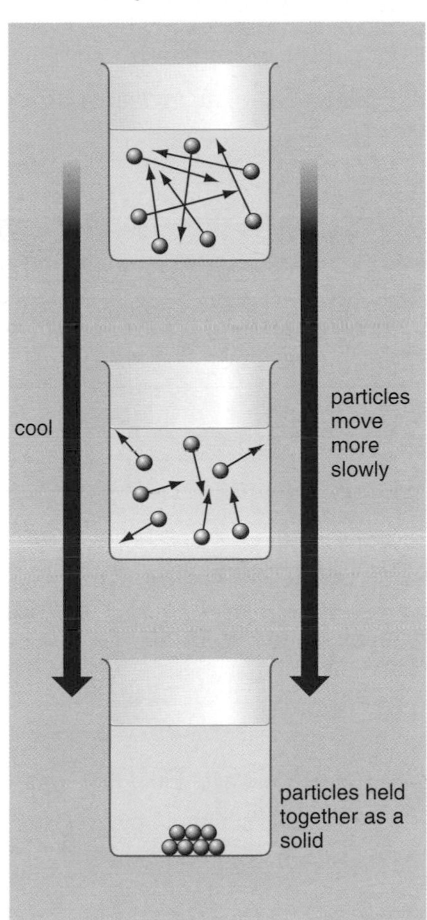

Figure 9 Crystallisation

If the solution and crystals are warmed, the process is reversed and the crystals will dissolve again.

Key terms

Check that you understand and can explain the following terms:

- ★ dissolve
- ★ solution
- ★ mixture
- ★ solute
- ★ solvent
- ★ soluble
- ★ insoluble
- ★ filter
- ★ evaporation
- ★ distillation
- ★ chromatography
- ★ saturated solution
- ★ solubility
- ★ crystallisation

Questions

1. How could you obtain salt from seawater?
2. How could you obtain pure water from seawater?
3. How could you obtain pure salt from rock salt?
4. Explain using the particle theory what happens when **a)** a solid dissolves in a liquid, and **b)** crystallisation takes place in a solution.
5. Sketch a graph to show how the solubility of a substance changes with temperature.

Chapter 7 The particle model:

What you need to know

1. Solids, liquids and gases have different physical properties, for example shape, strength and ability to flow.
2. These different properties can be explained by the arrangement and movement of their particles.
3. When a substance melts or boils, the arrangement and movement of its particles change.
4. When a solute is dissolved in a solvent, a solution is produced.
5. Solutions are mixtures.
6. The ingredients of a solution may be separated using techniques such as filtering, evaporating, distilling and chromatography.
7. The solubility of a solute increases with temperature.

How much do you know?

1 This question is about the different properties of solids, liquids and gases.

- ★ Iron is a solid at room temperature
- ★ Water is a liquid at room temperature
- ★ Oxygen is a gas at room temperature

a) In which substance are the particles completely free to move around at room temperature?

____

1 mark

b) In which substance are the particles arranged in a regular pattern at room temperature?

____

1 mark

c) In which substance are the forces of attraction between the particles strongest at room temperature?

____

1 mark

d) Which substance may condense if it is cooled down?

____

1 mark

e) Which substance may melt if its temperature is increased?

____

1 mark

f) Which two substances can be poured?

____

2 marks

2 This question is about the melting points and boiling points of five substances.

Substance	Melting point °C	Boiling point °C
A	0	100
B	220	78
C	1000	2000

Using the information in the table above answer the following questions.

a) Is substance A a solid, liquid or gas when it is at a temperature of 50 °C?

____

1 mark

b) Is substance B a solid, liquid or gas when it is at a temperature of 10 °C?

____

1 mark

c) Is substance C a solid, liquid or gas when it is at a temperature of 50 °C?

____

1 mark

d) Is substance D a solid, liquid or gas when it is at a temperature of 50 °C?

____

1 mark

e) Is substance E a solid, liquid or gas when it is at a temperature of 50 °C?

____

1 mark

f) Which two substances have a regular structure at 10 °C?

____

2 marks

How much do you know? *continued*

3 John adds a teaspoon of coffee to some hot water and stirs it until it has dissolved. Describe how John could now separate the coffee from the water.

___

___

2 marks

4 A mixture of sand and salt can be separated by adding water and then filtering.

a) Where is the sand after the mixture has been filtered?

___

1 mark

b) Where is the salt after it has been filtered?

___

1 mark

c) Explain why it would not be possible to separate a mixture of salt and sugar by adding them to water and then filtering.

___

___

2 marks

5 The diagram below shows the apparatus that is used to separate two liquids such as water and alcohol.

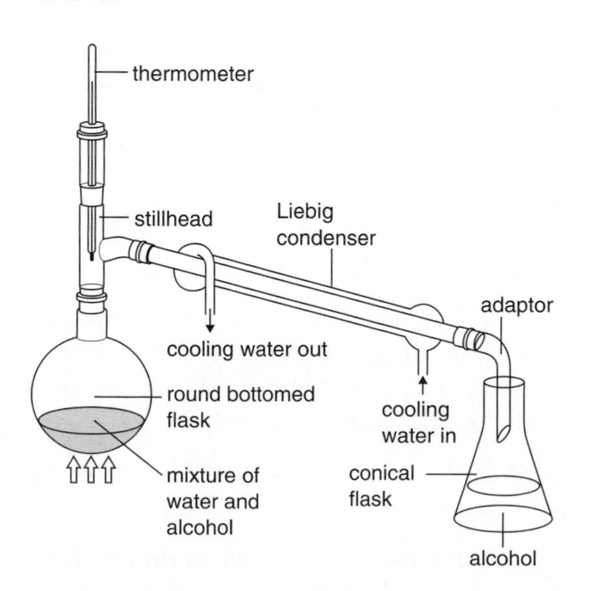

Water has a boiling point of 100 °C. Alcohol has a boiling point of 78 °C.

a) What happens to the alcohol when the mixture is heated to a temperature of 78 °C?

___

1 mark

b) What happens to any alcohol vapour which passes through the condenser?

___

1 mark

6 A forensic scientist is trying to discover which pen was used to write a false cheque. Using chromatography he was able to compare the ink from the cheque with that from three pens. The results are shown below.

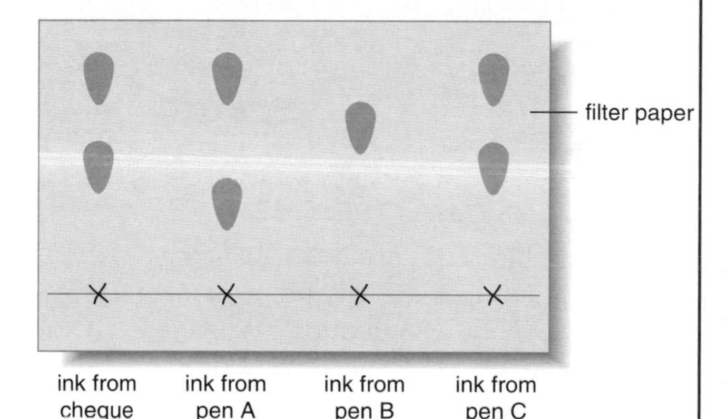

a) Which pen was used to write the cheque?

___

1 mark

b) Explain how the scientist knew this.

___

___

2 marks

8.1 Acids and alkalis

Acids and **alkalis** are two important groups of chemicals.

Acids

All the objects in Figure 1 contain acid.

Figure 1

In everyday life there are acids all around us. They are found not only in chemical laboratories and factories, but also in our homes and even inside our bodies. The sharp taste of a fruit, such as an orange or lemon, is caused by citric acid. The taste of sour milk is caused by lactic acid. In fact the word acid comes from the Latin word acidus which means sour. A car battery contains sulfuric acid. Our stomachs contain hydrochloric acid to help us digest our food.

Alkalis

All the objects in Figure 2 contain alkalis.

Figure 2

Alkalis are the chemical opposites of acids. They are **antacids** and can be used to **neutralise** the effect of acids. For example, if you ate too many green apples, which contain lots of malic acid, you might get indigestion or stomach ache. This happens because there is too much acid in your stomach. To neutralise the effects of the excess acid and stop the pain you could take some indigestion tablets or stomach powders which contain antacids.

The sting of a wasp is **alkaline** so the pain it causes can be neutralised by vinegar which is **acidic**. The sting of a bee, however, is acidic so the pain it causes can be neutralised using an antacid such as baking soda.

Figure 3 Pollution causes acid rain, which has made some lakes and rives acidic, an antacid (lime) is added to the water to neutralise it

Working safely with acids and alkalis

Handling chemicals, like some acids and alkalis, can be dangerous and should be done with great care. Hazard warning labels indicate the type of danger the chemicals may pose.

REMEMBER Goggles should always be worn when working with acids or alkalis.

Adding water to an acid or alkali dilutes it and makes it less hazardous.

Figure 4 Hazard warning labels

How can we tell an acid from an alkali?

Some **dyes** are different colours in acid or alkaline solution. **Litmus** is a good example of this. In an acid solution litmus is red, but in an alkaline solution it is blue. A dye which changes colour in this way and can be used to tell an acid from an alkali is called an **indicator**.

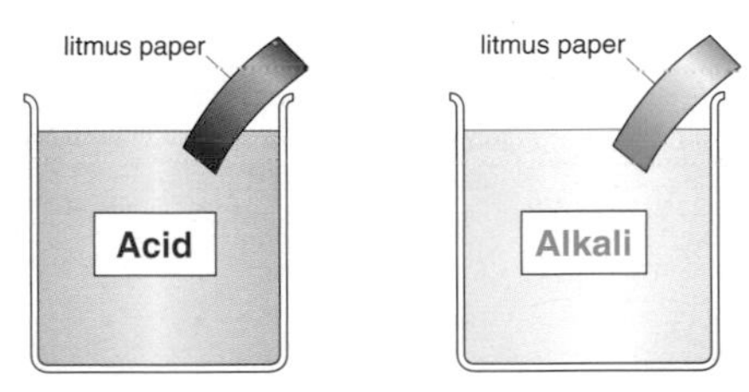

Figure 5 Using litmus to distinguish between an acid and an alkali

The table below shows the colours of some of the more common indicators in acid and alkaline solutions.

Indicator	Acid	Neutral	Alkaline
red litmus	red	red	blue
blue litmus	red	blue	blue
methyl orange	red	yellow	yellow
phenolphthalein	colourless	colourless	pink

One of the most useful indicators is **universal indicator**. This is a mixture of dyes and can produce a range of colours. Its colour indicates how acidic or how alkaline a solution is.

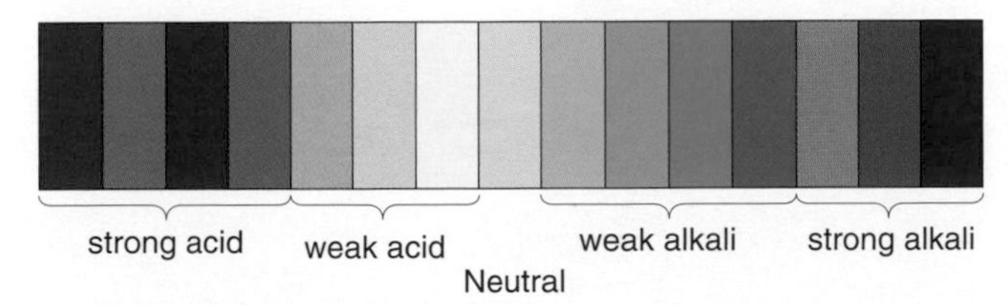

Figure 6 The different colours of universal indicator

The pH scale

Scientists also use a numbered scale, known as the **pH scale**, to describe the acidity or alkalinity of solutions.

★ A neutral solution has a pH value of 7.

★ Solutions which are acidic have a pH value of less than 7.

★ Solutions which are alkaline have a pH value of more than 7 (up to a maximum of pH 14).

litmus
strong acid weak acid neutral weak alkali strong alkali
universal indicator

Figure 7 The pH scale

Key terms

Check that you understand and can explain the following terms:

★ acid
★ alkali
★ antacid
★ neutralise
★ alkaline
★ acidic
★ dye
★ litmus
★ indicator
★ universal indicator
★ pH scale

Questions

1 Give two examples of everyday materials which **a)** contain acid and **b)** contain alkali.

2 Explain why indigestion powder eases the pain you might experience if you eat too many unripe, green apples.

3 Explain why the discomfort caused by a wasp sting can be eased by rubbing vinegar onto the sting.

4 Explain how we could distinguish an acid from an alkali.

5 All acids have a sour taste. Why is it a very bad idea to use taste as a way of testing to see if a substance is an acid?

6 Match the descriptions of the solutions in the first column with their pH values.

Neutral solution	pH 6
Slightly alkaline	pH 1
Very acidic	pH 7
Slightly acidic	pH 14
Very alkaline	pH 8

7 What is the pH value of a solution which causes universal indicator to turn **a)** red, **b)** yellow and **c)** blue?

8.2 Simple chemical reactions

Making new substances

Chemical reactions are taking place all around us. Chemical reactions within the cells of your body keep you alive. Chemical reactions take place when you cook food. Chemical reactions take place when you drive a car or light a fire.

Sometimes these reactions are rapid and spectacular like those in Figure 1, but often they take place much more slowly like the rusting of a car or the rotting of a piece of fruit. A chemical reaction takes place when substances come together to form new substances. The substances that are reacting are called the **reactants**. The substances that are produced by the chemical reaction are called the **products**.

We can summarise what is happening in a reaction by using a word equation:

Reactants → Products

There are several ways in which we might spot that a chemical reaction is taking place:

★ The reactants may look totally different from the products.

★ Bubbles in a liquid may indicate that a gas is being produced by the reaction.

★ One of the products might have a 'new' smell.

★ Energy may be released during the reaction and the substances may become warmer, or may glow. In some reactions energy is taken in and the products feel colder.

★ With vigorous reactions there may even be fizzing or popping noises.

Testing for some of the products of a chemical reaction

Testing for carbon dioxide

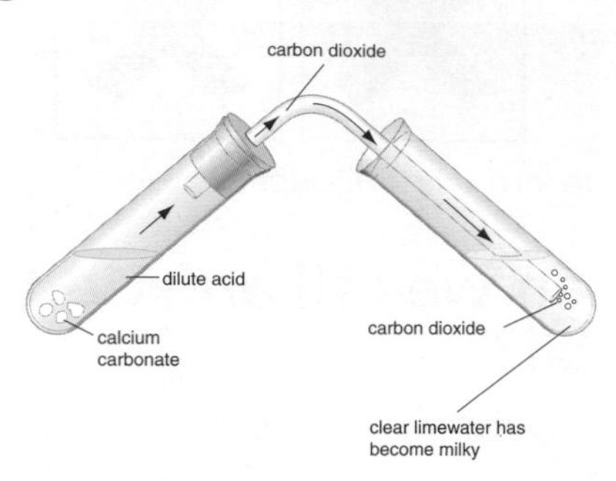

Figure 1 Limewater turns milky if carbon dioxide is bubbled through it for a few minutes

Testing for hydrogen

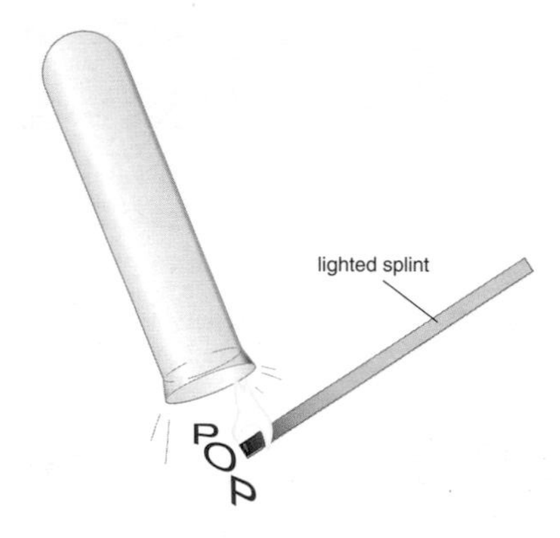

Figure 2 Hydrogen gives off a squeaky pop with a lighted splint

Testing for oxygen

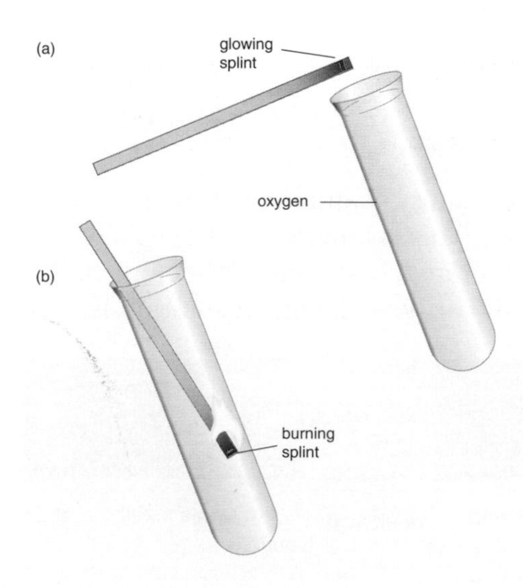

Figure 3 A glowing splint will relight when placed in a test tube filled with oxygen

Chemical reactions between acids and reactive metals

Acids can react with both metals and non-metals. The reaction between an acid and a metal can be described by the equation:

acid + metal → salt + hydrogen

★ Salts made from hydro**chlori**c acid are called **chlori**des.

★ Salts made from **nitr**ic acid are called **nitr**ates.

★ Salts made from **sulf**uric acid are called **sulf**ates.

Most acids will not react with less reactive metals such as copper, gold or silver.

Examples

hydrochloric acid (acid) + magnesium (metal) → magnesium chloride (salt) + hydrogen

sulfuric acid (acid) + calcium (metal) → calcium sulfate (salt) + hydrogen

nitric acid (acid) + potassium (metal) → potassium nitrate (salt) + hydrogen

Magnesium chloride, calcium sulfate and potassium nitrate are all examples of **salts**.

Chemical reactions between acids and carbonates

The reaction between an acid and a carbonate can be described by the equation:

acid + carbonate → salt + carbon dioxide + water

Examples

hydrochloric acid + magnesium carbonate → magnesium chloride (salt) + carbon dioxide + water

sulfuric acid + calcium carbonate → calcium sulfate (salt) + carbon dioxide + water

nitric acid + potassium carbonate → potassium nitrate (salt) + carbon dioxide + water

Chemical reactions between acids and alkalis

The reaction between an acid and an alkali can be described by the equation:

acid + alkali → salt + water

Examples

hydrochloric acid (acid)	+	sodium hydroxide (alkali)	→	sodium chloride (salt) + water
nitric acid (acid)	+	potassium hydroxide (alkali)	→	potassium nitrate (salt) + water
sulfuric acid (acid)	+	calcium hydroxide (alkali)	→	calcium sulfate (salt) + water

In all the above examples we know that a chemical reaction has taken place because new materials have been produced.

Burning fuels

Fuels such as coal, oil, gas and wood are very useful sources of energy. We burn them in order to release the energy they contain. A chemical reaction which releases energy is called an **exothermic reaction**. The energy released by the burning of fuels can be used to produce electricity, cook food, provide warmth, or produce light. **Combustion** is another word for burning.

When fuels burn they react with the oxygen in the air to produce oxides.

Fossil fuels are rich in compounds containing the element carbon. Therefore when fossil fuels are burnt, they react with the oxygen in the air to produce the gas carbon dioxide. These reactions can be described by the word equation:

fossil fuel + oxygen → carbon dioxide + water

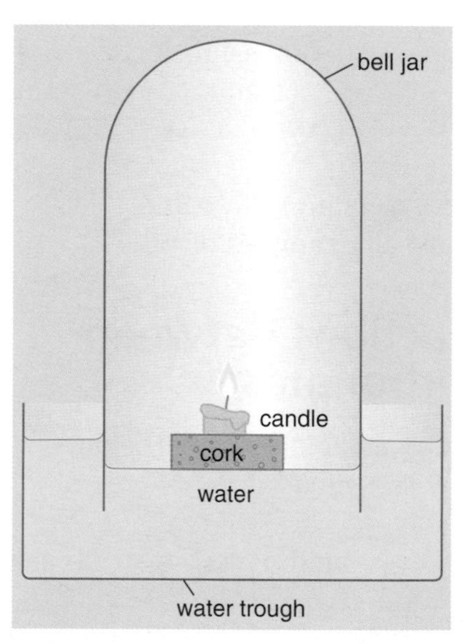

Figure 4 As the wax in this candle burns it uses up the oxygen in the air so the water level in the jar rises. When all the oxygen has been used up, the candle goes out

Although there are many advantages to using fossil fuels to provide the energy we need, there are also several disadvantages:

1. When fuels are burned they increase the amount of carbon dioxide in the atmosphere. Carbon dioxide is one of the gases which causes **global warming**.
2. Other gases, such as sulfur dioxide and nitrogen oxides, are sometimes released into the atmosphere from impurities in the fossil fuels. These gases can cause **acid rain**.

Acid rain

Metals, rocks, buildings and statues often suffer from the **corrosive** effects of acid rain. This is formed when gases such as carbon dioxide, sulfur dioxide and nitrogen dioxide dissolve in rain water. These gases are released into the atmosphere by cars, lorries, factories and power stations.

As acid rain runs down the hills it can affect the chemistry of the soil, killing trees and plants. When the acid rain reaches ponds and lakes it makes them acidic, killing many of the creatures that live in the water. Acid rain can also eat away the surfaces of statues and buildings.

Key terms

Check that you understand and can explain the following terms:

- ★ chemical reaction
- ★ reactant
- ★ product
- ★ salts
- ★ fuel
- ★ exothermic reaction
- ★ combustion
- ★ global warming
- ★ acid rain
- ★ corrosive

Questions

1. What is always produced in a chemical reaction?
2. Describe three ways in which you could detect that a chemical reaction is taking place.
3. What is always produced when an acid reacts with a reactive metal such as calcium? How would you test for this product?
4. Write a word equation which describes what happens when an acid reacts with a carbonate.
5. Describe a chemical reaction in which you could produce an oxide.
6. Describe how you would test a gas to see if it was **a)** carbon dioxide and **b)** oxygen.
7. Give one example of a fast spectacular chemical reaction.
8. Give one example of a slow chemical reaction.

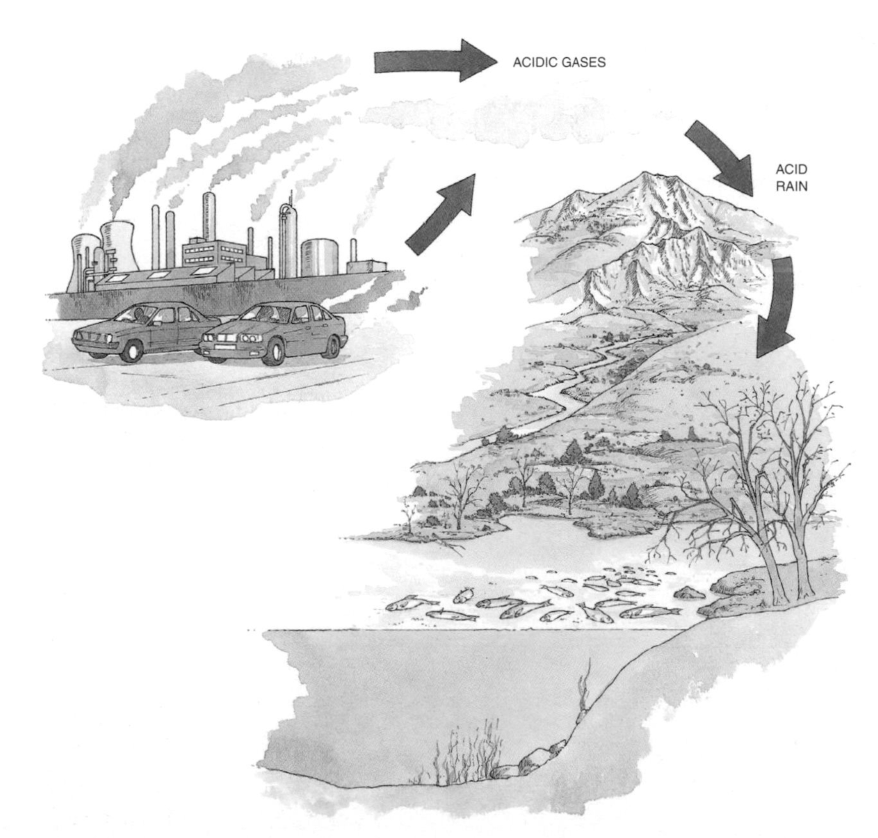

Figure 5 Formation and effects of acid rain

Chapter 8 Acids, alkalis and simple chemical reactions:

What you need to know

1. Acids and alkalis are two classes of chemicals. They can be identified from their chemical properties.
2. Acids have a pH of less than 7. Alkalis have a pH greater than 7. Neutral solutions have a pH of 7.
3. Indicators can be used to determine if a solution is acid, alkaline or neutral.
4. Acids and alkalis can be hazardous and so must be handled with care.
5. When chemical reactions take place new substances are formed.
6. Indications that a chemical reaction is taking place include a change in colour, a change in temperature and bubbles, i.e. a gas being produced.
7. When an acid reacts with a metal, a salt and hydrogen gas are produced.
8. When an acid reacts with a metal carbonate, a salt and carbon dioxide gas are produced.
9. When substances burn in oxygen (air), oxides are produced.
10. We burn fuels to release the energy they contain as heat.

How much do you know?

1 The chart below shows the pH of several household substances.

Substance	pH
Vinegar	2.9
Fresh milk	6.8
Oven spray	12.5
Brass cleaver	9.5
Kitchen cleaner	11.0

a) Which substance is very alkaline?

____

1 mark

b) Which substance is very acidic?

____

1 mark

c) Which substance is almost neutral?

____

1 mark

d) A wasp sting is slightly alkaline. Which of the substances in the chart could best be used to neutralise a wasp sting?

____

1 mark

2 The table below shows the colours of two indicators in solutions of different acidity.

pH	1	5	7	9	14
Methyl orange	red		orange		yellow
Litmus	red		purple		blue

a) What colour is litmus when added to a neutral solution?

____

1 mark

b) What colour is methyl orange when added to an acidic solution?

____

1 mark

c) To what type of solution must litmus be added for it to turn blue?

____

1 mark

d) What is the pH value of pure distilled water?

____

1 mark

How much do you know? *continued*

3 New materials are often made by heating one or more substances so that a chemical reaction takes place.

Tick the three boxes which describe a chemical change.

- ☐ Heating wax to make it melt.
- ☐ Heating the ingredients of a cake in an oven.
- ☐ Boiling a kettle filled with water.
- ☐ Grilling some bread to make toast.
- ☐ Burning gas on a stove.

3 marks

4 Bill collects samples of three different gases in six test tubes. Test tubes A and B contain oxygen, test tubes C and D contain carbon dioxide, and test tubes E and F contain hydrogen.

Bill then:

(i) adds a few drops of limewater to test tubes A, C and E

(ii) inserts a lighted splint into test tubes B, D and F.

Fill in the table below showing what happens in each case. If nothing happens write 'nothing happens'.

Test tube	A	B	C	D	E	F
Adding limewater						
Inserting lighted splint						

6 marks

5 The table below shows the products at the end of several different chemical reactions. Fill in the missing substances.

Reactant A	Reactant B	Product A	Product B
acid	alkali	salt	
acid		salt	hydrogen
acid		salt + water	carbon dioxide
	oxygen	water	carbon dioxide

4 marks

6 a) Explain how the gases released by cars and factories create acid rain.

____

____

____

____

3 marks

b) Describe the effects acid rain has on the environment.

____

____

2 marks

c) How can the effects acid rain on the environment be reduced?

____

____

____

2 marks

9.1 Atoms, elements and compounds

Figure 1 How many different materials can you see in this photograph?

Elements

Figure 1 shows lots of different objects made from a wide variety of materials, such as metals, non-metals, plastics, fabrics, wood and paper. Although these materials seem very different from the outside, they are actually all made from the same few building blocks, or **elements**. Elements are composed of tiny particles called **atoms**.

An element is made from one sort of atom. It is a substance that cannot be split into anything simpler.

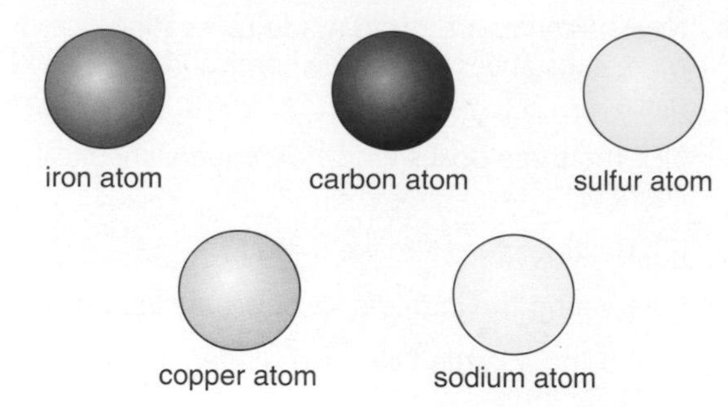

Figure 2 Models of several different elements

Other materials, such as glass, contain more than one type of atom joined together. We call glass a **compound**.

Altogether there are approximately 100 different elements. These can be grouped together in a special table called the **Periodic Table**.

Different elements have very different properties. They are arranged in the Periodic Table in such a way that elements which have similar properties are in the same part of the table.

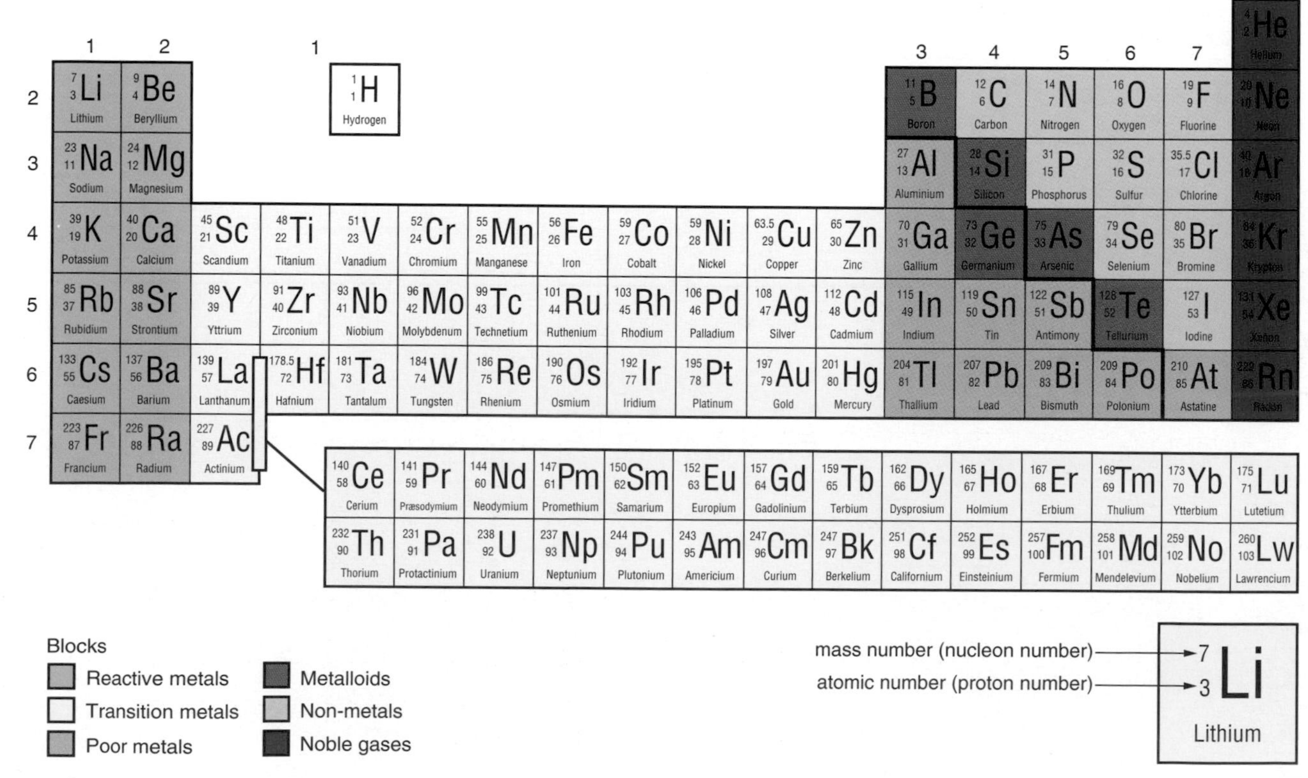

Figure 3 The Periodic Table

For example, the elements on the far right of the table – neon, argon, krypton, xenon and radon – are all gases and are very unreactive. The elements on the far left – including lithium, sodium and potassium – are the alkali metals and are very reactive.

Observations about the elements in the Periodic Table:

- ★ There are more metals than non-metals.
- ★ Most metals are non-magnetic.
- ★ Very few elements are liquids at room temperature.

The table lists some of the more common elements and their **chemical symbols**. You don't have to memorise them – the symbols are just a kind of shorthand for the names of the elements. Some of the symbols come from the first letter of the element, like H for hydrogen and S for sulfur. Other symbols may come from the Latin name for the element, like Fe for iron, which comes from the word ferrum. The first letter of a symbol is always a capital letter. If there is a second letter, this is always lower case.

Element	Symbol	Element	Symbol
hydrogen	H	sulfur	S
helium	He	chlorine	Cl
lithium	Li	calcium	Ca
carbon	C	iron	Fe
nitrogen	N	copper	Cu
oxygen	O	zinc	Zn
sodium	Na	silver	Ag
magnesium	Mg	lead	Pb

Compounds

Compounds are formed when atoms combine or join together. The smallest part of a compound that can exist is called a **molecule**. We can use chemical symbols to create a **chemical formula** for a compound.

Key terms

Check that you understand and can explain the following terms:

- ★ element
- ★ atom
- ★ compound
- ★ Periodic Table
- ★ chemical symbol
- ★ molecule
- ★ chemical formula

Examples

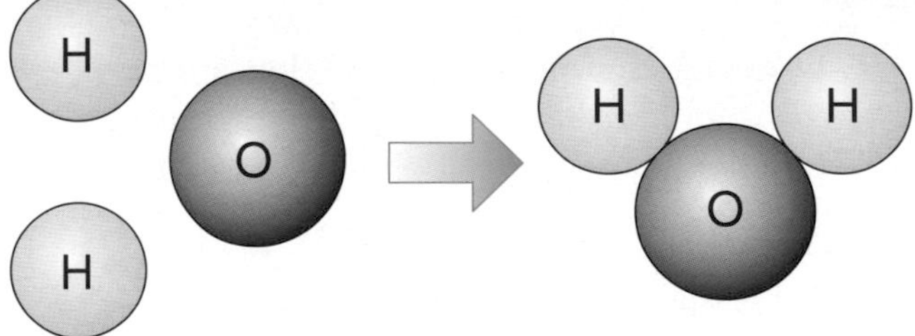

Figure 4 Two atoms of the element hydrogen join with one atom of oxygen to produce one molecule of the compound we call water

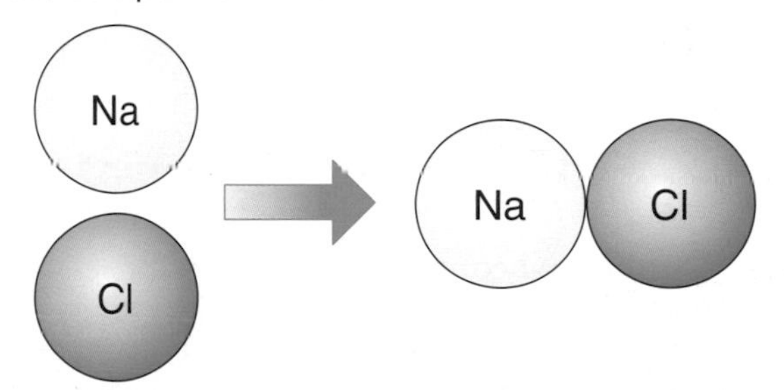

Figure 5 One atom of the element sodium joins with one atom of chlorine to produce one molecule of the compound sodium chloride (salt)

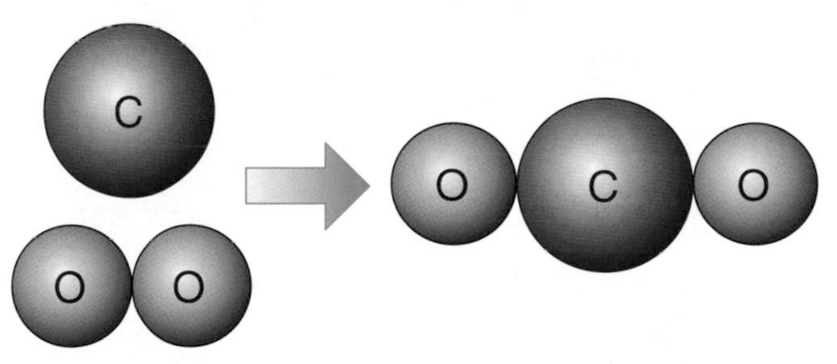

Figure 6 One atom of the element carbon joins with two atoms of the element oxygen to produce one molecule of the compound carbon dioxide

Questions

1. Explain the difference between an element and a compound.
2. Draw simple models and write down the chemical formulae for the following compounds:
   a) water, **b)** carbon dioxide, **c)** sodium chloride.
3. What are the chemical symbols for the following elements:
   a) iron, **b)** copper, **c)** lead, **d)** sodium, **e)** chlorine, **f)** silver, **g)** gold?
4. Where does the symbol for gold come from?
5. What are the names of the elements that have the following symbols:
   a) Sn, **b)** O, **c)** K, **d)** Al, **e)** Ca, **f)** C?

9.2 Compounds and mixtures

A **compound** contains elements that are **chemically combined**. A **mixture** contains substances that are not chemically combined.

Figure 1 This bowl contains the different ingredients needed to make a cake – including eggs, flour, margarine and sugar. These ingredients are not chemically combined to each other. It is a mixture

The air we breathe contains several different gases including nitrogen, oxygen and carbon dioxide. These gases are not chemically joined to each other. Air is a mixture of gases.

There are several ways in which we can tell the difference between a compound and a mixture:

1 In general it is much easier to separate the different substances in a mixture, than to separate the substances in a compound. Mixtures can be separated by filtration, evaporation, distillation or chromatography (see pages 71–72). Compounds cannot be separated by these methods. To separate the elements in a compound, you need much more complicated processes such as electrolysis.

2 The properties of mixtures tend to be similar to those of the substances which have been mixed together. The properties of a compound are often totally different from those of the substances from which it is made. Imagine the taste of the cake ingredients before cooking when they are not chemically combined and after cooking when chemical joining has taken place and compounds have formed.

An even more extreme example of these changes is seen when sodium and chlorine combine. Sodium is an extremely reactive metal. If it is placed in water it will react very violently. Any contact between sodium metal and skin is very dangerous. Chlorine is a yellowish, poisonous gas. But when sodium and chlorine join together, they form a solid, white, crystalline compound called sodium chloride. This compound is the common salt which we find dissolved in the sea and which we sprinkle on our food.

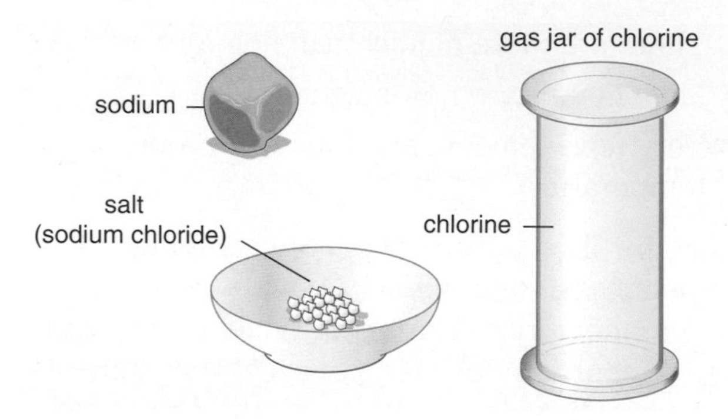

Figure 2 Sodium and chlorine react together to produce the compound sodium chloride

This joining together of substances to form a new compound is called a **chemical reaction**. This chemical change can be written as a **word equation**.

sodium + chlorine → sodium chloride (salt)

3 Compounds have a definite **composition**. Mixtures do not have a definite composition.

When you want to sweeten a cup of tea or coffee you add sugar. How much sugar you add depends upon your taste. You can do this because sugar dissolves in a liquid to form a **solution**. A solution is a kind of mixture. You can decide how much of each substance is present.

When substances react to produce compounds, the correct amount of each substance must join together. There is no choice.

When iron reacts with sulfur to produce the compound iron sulfide, there must be equal numbers of iron atoms and sulfur atoms.

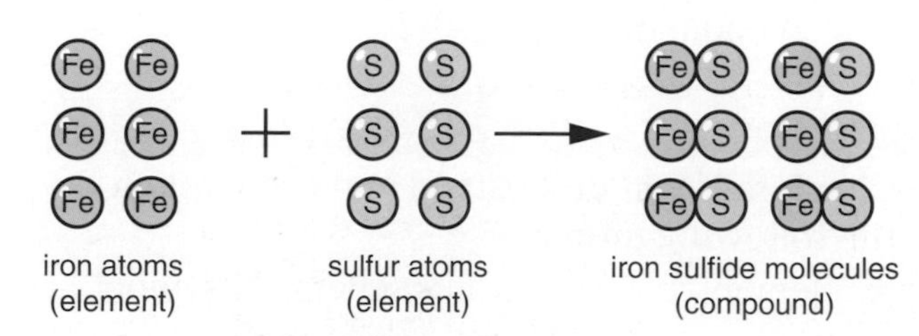

Figure 3 Equal numbers of iron and sulfur atoms combine to produce iron sulfide

When carbon reacts with oxygen to produce the compound carbon dioxide, two oxygen atoms combine with one carbon atom.

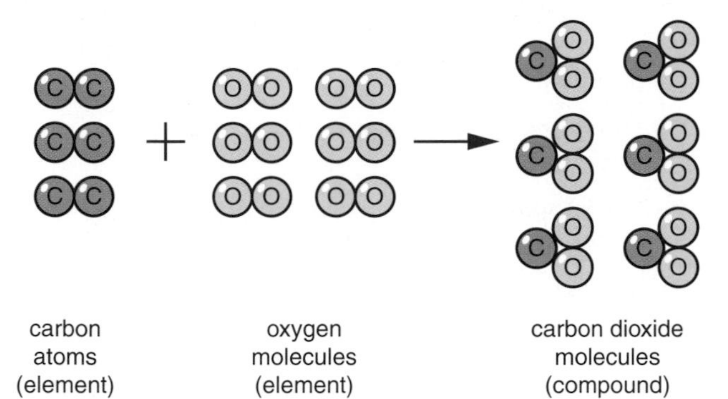

Figure 4 Carbon atoms join with twice as many oxygen atoms to produce carbon dioxide

Chemical formulae

Because compounds have a definite composition it is possible to describe them using a **chemical formula**. Some examples of compounds and their chemical formulae are shown in the table below.

Name	Composition	Chemical formula
sodium chloride	one sodium (Na) atom for each chlorine (Cl) atom	NaCl
iron sulfide	one iron (Fe) atom for each sulfur (S) atom	FeS
carbon dioxide	two oxygen (O) atoms for each carbon (C) atom	CO_2
water	two hydrogen (H) atoms for each oxygen (O) atom	H_2O
ammonia	three hydrogen (H) atoms for each nitrogen (N) atom	NH_3

Melting points and boiling points

Elements and compounds melt and boil at particular temperatures. For example we know that pure ice melts at 0 °C and pure water boils at 100 °C.

But this is not true for mixtures. We cannot say what the **freezing point** or **boiling point** of salt water is, as this depends upon how much salt is dissolved in the water. In other words, the **melting point** and boiling point of a mixture can be varied by changing the composition of the mixture.

Adding salt to pure water decreases the freezing point of the mixture. Increasing the amount of salt dissolved in the water lowers the freezing point even further.

Adding salt to pure water increases its boiling point. Increasing the amount of salt dissolved raises the boiling point of the mixture even further.

Key terms

Check that you understand and can explain the following terms:

- ★ compound
- ★ chemically combined
- ★ mixture
- ★ chemical reaction
- ★ word equation
- ★ composition
- ★ solution
- ★ chemical formula
- ★ freezing point
- ★ boiling point
- ★ melting point

Questions

1 Explain the difference between a mixture and a compound. Give one example of each.

2 Draw model diagrams to show how the compound iron sulfide is produced by a chemical reaction between iron atoms and sulfur atoms.

3 Complete the table below.

Name	Composition	Chemical formula
magnesium chloride		$MgCl_2$
calcium oxide	one calcium (Ca) atom for each oxygen (O) atom	
carbon monoxide	one oxygen (O) atom for each carbon (C) atom	
sulfur trioxide	three sulfur (S) atoms for each oxygen (O) atom	
aluminium chloride		$AlCl_3$

4 What practical use can you think of for the fact that adding salt to ice lowers its freezing point? (*Hint*: Think about snowy roads in winter.)

Chapter 9 Atoms, elements, compounds and mixtures:

What you need to know

1. The smallest particle of an element is an atom of that element.
2. An element is made from one sort of atom. There are only a small number of elements.
3. A compound is made from more than one type of atom. There are a very large number of compounds, with a huge range of properties.
4. Elements and compounds can be written using symbols and formulae.
5. A compound is created when elements combine together chemically.
6. A mixture contains elements or compounds which are not joined together chemically.
7. Elements have a definite composition, mixtures do not.
8. The ingredients of a compound cannot be easily separated.
9. The properties of a compound are often totally different from those substances from which it has been made.
10. Elements and compounds have definite melting points and boiling points, mixtures do not.

How much do you know?

1 Look carefully at the formulae below. Which of them represent elements and which of them represent compounds?

a) C ________________ *1 mark*

b) O_2 ________________ *1 mark*

c) CO_2 ________________ *1 mark*

d) S_8 ________________ *1 mark*

e) NO ________________ *1 mark*

2 Look carefully at the list below. Indicate for each substance whether it is a mixture or a compound.

a) Cup of coffee ________________ *1 mark*

b) Cup of pure water ________________ *1 mark*

c) A baked cake ________________ *1 mark*

d) Air ________________ *1 mark*

e) Common salt ________________ *1 mark*

3 The diagrams show atoms of elements which have combined chemically to produce compounds. Write the chemical formula for each of the compounds.

a) ________________ *1 mark*

Cl Ca Cl

b) ________________ *1 mark*

Cu O

c) ________________ *1 mark*

Fe S

d) ________________ *1 mark*

H H
O

How much do you know? *continued*

e) ________________

1 mark

Cl Ca Cl

4 Use the Periodic Table on page 84 to answer the following questions.

a) Write down the name of an element that is a metal.

____

1 mark

b) Write down the name of two elements that are gases at room temperature.

____

1 mark

c) Write down the name of an element that is a liquid at room temperature.

____

1 mark

d) Write down the names of the two elements that combine chemically to make water.

____

1 mark

e) Write down the name of an element that might be used to make expensive jewellery.

____

1 mark

5 a) What happens to the freezing point of water if a small amount of salt is added to it?

____

1 mark

b) What happens if a lot of salt is added to the water?

____

1 mark

c) Give one practical use for the effect you have described in part a)

____

1 mark

6 a) Describe the reaction between sodium metal and chlorine gas as a word equation.

____

1 mark

b) Give one reason why you know a chemical reaction has taken place between the sodium and the chlorine.

____

1 mark

10.1 Rocks and weathering

Rocks are usually made up of a mixture of mineral **grains**.

The chemical composition of these minerals, the sizes of the grains and how closely the grains are packed together, determine the properties of the different rocks. For example, the grains of sandstone are not packed closely together so air and water can pass between them. As a result sandstone is a **porous** rock. The grains in granite are packed very closely together. Air and water cannot pass through this structure so granite is described as a **non-porous** rock.

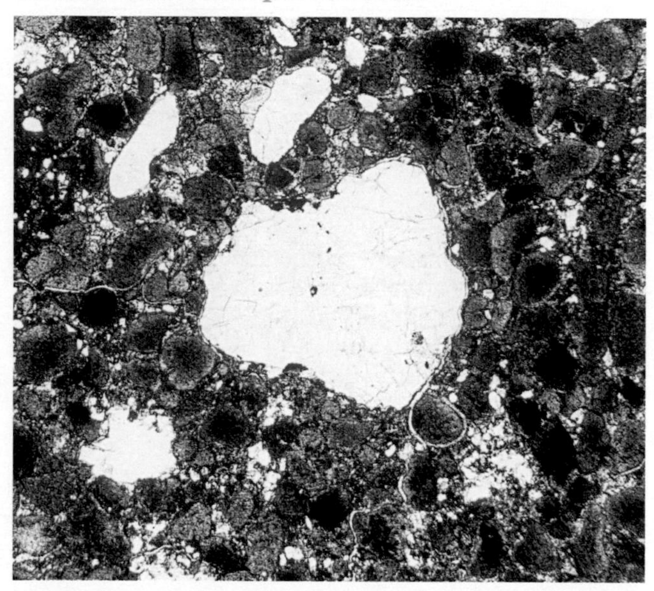

Figure 1 You can see that the grains in this piece of sandstone are large and loosely packed

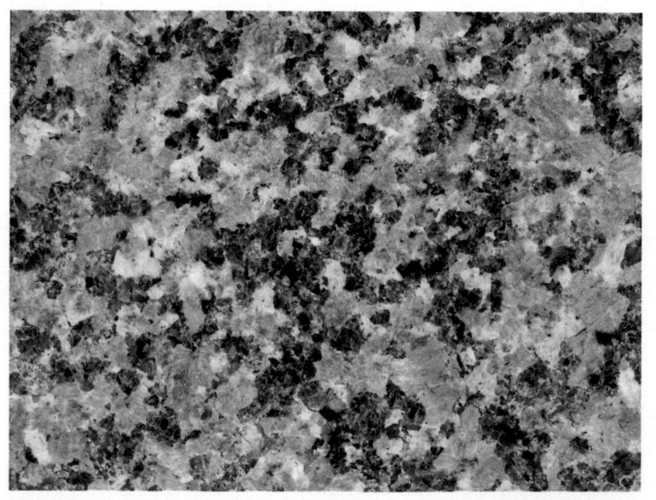

Figure 2 The grains in this sample of granite are much more tightly packed

Rocks have many important uses. They can be used for the walls and roofs of houses, for the floors and pillars of large buildings, and in the manufacture of plaster and cement. The different properties of the various kinds of rock will determine which is used for which purpose.

Rock	Description	Use
granite	very hard	building stone
sandstone	very hard	building stone
limestone	light colour	making cement
marble	hard, smooth	statues, floors
slate	hard, waterproof	roofing tiles

Weathering

Rocks at the surface of the Earth are continually being changed and broken down into smaller pieces through exposure to the environment. This **disintegration** of the rocks is called **weathering**. There are two main types of weathering:

1 **physical weathering**
2 **chemical weathering**

1 Physical weathering

In physical weathering, rocks are broken down by the action of water (Figure 3) and wind (Figure 4). One of the main methods of physical weathering is freeze and thaw.

Figure 3 Rain and running water can wear away rock surfaces. This process is called **erosion**

Figure 4 The surfaces of these rocks have been worn away by wind erosion

Figure 5 shows the processes involved in freeze and thaw.

a) Water seeps into a crack in the rock.

b) The water then freezes and expands, forcing the pieces of rock apart and making the crack larger.

c) The ice thaws and water seeps further into the crack.

d) This process continues and over a period of time the rock will break into two or more smaller pieces.

Figure 6 Scree slopes are created from fragments of rock that have broken from cliff faces by freeze and thaw weathering

Heating and cooling

If a rock is heated and then cooled, the minerals from which it is made expand and contract. If the different minerals expand and contract by different amounts, or different parts of the rock are at different temperatures, large enough forces may be created to cause the rock to crack and break into smaller pieces.

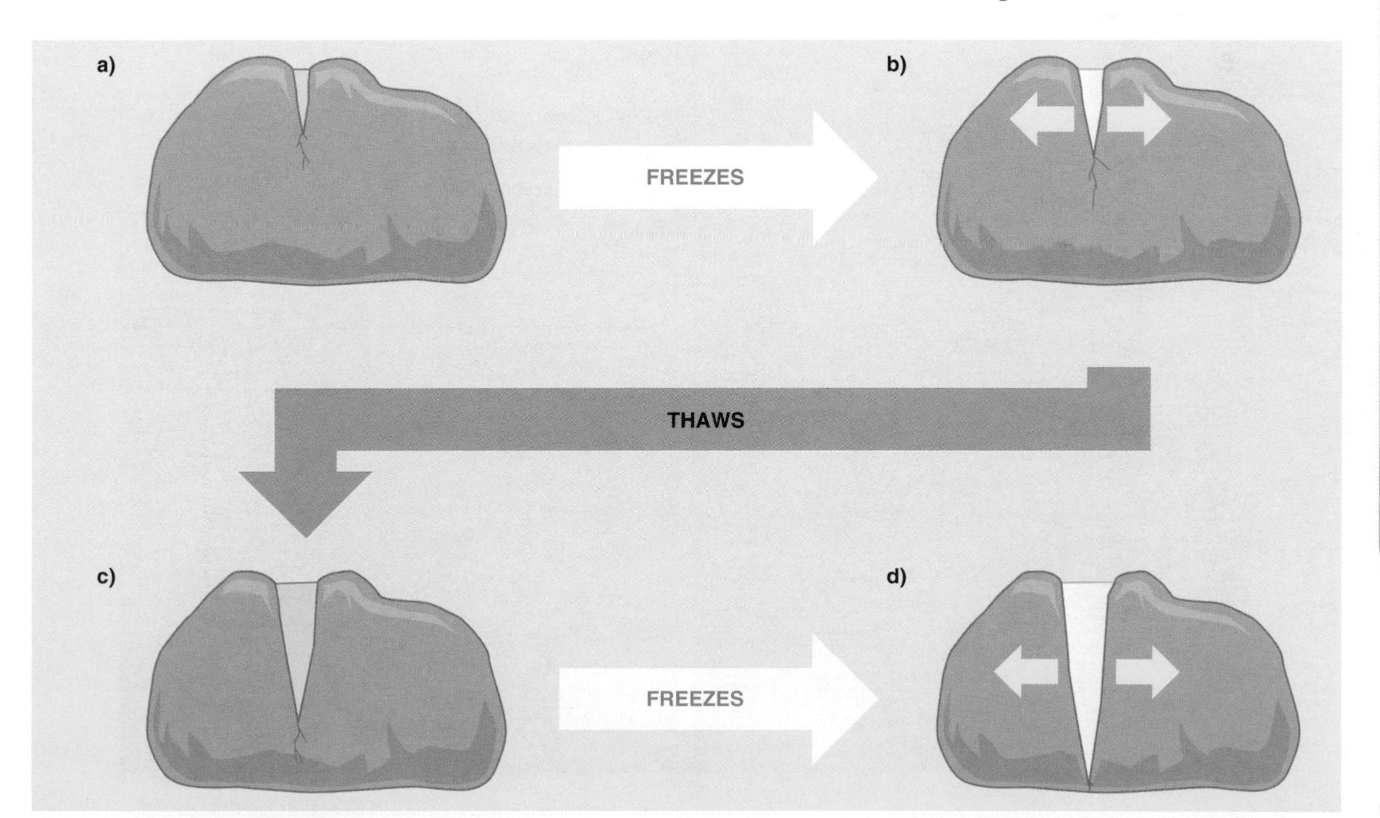

Figure 5 Freeze and thaw

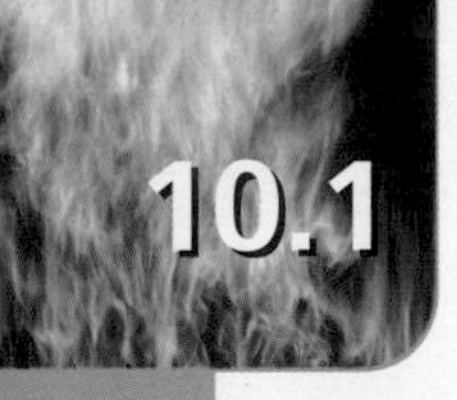

10.1 Rocks and weathering (*continued*)

2 Chemical weathering

As rainwater falls, some of the gases in the atmosphere become dissolved in it. Two of these gases, carbon dioxide and sulfur dioxide, make the water acidic. (See acid rain on pages 109–10). This **acid rain** reacts with the minerals in the rock, dissolving them and so chemically weathers the rock. Different types of rocks contain different minerals and so they are dissolved by the acid rain at different rates.

Limestone is very susceptible to attack from acid rain and over a long period of time will completely dissolve. However, granite contains several different minerals, some of which are quite resistant to attack from the acid rain.

What happens to weathered rocks?

Weathering gradually break rocks down into small particles. These smaller particles are then **transported** by rivers and winds and are called **sediments**. Eventually they come to rest and are **deposited** as layers of sediment. Smaller rock particles or grains are lighter and so are usually transported over greater distances.

During transportation, friction between the particles causes them to become more rounded. Larger particles are heavier and so tend to be transported over shorter distances and so have less rounded shapes.

Over a long period of time, other layers of particles settle on top of them. The lower layers become squashed and eventually form a new type of rock with a layered structure. This rock is called **sedimentary rock**. Limestone and sandstone are sedimentary rocks.

Some of the largest sediments found in these rocks are the fossilised remains of dead plants and animals that used to live in the sea. These **fossils** are very useful as they can be used to age a piece of rock.

Sedimentary layers can also be formed by evaporation of waters that contain dissolved salts.

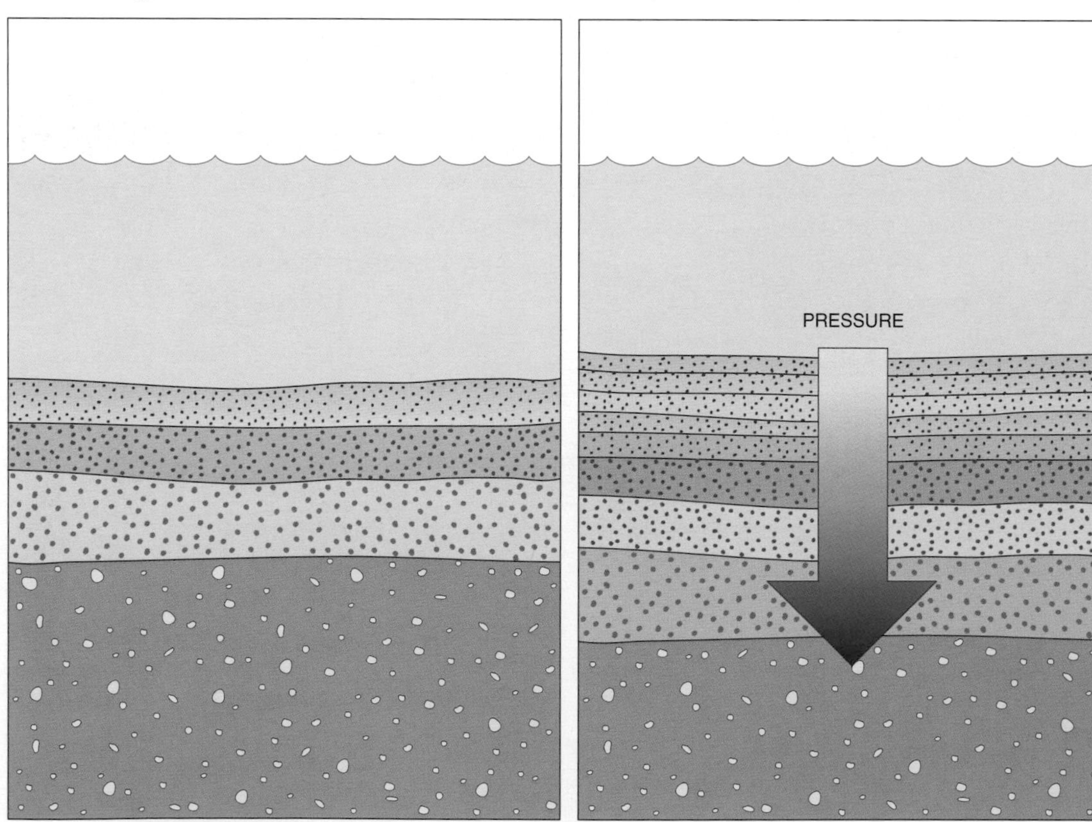

Figure 7 Formation of sedimentary rocks

Figure 8 The layers of sedimentary rock are clearly visible in this limestone quarry

Key terms

Check that you understand and can explain the following terms:

- ★ grain
- ★ porous
- ★ non-porous
- ★ disintegration
- ★ weathering
- ★ physical weathering
- ★ chemical weathering
- ★ erosion
- ★ acid rain
- ★ transported
- ★ sediment
- ★ deposit
- ★ sedimentary rock
- ★ fossil

Questions

1. Explain the structural difference between a porous rock and a non-porous rock.
2. Name three ways in which a rock may be physically weathered.
3. Explain how the presence of certain gases in the atmosphere leads to chemical weathering.
4. What is a sedimentary rock and how can the age of this rock be found?
5. Why are all the particles in one layer of a sedimentary rock approximately the same size?

10.2 Rocks and the rock cycle

There are three main groups of rocks.

★ Sedimentary
★ Igneous
★ Metamorphic

The different ways in which these rocks are formed affect their properties and therefore the purposes for which they are used.

Sedimentary rocks

Sedimentary rocks include sandstone, chalk and limestone. They have the following properties:

★ a crumbly texture
★ a layered structure
★ they are porous
★ they may contain fossils
★ they are mixtures with different compositions depending upon how, where and when they were formed
★ at least 50 % of limestone and chalk is calcium carbonate. We know this is present in the rocks because limestone and chalk react with acid to give carbon dioxide.

Figure 1 Limestone is a sedimentary rock

Igneous rocks

Igneous rocks include granite, basalt and obsidian. They are formed in the following way:

★ Under the surface of the Earth the temperature is so high that the rocks here have melted. This molten rock is called **magma**.
★ When magma escapes to the surface of the Earth, for example when there is a volcanic eruption, it is called **lava**.
★ When liquid rock cools and solidifies, igneous rocks are formed.
★ Igneous rocks are very hard and contain crystals.
★ If the magma solidifies and cools slowly, for example whilst it is still underground, the crystals have time to grow and are quite large. These are called **intrusive igneous rocks**. Granite is an example of an intrusive igneous rock.
★ If the magma cools rapidly, for example after reaching the Earth's surface, the crystals will be quite small. These are called **extrusive igneous rocks**. Obsidian and basalt are extrusive igneous rocks.
★ The strength and hardness of igneous rocks depends upon the size of their crystals.

Figure 2 Three igneous rocks – (a) granite, (b) basalt and (c) obsidian

Metamorphic rocks

Metamorphic rocks include marble and slate. They are formed in the following way:

★ Rocks which are deep in the ground, or are close to erupting volcanoes, may change in structure due to the high pressures and temperatures they experience. These new structures are called metamorphic rocks.

★ Metamorphic rocks are usually hard.
★ They tend to be less porous than sedimentary rocks, but may contain some fossils.
★ When limestone experiences high temperatures and pressures it changes into marble. Under similar conditions, mudstone will change into slate.

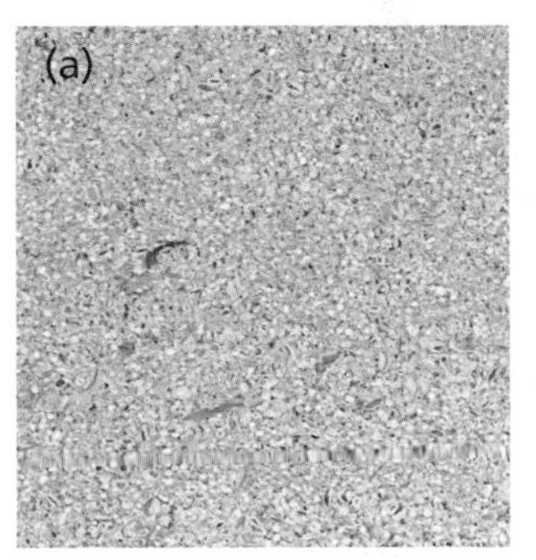

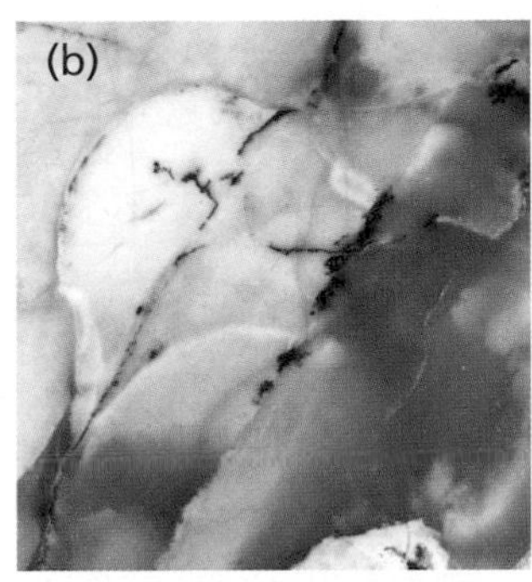

Figure 3 Under conditions of high temperature and pressure the sedimentary rock limestone (a) changes into the metamorphic rock marble (b)

The rock cycle

Within the crust of the Earth all three types of rock are being continuously created, destroyed and then recreated over millions of years. These geological changes are called the **rock cycle**.

Key terms

Check that you understand and can explain the following terms:

★ sedimentary rock
★ igneous rock
★ magma
★ lava
★ intrusive igneous rock
★ extrusive igneous rock
★ metamorphic rock
★ rock cycle

Questions

1. Describe the structure of a typical sedimentary rock such as limestone.
2. Explain under what conditions limestone may change into marble. What kind of rock is marble?
3. Explain the difference between magma and lava.
4. Explain, using the particle theory, why slow-cooling molten rock will form a solid containing large crystals.
5. Explain in your own words what you understand by the phrase 'the rock cycle'.

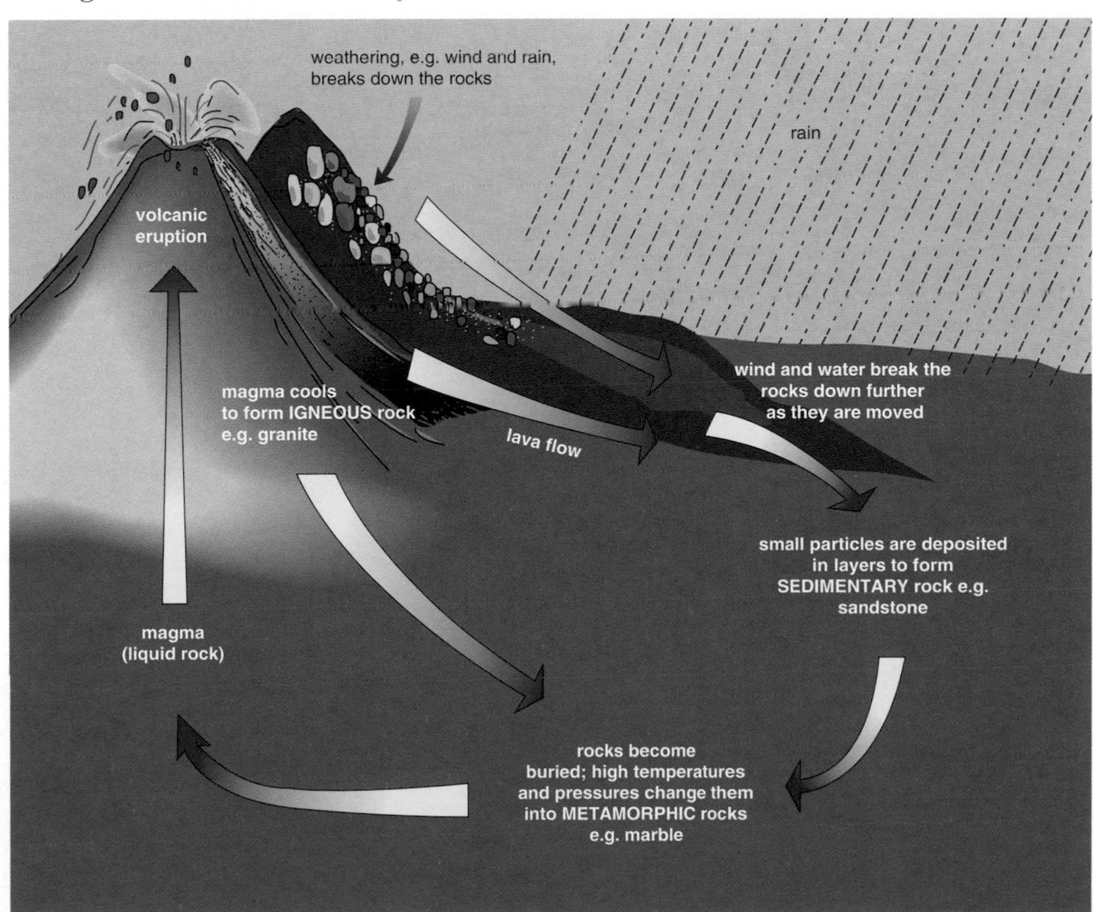

Figure 4 The rock cycle

Chapter 10 Rocks:

What you need to know

1. Rocks are usually made up of mixtures of mineral grains.
2. The sizes of the grains affect the properties of the rocks.
3. Rocks at the Earth's surface are broken down by physical and chemical weathering.
4. Weathered rock can be transported by wind and water. This is called erosion.
5. Sedimentary rock is formed from small particles of weathered rocks.
6. Metamorphic rocks are formed by the action of high temperatures and pressures on other types of rock.
7. Igneous rocks are formed from molten magma as it cools and solidifies.
8. The rock cycle describes how rocks are continuously being created and destroyed within the Earth's crust.

How much do you know?

1 Name three ways in which a rock might be physically broken down into very small particles by the weather.

____

____

____

3 marks

2 a) Explain what is meant by the term 'chemical weathering'.

____

____

2 marks

b) Name one rock which readily undergoes chemical weathering.

____

1 mark

3 Rocks can be classified into three groups – igneous rocks, sedimentary rocks and metamorphic rocks.

Look carefully at the information below and decide which group each of the four different rocks A, B, C and D belongs in.

Rock	Description	Group
A	Has a layered structure and contains pieces of shell	
B	Very hard, glassy, smooth appearance	
C	Has a crystalline structure but no layers can be seen	
D	Hard but is easily split into thin layers	

4 marks

4 a) The diagram below shows the rock cycle.

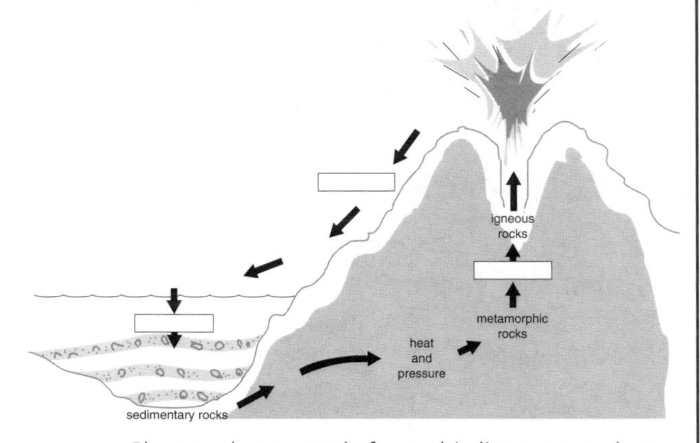

Choose three words from this list to complete the labelling of the diagram.

weathering deposition transport melting

3 marks

b) What two conditions are necessary to change a sedimentary rock into a metamorphic rock?

____

____

2 marks

5 a) Explain the difference between extrusive igneous rock and intrusive igneous rock.

____

____

2 marks

b) Explain how this difference affects the strength and hardness of the rock.

____

____

2 marks

How much do you know? *continued*

6 The diagram below shows the path followed by a river.

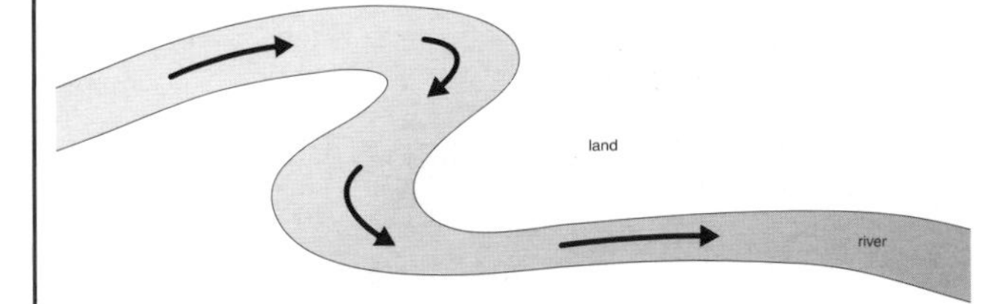

a) Mark on the diagram with an X a place where sediment being carried by the river is likely to be deposited.

1 mark

b) Explain why you have chosen this place.

____

____

2 marks

c) What kind of rock may eventually be formed here?

____

1 mark

d) Explain why the rock here may eventually contain fossils.

____

____

2 marks

7 Explain the meaning of the following words.

a) weathering

____

____

1 mark

b) transport

____

____

1 mark

c) deposit

____

____

1 mark

The diagram below shows physical weathering of a rock.

(a)

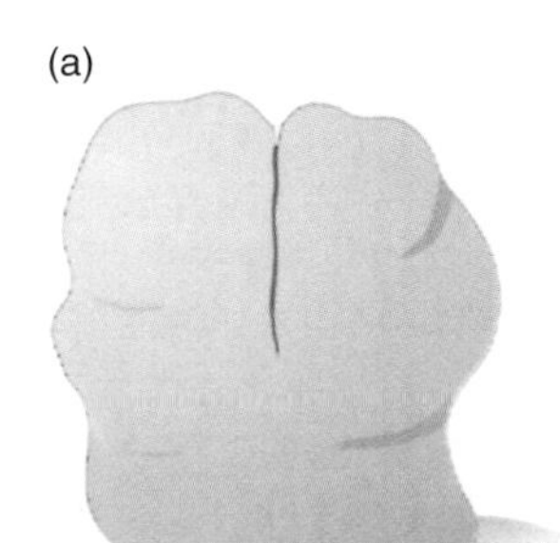

(b)

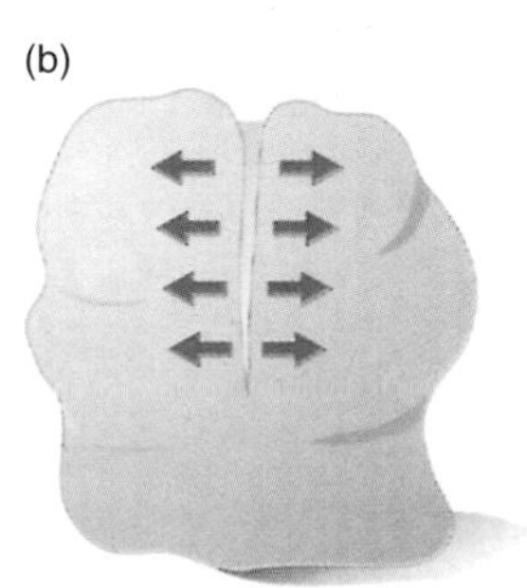

(c)

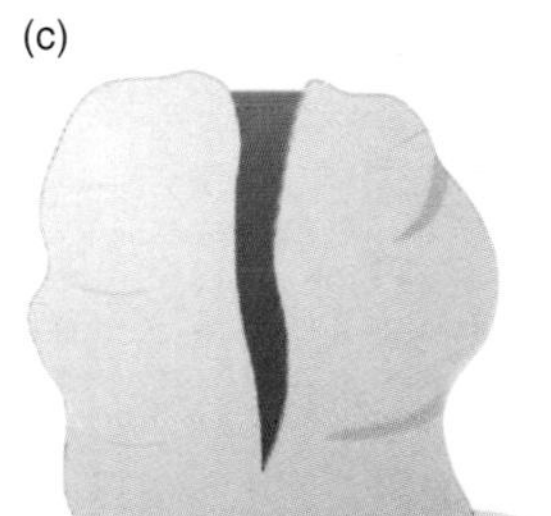

(d)

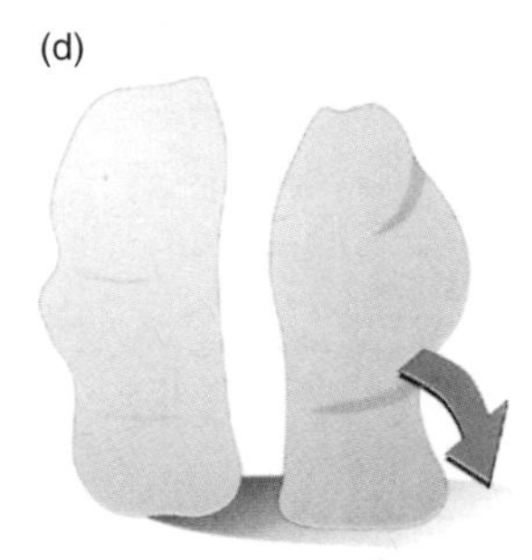

c) Explain in detail how the rock is weathered.

____

____

____

____

____

4 marks

11.1 Reactions of metals and non-metals

Metals and non-metals

All the elements in the Periodic Table are either **metals** or **non-metals**. The **physical** and **chemical properties** of an element determine whether it is a metal or a non-metal.

Physical properties of metals

★ All metals are good conductors of heat and electricity. This makes them useful in wires and pipes.

★ Most metals make a ringing sound when they are struck. This is why metals are used to make bells.

★ Most metals have a shiny appearance, so can be used to make mirrors.

★ Most metals are strong, so they can be used to make girders or cables to support large structures.

★ Metals are flexible and **ductile**, so are handy for making springs and wires.

★ Most metals have a high melting point and boiling point. The metal tungsten is used to make the filament in a light bulb.

	1	2											3	4	5	6	7	0
						1												
						$^{1}_{1}H$ Hydrogen												$^{4}_{2}He$ Helium
2	$^{7}_{3}Li$ Lithium	$^{9}_{4}Be$ Beryllium											$^{11}_{5}B$ Boron	$^{12}_{6}C$ Carbon	$^{14}_{7}N$ Nitrogen	$^{16}_{8}O$ Oxygen	$^{19}_{9}F$ Fluorine	$^{20}_{10}Ne$ Neon
3	$^{23}_{11}Na$ Sodium	$^{24}_{12}Mg$ Magnesium											$^{27}_{13}Al$ Aluminium	$^{28}_{14}Si$ Silicon	$^{31}_{15}P$ Phosphorus	$^{32}_{16}S$ Sulphur	$^{35.5}_{17}Cl$ Chlorine	$^{40}_{18}Ar$ Argon
4	$^{39}_{19}K$ Potassium	$^{40}_{20}Ca$ Calcium	$^{45}_{21}Sc$ Scandium	$^{48}_{22}Ti$ Titanium	$^{51}_{23}V$ Vanadium	$^{52}_{24}Cr$ Chromium	$^{55}_{25}Mn$ Manganese	$^{56}_{26}Fe$ Iron	$^{59}_{27}Co$ Cobalt	$^{59}_{28}Ni$ Nickel	$^{63.5}_{29}Cu$ Copper	$^{65}_{30}Zn$ Zinc	$^{70}_{31}Ga$ Gallium	$^{73}_{32}Ge$ Germanium	$^{75}_{33}As$ Arsenic	$^{79}_{34}Se$ Selenium	$^{80}_{35}Br$ Bromine	$^{84}_{36}Kr$ Krypton
5	$^{85}_{37}Rb$ Rubidium	$^{88}_{38}Sr$ Strontium	$^{89}_{39}Y$ Yttrium	$^{91}_{40}Zr$ Zirconium	$^{93}_{41}Nb$ Niobium	$^{96}_{42}Mo$ Molybdenum	$^{99}_{43}Tc$ Technetium	$^{101}_{44}Ru$ Ruthenium	$^{103}_{45}Rh$ Rhodium	$^{106}_{46}Pd$ Palladium	$^{108}_{47}Ag$ Silver	$^{112}_{48}Cd$ Cadmium	$^{115}_{49}In$ Indium	$^{119}_{50}Sn$ Tin	$^{122}_{51}Sb$ Antimony	$^{128}_{52}Te$ Tellurium	$^{127}_{53}I$ Iodine	$^{131}_{54}Xe$ Xenon
6	$^{133}_{55}Cs$ Caesium	$^{137}_{56}Ba$ Barium	$^{139}_{57}La$ Lanthanum	$^{178.5}_{72}Hf$ Hafnium	$^{181}_{73}Ta$ Tantalum	$^{184}_{74}W$ Tungsten	$^{186}_{75}Re$ Rhenium	$^{190}_{76}Os$ Osmium	$^{192}_{77}Ir$ Iridium	$^{195}_{78}Pt$ Platinum	$^{197}_{79}Au$ Gold	$^{201}_{80}Hg$ Mercury	$^{204}_{81}Tl$ Thallium	$^{207}_{82}Pb$ Lead	$^{209}_{83}Bi$ Bismuth	$^{209}_{84}Po$ Polonium	$^{210}_{85}At$ Astatine	$^{222}_{86}Rn$ Radon
7	$^{223}_{87}Fr$ Francium	$^{226}_{88}Ra$ Radium	$^{227}_{89}Ac$ Actinium															

Blocks

☐ Metals

☐ Non-metals

Figure 1 The Periodic Table. All the metals are coloured yellow and all the non-metals are coloured green

★ Most metals can be hammered into shape. We say they are **malleable**. This makes them useful for horseshoes.

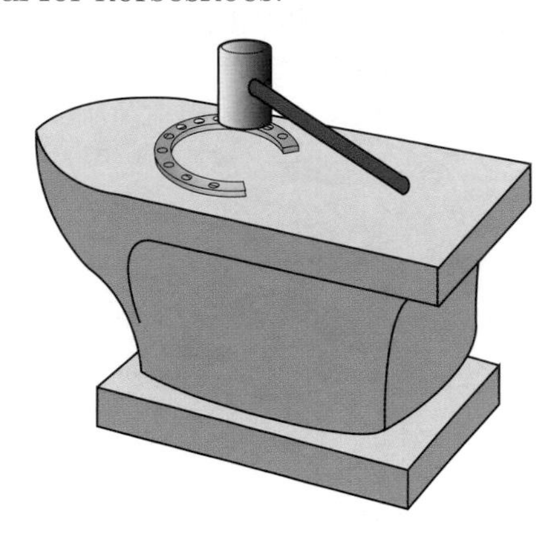

★ Most metals are dense, so a small amount of them weighs a lot. Lead is a very heavy metal used to line diving boots.

Physical properties of non-metals

The physical properties of non-metals vary widely and, unlike the metals, show few trends. For example at room temperature some non-metals are gases, some are solids and one is a liquid. But all non-metals are poor conductors of heat and electricity, except for graphite which is a good conductor of electricity.

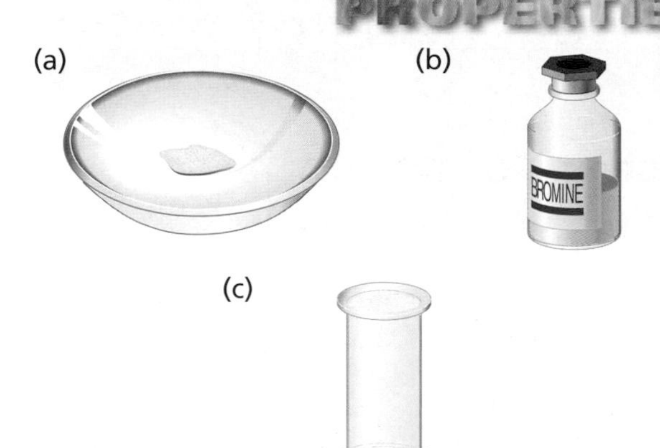

Figure 2 Non-metals can be solids like sulfur (a), liquids like bromine (b) or gases like chlorine (c)

Obtaining metals

Most metals come from rocks in the ground. Some very unreactive metals, such as gold, can be found naturally, but most react with other substances in the air or the ground and are found as compounds called **ores**. In order to obtain a metal, the ore needs to be chemically changed.

Getting iron from iron ore

Haematite is an ore which contains iron. It is a compound of iron and oxygen. The oxygen is removed from the iron by heating the ore with coke (carbon) in a blast furnace.

The **word equation** for this reaction is:

carbon + iron → carbon + iron
monoxide oxide dioxide

(The carbon monoxide is made by heating the carbon (coke) in the furnace.)

Getting copper from copper ore

There are several ores which contain copper. One of these is copper oxide. The oxygen can be removed by heating the copper oxide ore with carbon in a furnace. The word equation for this reaction is:

copper + carbon → carbon + copper
oxide dioxide

Getting aluminium from aluminium ore

The main ore which contains aluminium is called bauxite. It is a compound of aluminium and oxygen. These two elements cannot be separated by heating with carbon. Instead they are separated by heating the ore until it melts then passing an electric current through it. The process is called **electrolysis**.

Reactions of metals

Reactions of metals and acids

If a piece of metal such as zinc is placed in a solution of hydrochloric acid, we can see that a chemical reaction takes place because the metal dissolves in the acid and a gas is given off. If the gas is tested with a lighted splint, it gives a squeaky pop, so we know that it is hydrogen. We can describe this chemical reaction with the following word equation:

zinc + hydrochloric acid → zinc chloride + hydrogen

We can also write this as a **formula equation**:

$$Zn + 2HCl \rightarrow ZnCl_2 + H_2$$

This equation shows us that during a chemical reaction atoms join together in different ways. In this example we can see that after the reaction, the zinc has combined with the chlorine and the hydrogen atoms have formed pairs.

Remember: A hydrogen atom nearly always combines with another hydrogen atom to make one molecule of hydrogen gas (H_2).

Metals that react with acids always produce compounds called **salts** plus hydrogen. The salt in this case is zinc chloride. The general equation for this type of reaction is written as:

metal + acid → salt + hydrogen

Other examples of this kind of reaction include

iron + sulfuric acid → iron sulfate (salt) + hydrogen

$$2Fe + 3H_2SO_4 \rightarrow Fe_2(SO_4)_3 + 3H_2$$

and

magnesium + nitric acid → magnesium nitrate (salt) + hydrogen

$$Mg + 2HNO_3 \rightarrow Mg(NO_3)_2 + H_2$$

Notice in each case the change in how the atoms are combined after the reaction, compared with before the reaction.

Reactions of metals and carbonates

If we add an acid to a metal carbonate, a chemical reaction will take place. We know this is true because:

★ a gas which turns limewater milky (carbon dioxide) is given off

★ there is a rise in temperature

★ there is usually a colour change.

The general word equation for this type of reaction is:

metal carbonate + acid → salt + carbon dioxide + water

The word equations and formula equations below describe some of these reactions and show how the atoms have joined in different ways after the reaction.

copper carbonate + sulfuric acid → copper sulfate (salt) + carbon dioxide + water

$$CuCO_3 + H_2SO_4 \rightarrow CuSO_4 + CO_2 + H_2O$$

zinc carbonate + hydrochloric acid → zinc chloride (salt) + carbon dioxide + water

$$ZnCO_3 + 2HCl \rightarrow ZnCl_2 + CO_2 + H_2O$$

magnesium carbonate + nitric acid → magnesium nitrate (salt) + carbon dioxide + water

$$MgCO_3 + 2HNO_3 \rightarrow Mg(NO_3)_2 + CO_2 + H_2O$$

Reactions of metal oxides and acids

If we add an acid to a metal oxide, a chemical reaction will take place. We know that this is true because:

★ there is a rise in temperature

★ there is usually a colour change.

The general word equation for this type of reaction is:

metal oxide + acid → salt + water

The word equations and formula equations below describe some of these reactions.

copper oxide + sulfuric acid → copper sulfate (salt) + water

$$CuO + H_2SO_4 \rightarrow CuSO_4 + H_2O$$

zinc oxide + hydrochloric acid → zinc chloride (salt) + water

$$ZnO + 2HCl \rightarrow ZnCl_2 + H_2O$$

magnesium oxide + nitric acid → magnesium nitrate (salt) + water

$$MgO + 2HNO_3 \rightarrow Mg(NO_3)_2 + H_2O$$

Making salts

All the above reactions have produced compounds we call salts. We can also produce salts by reacting an acid with an alkali.

The general word equation for this type of reaction is:

acid + alkali → salt + water

Example:

hydrochloric acid + sodium hydroxide → sodium chloride (salt) + water

$$HCl + NaOH \rightarrow NaCl + H_2O$$

All acids have a pH value of less than 7. All alkalis have a pH value of more than 7. If the reaction between an acid and an alkali produces a solution of pH 7, like water, we say that **neutralisation** has taken place. To make a salt by neutralisation we need to add exactly the right amounts of acid and alkali. We can do this using a method called **titration**.

The colour of the indicator added to the acid in the conical flask shows that the solution is acidic. Alkali is now added to this a drop at a time from the burette until the indicator just begins to change colour. At this point, the solution is neutral, but adding more alkali will make it an alkaline solution.

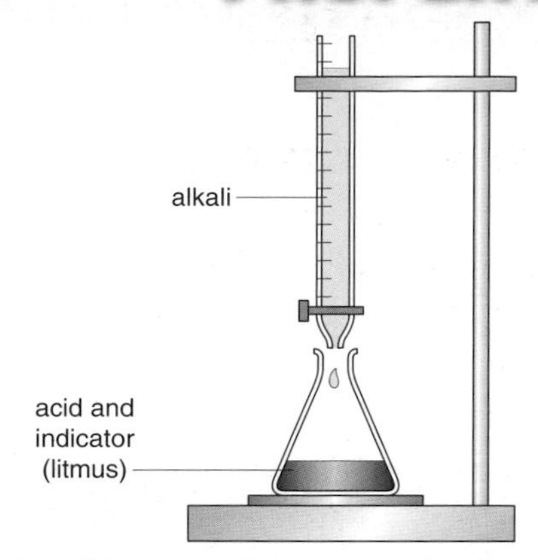

Figure 3 Titrating to make a neutral solution

Uses of salts

The different salts made in these reactions can be very useful to us. For example:

- ★ sodium chloride is used for seasoning food and gritting icy roads
- ★ potassium sulfate is used to make fertilisers
- ★ calcium carbonate is used to make cement
- ★ sodium carbonate is used in washing soda (as a water softener)
- ★ calcium sulfate is used to make plaster for buildings.

Key terms

Check that you understand and can explain the following terms:

- ★ metal
- ★ non-metal
- ★ physical properties
- ★ chemical properties
- ★ ductile
- ★ malleable
- ★ ore
- ★ word equation
- ★ electrolysis
- ★ formula equation
- ★ salt
- ★ neutralisation
- ★ titration

Questions

1. List four physical properties of metals.
2. Explain briefly how copper can be extracted from copper ore.
3. Write down the word equation which describes the reaction between an acid and a metal carbonate.
4. Suggest three ways in which you might recognise that a chemical reaction is taking place between an acid and a metal carbonate.
5. Name four salts and give one use for each.

11.2 Patterns of reactivity

The gold necklace shown in Figure 1 is hundreds of years old and yet it still has a bright **untarnished** appearance. The gate in Figure 2 is only a few years old but the **corrosive** effect of air and water is very clear to see. We describe this difference in behaviour by saying that iron is **more reactive** than gold.

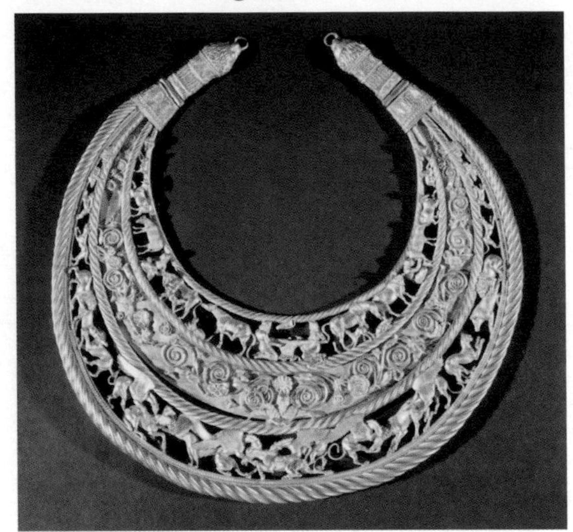

Figure 1 This Roman gold necklace is very old, but is still shiny

Figure 2 This gate is fairly new, but has rusted badly

It is important that we have a good understanding of how reactive different metals are so that we can choose the right metal for the right job. By comparing several different types of reactions we can list the metals in order of **reactivity**. This is called the **reactivity series**.

Reactions between metals and oxygen

Most metals will react with oxygen when heated to form a metal oxide.

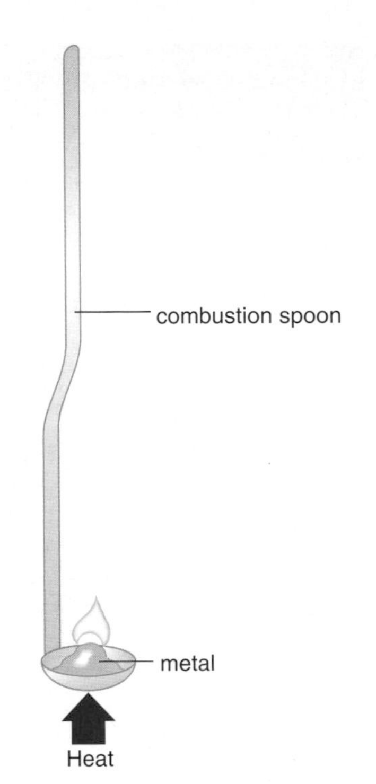

Figure 3 Reacting a metal with oxygen

Metal	Reaction with oxygen
Sodium	Burns brightly in air to form a white powder, sodium oxide $4Na + O_2 \rightarrow 2NaO$
Magnesium	Catches fire after being heated and burns with a bright white flame. A white powder, magnesium oxide, is formed by the reaction $2Mg + O_2 \rightarrow 2MgO$
Iron filings	Do not burn, but glow brightly if heated strongly. They react with the oxygen to form black iron oxide $2Fe + 3O_2 \rightarrow 2Fe_2O_3$
Copper foil	Does not burn, but if it is heated strongly a coating of black copper oxide forms $2Cu + O_2 \rightarrow 2CuO$

Reactions between metals and water

Many metals will react with water to form a metal oxide or hydroxide.

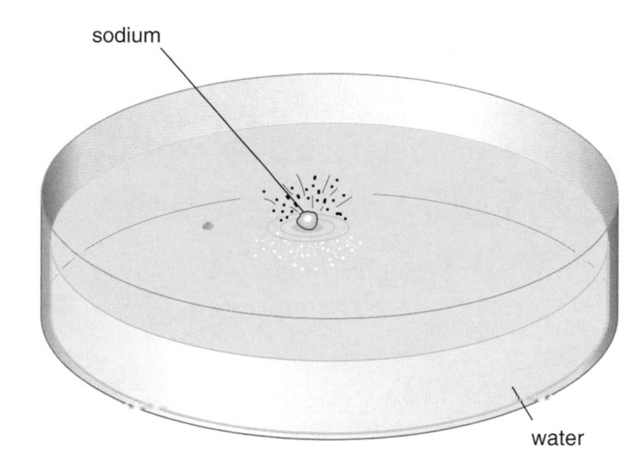

Figure 4 Reacting a metal with water

Metal	Reaction with water
Sodium	Reacts violently with cold water, becoming so hot that it melts and whizzes around on a layer of hydrogen. The reaction produces the compound sodium hydroxide $2Na + 2H_2O \rightarrow 2NaOH + H_2$
Magnesium	Reacts very slowly with cold water forming magnesium oxide and hydrogen $Mg + H_2O \rightarrow MgO + H_2$
Iron	Reacts extremely slowly with cold water (days or weeks) to produce iron oxide (rust) and hydrogen $2Fe + 3H_2O \rightarrow Fe_2O_3 + 3H_2$
Copper	Does not react with water and so it is often used for water pipes.

Reactions between metals and an acid (hydrochloric acid)

Metal	Reaction with acid
Sodium	The reaction between dilute hydrochloric acid and sodium is extremely violent and is too dangerous to carry out. The reaction would produce sodium chloride and hydrogen $2Na + 2HCl \rightarrow 2NaCl + H_2$
Magnesium	Reacts vigorously to produce magnesium chloride and hydrogen $Mg + 2HCl \rightarrow MgCl_2 + H_2$
Iron	Reacts very slowly to produce iron chloride and hydrogen $4Fe + 6HCl \rightarrow 2Fe_2Cl_3 + 3H_2$
Copper	Does not react with hydrochloric acid even if the acid is concentrated

If we use experiments like these for other metals, a reactivity series for metals similar to that shown below can be constructed.

Potassium (K)
Sodium (Na)
Calcium (Ca)
Magnesium (Mg)
Aluminium (Al)
Zinc (Zn)
Iron (Fe)
Lead (Pb)
Copper (Cu)
Silver (Ag)
Gold (Au)

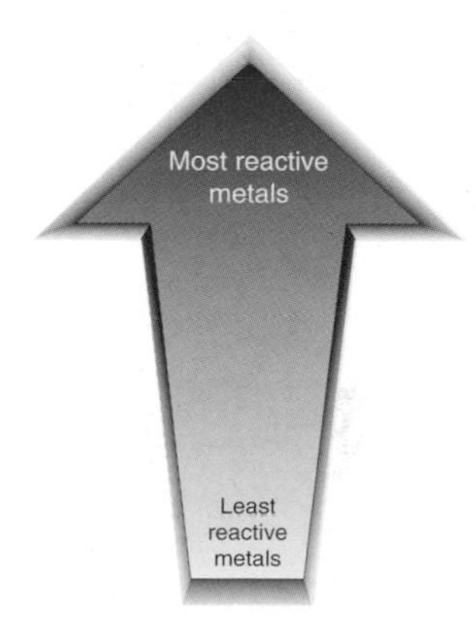

This sentence might help you to remember the order of the metals in the series:

Poor **S**ally **C**an't **M**anage **A**ny **Z**eal **I**n **L**atin **C**os **S**he's **G**lum

Figure 5

Patterns of reactivity (*continued*)

Displacement reactions

The more reactive a metal is, the more it *wants* to form compounds. We can use this idea to predict what will happen in a **displacement reaction**.

In a displacement reaction, a more reactive metal will displace a **less reactive** metal from a salt solution.

This reaction is described by the equation:

magnesium + copper sulfate → magnesium sulfate + copper

$$Mg + CuSO_4 \rightarrow MgSO_4 + Cu$$

The more reactive magnesium goes into solution and displaces the less reactive copper. The copper is forced out of solution and forms a coating on the magnesium ribbon (Figure 6).

A different displacement reaction takes place if copper metal is added to a silver salt solution.

This reaction is described by the equation:

copper + silver nitrate → copper nitrate + silver

$$Cu + 2AgNO_3 \rightarrow Cu(NO_3)_2 + 2Ag$$

The more reactive copper **displaces** the less reactive silver (Figure 7).

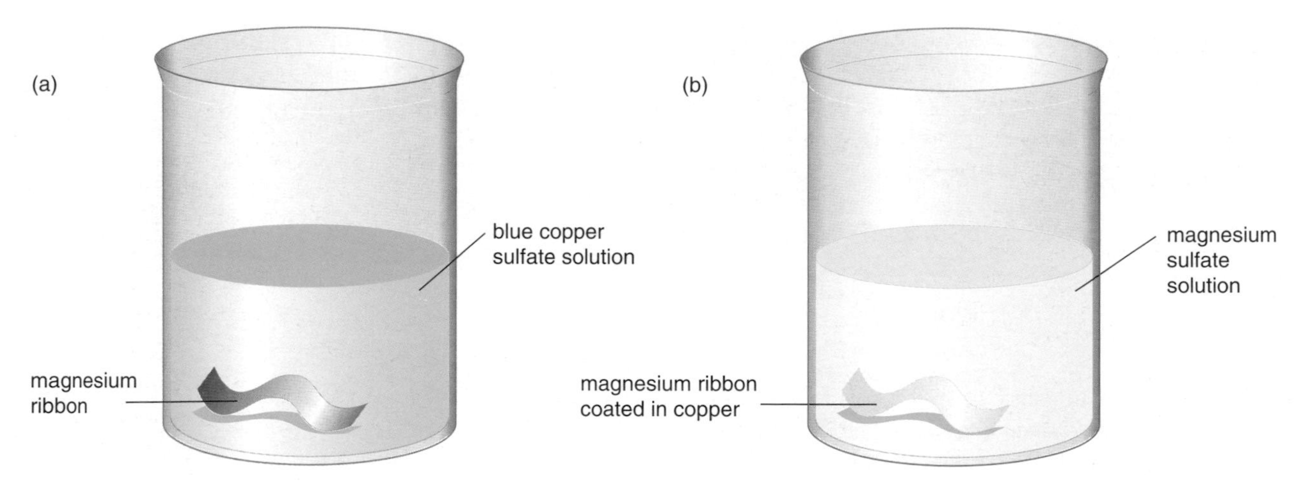

Figure 6 The displacement of copper by magnesium

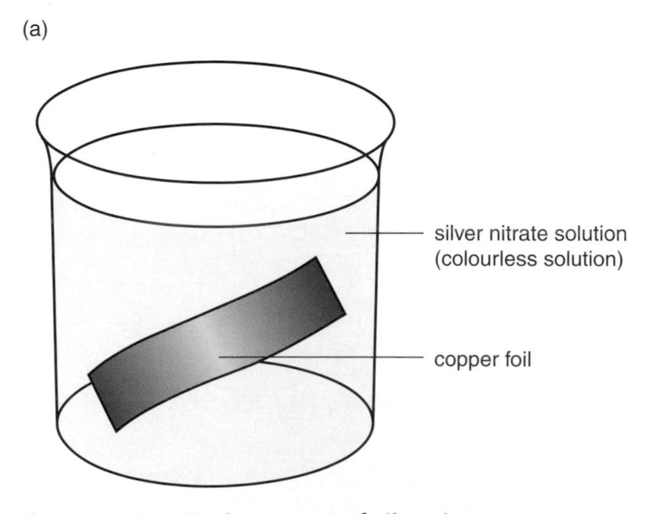

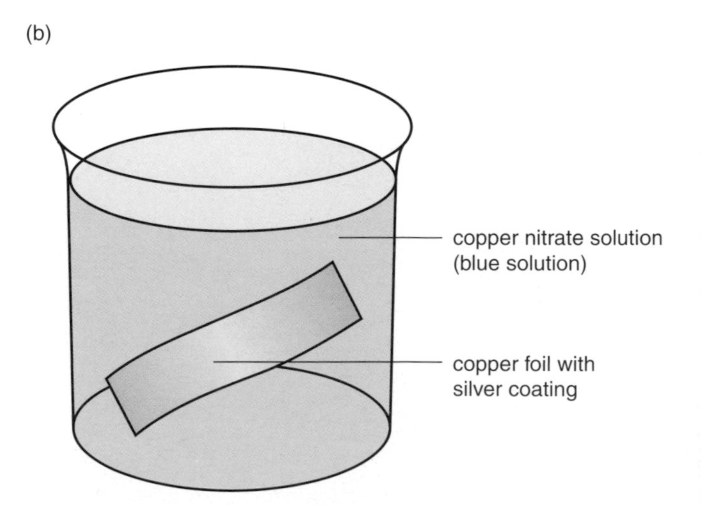

Figure 7 The displacement of silver by copper

Competing for oxygen

If a mixture of iron powder and copper oxide is heated, a displacement reaction takes place. The iron displaces the copper because it is higher in the reactivity series.

iron + copper oxide → iron oxide + copper

$$2Fe + 3CuO \rightarrow Fe_2O_3 + 3Cu$$

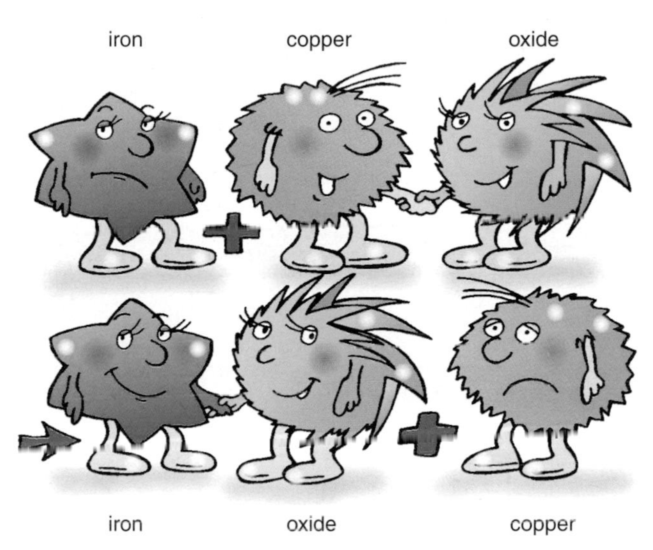

Figure 8 The displacement of copper by iron

The more reactive iron displaces the less reactive copper.

If a mixture of aluminium powder and iron oxide is heated, a displacement reaction takes place. The aluminium displaces the iron because it is higher in the reactivity series.

aluminium + iron oxide → aluminium oxide + iron

$$2Al + Fe_2O_3 \rightarrow Al_2O_3 + 3Fe$$

Figure 9 The displacement of iron by aluminium

The more reactive aluminium displaces the less reactive iron.

This reaction is so violent that the temperature of the reactants and products exceeds the melting point of iron (1500 °C). As a result the iron produced by the reaction is molten (liquid). This reaction is called the **thermit reaction** and is often used to mend railway lines.

Key terms

Check that you understand and can explain the following terms:

★ untarnished
★ corrosive
★ more reactive
★ reactivity
★ reactivity series
★ displacement reaction
★ less reactive
★ displace
★ thermit reaction

Questions

1 Explain what is meant by the reactivity series.

2 Use the reactivity series to explain the following:
a) Sodium metal is not used to make knives, forks and spoons.
b) Magnesium is not used for car bodies.
c) Gold jewellery from ancient civilisations remains unaffected by air and water.
d) It is unsafe to add an acid to potassium.
e) Iron metal is never found naturally in the ground.

3 What kind of reaction is the thermit reaction? Give one example of a use for this reaction.

4 Write down a word equation and a formula equation for each of the following:
a) sodium metal heated in air.
b) iron reacting with water.
c) magnesium reacting with hydrochloric acid.

Chapter 11 Reactions and reactivity:

What you need to know

1. Elements in the Periodic Table can be divided into two groups – metals and non-metals.
2. Metals have similar properties e.g. good conductors, strong, ductile. Non-metals have very varied properties.
3. Some metals such as gold are found naturally in the ground. Others such as iron and copper are not and need to be extracted from their ores.
4. Most metals will react with acids to give a salt and hydrogen.
5. Metal carbonates react with acids to give a salt, carbon dioxide and water.
6. Metal oxides react with acids to give a salt and water.
7. Acids and alkalis neutralise each other to give a salt and water.
8. Some metals are more reactive than others.
9. By observing their reactions, metals can be placed in a reactivity series.
10. It is possible to predict how metals will react if their position in the reactivity series is known.
11. Displacement reactions may take place between metals and solutions of other metals.

How much do you know?

1 The diagram below shows four test tubes containing equal amounts of a metal powder and 20 cm^3 of dilute hydrochloric acid. The number of bubbles of hydrogen being given off indicates the speed of the reaction.

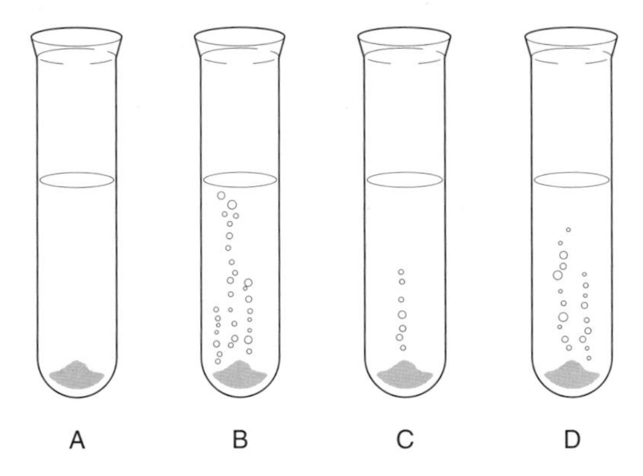

Place the four metals in order of reactivity.

Most reactive ______________

____

____

Least reactive ______________

4 marks

2 The list below shows four metals in order of their reactivity.

Most reactive sodium
calcium
magnesium
Least reactive copper

a) Using this information complete the table below.

Metal	What happened when added to water
sodium	violent reaction, lots of bubbles of gas released
calcium	
magnesium	very slow reaction with cold water, very few bubbles
copper	

2 marks

b) Name the gas released when sodium reacts with water?

____

1 mark

c) Why are copper pipes used in houses to carry water?

____

1 mark

d) Name the gas released when sulfuric acid reacts with calcium carbonate.

____

1 mark

e) Name the salt produced when sulfuric acid reacts with calcium carbonate.

____

1 mark

How much do you know? *continued*

3 An iron nail is placed in a solution of copper sulfate. After 24 hours it is found that the nail is coated with a reddish-brown material.

a) What is the reddish-brown coating on the nail?

____

1 mark

b) What kind of reaction has taken place?

____

1 mark

c) Write a word equation which describes this reaction.

____

4 marks

d) Which is higher in the reactivity series, iron or copper?

____

1 mark

4 Put each of the properties listed below in the correct column of the table.

strong

brittle

flexible

shiny

low melting point

good conductor of heat

low density

poor conductor of electricity

Property of a metal	Property of a non-metal

8 marks

5 Name the salt produced in each of the following reactions.

a) sodium carbonate + nitric acid

____

1 mark

b) calcium oxide + hydrochloric acid

____

1 mark

c) magnesium + sulfuric acid

____

1 mark

d) Name the gas produced by reaction a).

____

1 mark

e) Describe how you would test the gas produced by reaction a).

____

____

2 marks

6 Using the apparatus shown below describe how you would produce a neutral salt solution from an acid and an alkali.

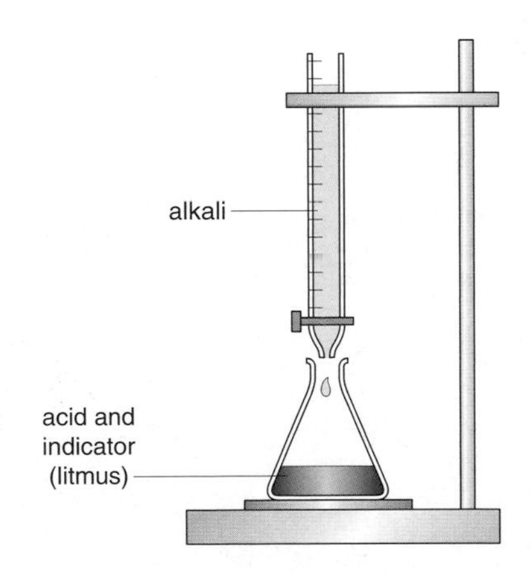

____

____

____

____

4 marks

12.1 Environmental chemistry

Our environment is being continually affected by natural processes and the activities of humans. It is important that we understand these processes and monitor the effects of any changes.

Soils

Soils are made up of:

★ fragments of rocks which have been broken down by weathering
★ organic matter called **humus** made from the decomposed remains of plants and animals
★ water
★ air.

The properties of a soil will vary from place to place. They will depend on several factors including:

★ the proportions of each of the above found in a soil
★ the type of rock which has been weathered
★ the type of plant and animal remains found there
★ the climatic conditions.

Acid and alkaline soils

Most plants will grow best if the soil is very slightly acidic, around pH 6.5. But if the soil is too acidic or is alkaline, important **nutrients** needed by the plants, such as nitrogen and phosphorus, will not be absorbed and growth may be restricted. It is important that we monitor the pH of a soil to ensure we get strong healthy plant growth. We can do this using pH soil testing kits.

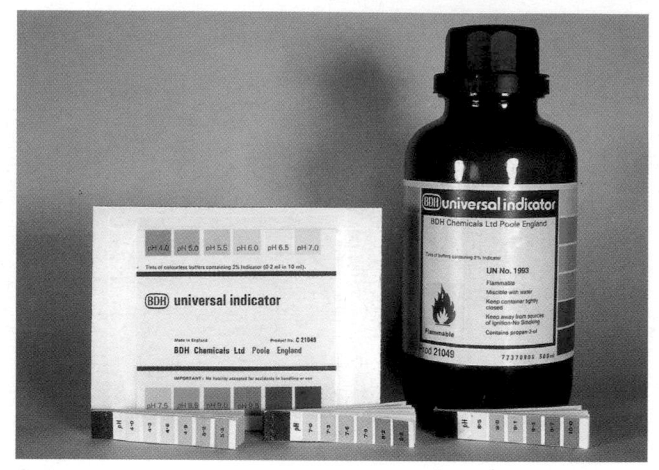

Figure 2 A soil testing kit and pH card

(a)

(b)

Figure 1 The different properties of soils affect the kinds of plants and trees that can grow

Figure 3 Acid rain has dissolved away the limestone creating caves like the one shown. Acid rain has also damaged this limestone statue

A small amount of soil is shaken with **distilled water**. The mixture is left for a few minutes so that the soil can settle. A few drops of **universal indicator** are added and the colour of the solution is compared with a colour pH chart.

Most soils over a period of time become more acidic because of the effects of **acid rain**. To reduce this acidity calcium hydroxide (lime) is added to the soil.

Acid rain and chemical weathering

As rain falls, some of the gases in the atmosphere dissolve in the water making it acidic.

★ Carbon dioxide dissolves in water to produce carbonic acid

★ Sulfur dioxide dissolves in water to produce sulfurous acid

★ Nitrogen dioxide dissolves in water to produce nitric acid

Limestone reacts with acids. Therefore rocks and buildings made from limestone will be attacked over a period of time and **weathered** (damaged) by acid rain.

The rate at which this weathering occurs will be most affected by:

★ the annual rainfall, i.e. the higher this is, the greater the rate of weathering

★ the pH of the rain, i.e. the greater the acidity (the lower the pH), the greater the rate of weathering.

The acidity of the rain is greatly affected by the amount of carbon dioxide, sulfur dioxide and nitrogen dioxide present in the atmosphere. Burning fossil fuels puts more of these gases into the atmosphere.

There are several ways in which we can reduce the emissions of these gases:

★ burn fewer fossil fuels by using alternative sources of energy, like wind and water power

★ use low sulfur petrol and diesel in cars and lorries

★ fit **catalytic converters** to all cars to reduce the levels of pollutants such as nitrogen dioxide emitted from the exhaust.

Other effects of acid rain

Acid rain does not just affect buildings and rocks. It also affects other things in our environment.

- ★ It affects the chemistry of the soil, killing trees and plants.
- ★ It damages the leaves of plants.
- ★ It makes ponds and lakes more acidic, killing many of the creatures that live in the water.
- ★ It can increase the rate of corrosion of objects made from metals such as iron.

Figure 4 Acid rain has had a devastating effect on this forest in the Czech Republic

It is important to monitor the level of all **pollutants** both in the air and in water. We can do this using **indicator organisms**. These are plants or animals that are very susceptible to pollutants and therefore die or disappear if pollution levels increase. For example, the presence of salmon or trout in a river indicates that the water is clean. If these fish disappear or die, this may indicate that the river is becoming polluted.

Alternatively, static monitoring stations similar to the one in Figure 5 could be used. This station measures the levels of sulfur dioxide, nitrogen dioxide and other pollutants in the air.

Figure 5 A static monitoring station

Global warming

Most countries in the world seem to be experiencing warmer conditions than normal. It is possible that this **global warming** is being caused by a phenomenon called the **greenhouse effect**.

Heat energy from the Sun travels to the Earth as rays called **infrared rays**. These rays pass through the atmosphere and are absorbed by the Earth's surface. The surface of the Earth becomes warm and emits infrared rays (all warm objects give off infrared rays, even though you can't see them). Carbon dioxide in the atmosphere forms a 'blanket' around the Earth, preventing some of this heat from escaping. The more carbon dioxide there is in the atmosphere, the less heat can escape. By increasing the amount of carbon dioxide in the atmosphere, we are increasing the amount of heat which is unable to escape, making the Earth and its atmosphere hotter. This increase in temperature is called the greenhouse effect, because this effect is similar to the way in which a glass greenhouse keeps its heat.

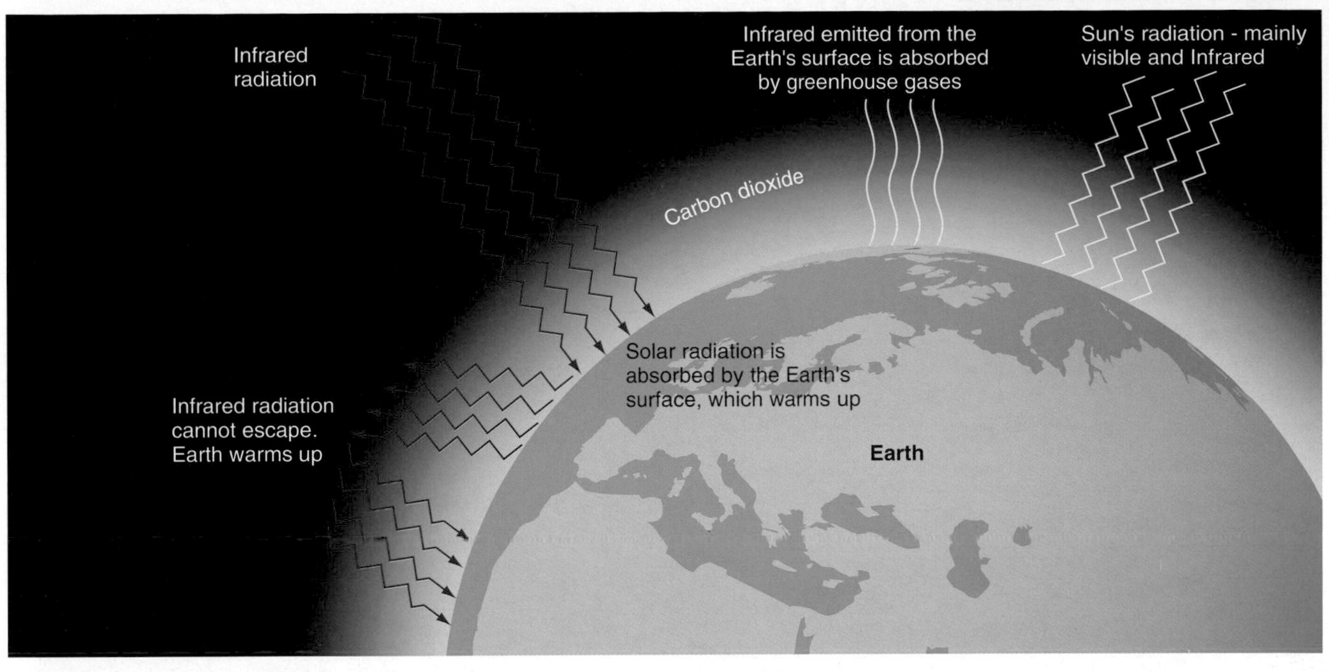

Figure 6 How the greenhouse effect occurs

Scientists can't be sure that the changes we are experiencing are truly caused by the greenhouse effect, but we do know that burning fossil fuels increases the amount of carbon dioxide in the atmosphere. It is therefore a wise precaution to reduce the amount of carbon dioxide we are releasing into the atmosphere.

Key terms

Check that you understand and can explain the following terms:

- ★ soil
- ★ humus
- ★ nutrients
- ★ distilled water
- ★ universal indicator
- ★ acid rain
- ★ weathering
- ★ catalytic converter
- ★ pollutants
- ★ indicator organisms
- ★ global warming
- ★ greenhouse effect
- ★ infrared rays

Questions

1 What are the four main constituents of soil?

2 Describe how you would test the pH of a soil.

3 Name three gases that dissolve in rain water to make it acidic.

4 Name three effects acid rain has on the environment.

5 Give three ways in which we could reduce the effects of acid rain.

6 Find out why planting more trees may help to reduce the problem of carbon dioxide in the atmosphere?

12.2 Using chemistry

There are lots of different kinds of chemical reactions. Some of these we use because of the energy they release. Others we use because of the new materials produced by the reactions.

Energy from chemical reactions

A **fuel** is a substance which releases energy when burnt. Three of the most important fuels we use are the fossil fuels, coal, oil and natural gas.

Figure 1 The fuel burnt in this car is petrol

Oil and natural gas are mixtures of compounds. These compounds all contain hydrogen atoms and carbon atoms. They are called **hydrocarbon fuels**. Petrol is a hydrocarbon.

When hydrocarbon fuels are burnt, water, carbon dioxide and sometimes carbon monoxide are formed. For example if we open the air hole of a Bunsen burner so that there is a good supply of oxygen, **complete combustion** occurs and the products of the reaction are water and carbon dioxide.

natural gas (methane) + oxygen → water + carbon dioxide

$$CH_4 + 2O_2 \rightarrow 2H_2O + CO_2$$

If the air hole is not completely opened and there is not a good supply of oxygen, **incomplete combustion** takes place and water and carbon are produced.

methane + oxygen → water + carbon (soot)

$$CH_4 + O_2 \rightarrow 2H_2O + C$$

Sometimes gas heaters in the home can produce the poisonous gas carbon monoxide if they are not regularly checked and serviced. This again is caused by incomplete combustion.

methane + oxygen → water + carbon monoxide

$$2CH_4 + 3O_2 \rightarrow 4H_2O + 2CO$$

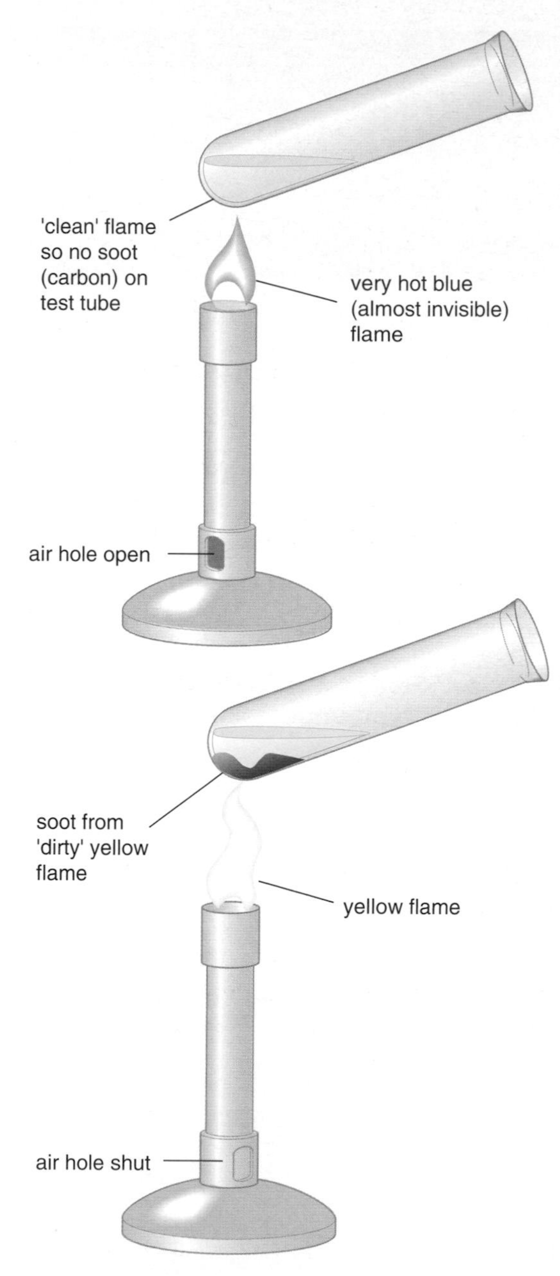

Figure 2 The Bunsen burner at the top has its air hole completely open. The Bunsen at the bottom has its air hole only partially open, so combustion is incomplete. The soot on the outside of the test tube is the carbon formed in this reaction

There are many other hydrocarbons which we use as fuels. These include propane, butane and diesel.

A good fuel should:

★ give out lots of energy when burnt
★ produce as little pollution as possible
★ be plentiful
★ be cheap (economic).

Hydrogen may become an important fuel of the future. It has several advantages:

★ when burnt it produces no carbon dioxide, only water
★ it weighs less than other fuels.

But there are disadvantages:

★ the reaction with air when burnt can be explosive
★ it would need to be compressed for storage
★ at present it is more expensive than petrol.

Other chemical reactions that can be used as sources of energy

As we saw on page 105, if aluminium powder and iron oxide are mixed and ignited in the thermit reaction, a displacement reaction takes place in which a large amount of heat energy is released.

aluminium + iron oxide → aluminium oxide + iron + HEAT

$$2Al + Fe_2O_3 \rightarrow Al_2O_3 + 2Fe + HEAT$$

The energy released is so great that the temperature of the reactants and products rises above the melting point of iron (1500 °C).

If zinc powder is added to a solution of copper sulfate, a different displacement reaction takes place and heat energy is released (Figure 3).

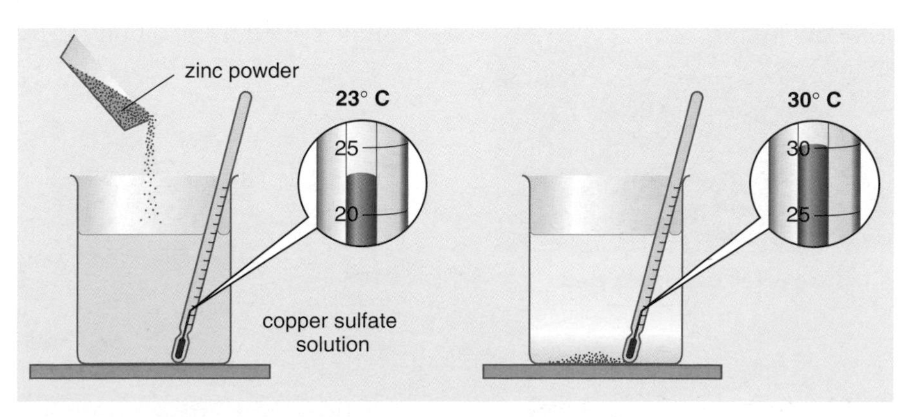

Figure 3 Displacement reaction between zinc and copper sulfate

zinc + copper sulfate → zinc sulfate + copper + HEAT

$$Zn + CuSO_4 \rightarrow ZnSO_4 + Cu + HEAT$$

This reaction does not release as much energy as the thermit reaction and the solution is likely to increase in temperature by only 5–6 °C.

Voltaic cells

If two **dissimilar metals** are placed in a beaker of acid, a reaction takes place and energy is released in the form of electricity. We call this a **voltaic cell**.

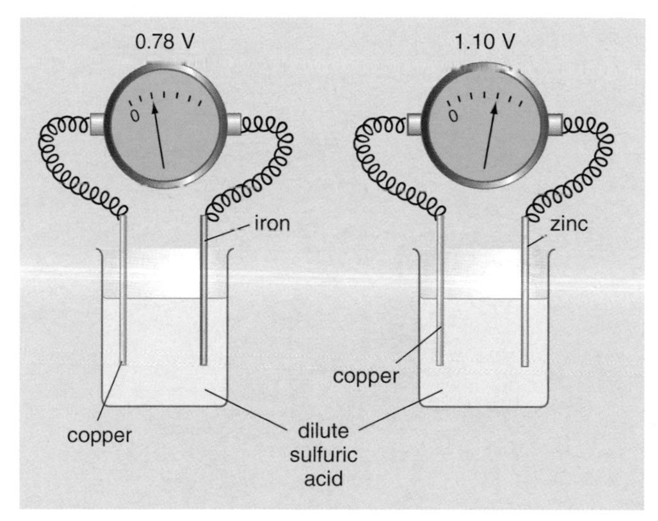

Figure 4 Examples of voltaic cells

The further apart the metals are in the **reactivity series**, the greater the voltage produced by the reaction.

Oxidation reactions

Flash camera!

If magnesium powder is heated in air it reacts with oxygen. This is a very vigorous reaction which gives off a lot of light. The reaction was used by early photographers for flashlights.

magnesium + oxygen → magnesium oxide

$$2Mg + O_2 \rightarrow 2MgO$$

This kind of reaction is called an **oxidation reaction**. An **oxide** is produced.

Figure 5

Oxidation of metals

Some oxidation reactions are not useful. For example, iron will react with the oxygen and water in the air producing iron oxide or rust. This kind of chemical change is called **corrosion**.

Figure 6 This car has rusted after exposure to air and water

The simple experiment illustrated in Figure 7 shows that both air and water are needed if an object is to rust.

★ Tube 1 has nails in dry air – the anhydrous calcium chloride absorbs any water from the air.

★ Tube 2 has nails in water which has been boiled to remove all the air. The layer of oil stops any more air from entering the water.

★ Tube 3 has nails exposed to both air and water. Only in this tube have the nails rusted.

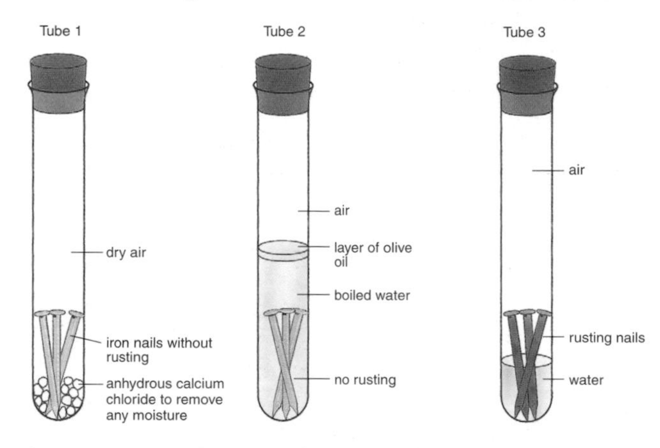

Figure 7 Investigating the conditions necessary for rusting to occur

To stop a surface from corroding it can be coated with paint, grease, plastic or another metal, such as tin, which does not corrode.

Oxidation of food

Some foods also react with the oxygen in the air. This may affect their taste. There are several ways of avoiding this:

★ food can be vacuum packed so there is no contact between it and the air e.g. bacon

★ food can be kept in the refrigerator to slow down any reaction that may take place e.g. fats such as lard and butter.

New materials from chemical reactions

Chemical reactions *always* create new materials, some of which are very useful. For example if copper oxide is heated with carbon, the following reaction takes place.

copper oxide + carbon → copper + carbon dioxide

$$2CuO + C \rightarrow 2Cu + CO_2$$

The oxygen has been removed from the copper oxide to make copper metal. This kind of reaction is called a **reduction reaction**. We say that the copper oxide has been reduced.

Similarly if we heat iron oxide with carbon monoxide we can produce iron.

iron oxide + carbon monoxide → iron + carbon dioxide

$$Fe_2O_3 + 3CO \rightarrow 2Fe + 3CO_2$$

Conservation of mass

Whether we use a reaction for the energy it releases or for the new substances formed, the atoms of the reactants are always rearranged during the reaction. For example:

hydrogen + oxygen → water

$$2H_2 + O_2 \rightarrow 2H_2O$$

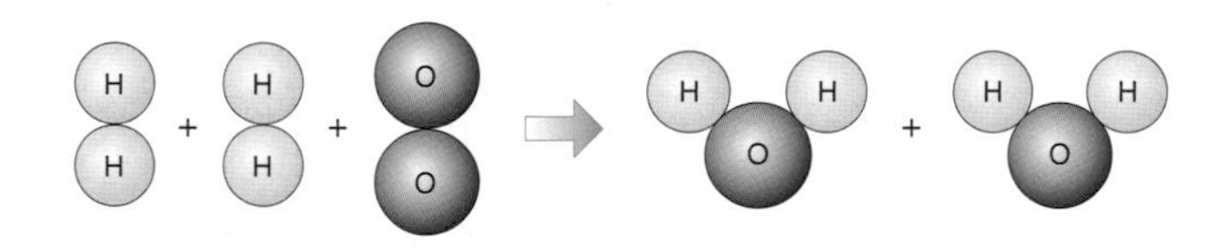

Figure 8

carbon + oxygen → carbon dioxide

$$C + O_2 \rightarrow CO_2$$

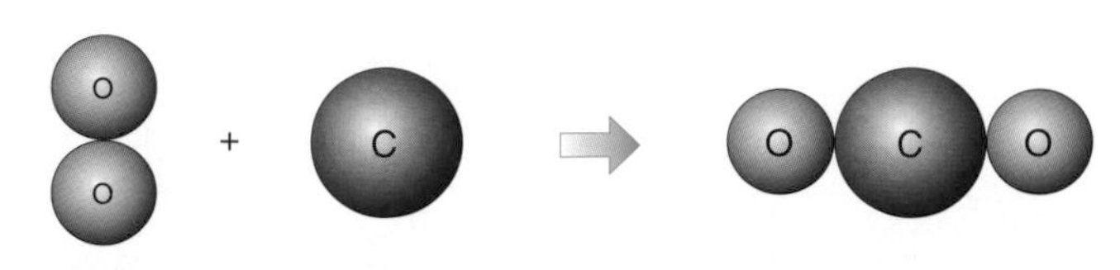

Figure 9

We can see from these examples that at the beginning and end of a reaction, there are the same number of atoms of each element. In other words, the mass of the reactants is equal to the mass of the products. This principle of the **conservation of mass** is very important.

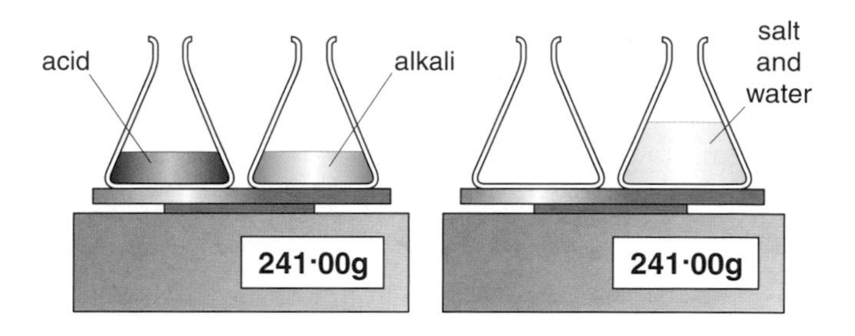

Figure 10 The mass of the reactants always equals the mass of the products

When one of the products is a gas, it may appear that after the reaction the mass of the products is less than the mass of the reactants. This happens because the gas has been lost to the air, so its mass has not been measured.

Key terms

Check that you understand and can explain the following terms:

★ fuel
★ hydrocarbon fuel
★ complete combustion
★ incomplete combustion
★ dissimilar metals
★ voltaic cell
★ reactivity series
★ oxidation reaction
★ oxides
★ corrosion
★ reduction reaction
★ conservation of mass

Questions

1 What are hydrocarbon fuels? Give three examples of hydrocarbon fuels.
2 Why might the incomplete combustion of methane in the home be dangerous?
3 Give one advantage and one disadvantage of using hydrogen as a fuel.
4 Give one example of an oxidation reaction and its use.
5 Give one example of an oxidation reaction which is a disadvantage.
6 Give one example of a reduction reaction and its use.

Chapter 12 Importance of chemistry:

What you need to know

1. Acid rain is formed when certain gases in the atmosphere dissolve in rain.
2. Acid rain affects rocks, building materials and living things.
3. Increasing the amount of carbon dioxide in the atmosphere may lead to global warming.
4. It is important to monitor pollution in the air and water.
5. Chemical reactions can be used to make new materials.
6. Chemical reactions can be used as a source of energy.
7. During a chemical reaction there is no loss or gain in mass.
8. Chemical reactions can be described by word equations or formula equations.

How much do you know?

1 a) What is the ideal pH of a soil which will allow strong healthy growth of most plants?

____

1 mark

b) Give one reason why a soil may become acidic.

____

1 mark

c) How could the acidity of a soil be reduced.

____

1 mark

d) How would you check the pH of your soil?

____

____

____

3 marks

2 a) Suggest three ways in which we could reduce the emissions of gases into the atmosphere which cause acid rain.

____

____

____

3 marks

b) Suggest four ways in which acid rain could damage the environment.

____

____

____

____

4 marks

c) What are indicator organisms?

____

____

2 marks

d) Name one indicator organism and explain how it could be used to monitor pollution.

____

____

____

3 marks

3 a) How does heat energy travel from the Sun to the Earth?

____

1 mark

b) Explain what happens when this heat reaches the Earth's surface.

____

1 mark

c) Explain why increasing the amount of carbon dioxide in the atmosphere will lead to global warming.

____

____

____

____

4 marks

4 a) What is a fuel?

____

1 mark

How much do you know? *continued*

b) Give one example of a fuel and write down the word equation which describes what happens when it is used.

______

______

2 marks

c) Name three properties that a good fuel should have.

______

______

______

3 marks

5 The diagram below shows a beaker containing a solution of copper sulfate and a thermometer.

zinc powder

copper sulfate solution

a) What kind of chemical reaction takes place if zinc powder is added to the solution.

______

1 mark

b) What happens to the reading on the thermometer after the zinc has been added?

______

1 mark

c) State one way in which the reaction would change if a metal higher in the reactivity series than zinc, e.g. magnesium, was added to the solution instead of the zinc powder.

______

1 mark

6 a) Write a word equation which describes the reaction between iron and oxygen.

______

1 mark

b) What is rust?

______

1 mark

c) Describe three ways in which we could prevent an iron object from rusting.

______

______

______

3 marks

7 The diagrams below show the mass of an acid (hydrochloric acid) and a metal carbonate (magnesium carbonate) before they react and after they react.

hydrochloric acid

magnesium carbonate

310·00g

308·00g

a) Write down a word equation which describes the chemical reaction which takes place when the two solutions are added together.

______

2 marks

b) Write down the formula equation for this reaction.

______

6 marks

c) Why in the above diagram does it appear as if the mass of the products is less than the mass of the reactants?

______

______

2 marks

13.1 Energy and energy changes

The need for energy

You need **energy** every day of your life to run, to walk, to play and even to sleep. You receive all this energy from the food you eat.

Figure 1 How you get energy

Trees and plants need energy to grow and reproduce. They get most of their energy from the Sun.

Figure 2 How green plants and trees get their energy

Machines need energy to be able to work. There are many ways in which machines can be given this energy.

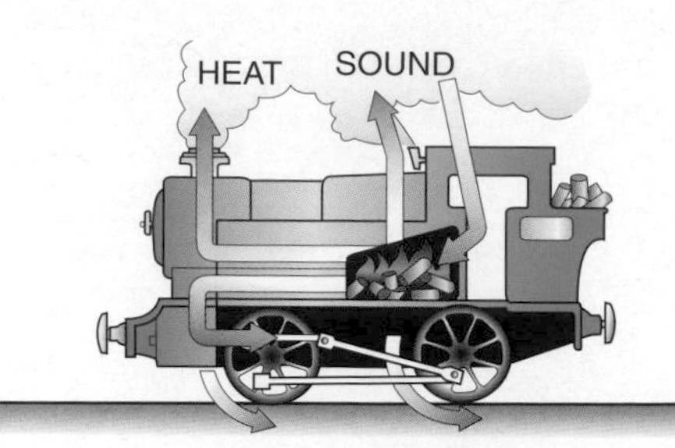

Figure 3 a) The energy needed to move this train comes from burning coal

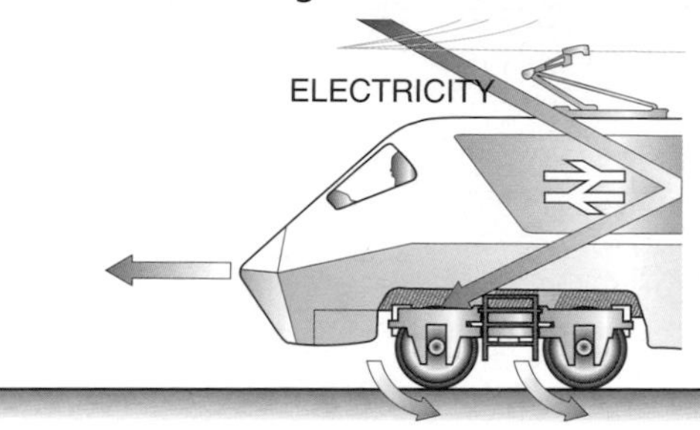

b) The energy needed to move this train comes from overhead electric cables

c) The energy needed to move this car comes from the wound-up spring

Different kinds of energy

There are many different kinds of energy.

★ **Light energy** which we obtain from luminous objects, such as the Sun, the stars, fires and light bulbs.

★ **Heat (thermal) energy** which we obtain from hot objects, such as the Sun and fires.

★ **Sound energy** which we obtain from objects that are vibrating, such as the strings of a guitar or the vocal cords in your throat.

★ **Electrical energy** which is available to us every time current flows.

★ **Kinetic energy** which is the energy an object has because it is moving.

★ **Elastic** or **strain potential energy** which is the energy an object has when it is stretched or twisted out of shape.

★ **Gravitational potential energy** which is the energy an object has because of its position. The higher it is, the more gravitational potential energy it has.

★ **Nuclear energy** which we obtain from reactions that take place in the centre of atoms.

Figure 4 The strain energy stored in this bow is going to be used to shoot an arrow. Once the arrow has been released and is flying through the air, it has kinetic energy because of its movement. As the arrow climbs higher and higher it gains gravitational potential energy

Energy transfers

When energy is used it does not disappear. It changes into a different form of energy.

Energy at the start	Energy changer	Energy after
chemical (in the wax)	candle	heat and light
electrical energy	radio	sound
electrical energy	electric fire	light and heat
strain energy	catapult	kinetic energy
kinetic energy	dynamo	electrical energy
electrical energy	electric motor	kinetic energy
light energy	green plants	chemical energy
chemical energy	battery	electrical energy

Although energy is never lost, it often changes into a less **concentrated** form which is not easy to use again. For example, the wax of a candle is a concentrated source of **chemical energy**, but the light and heat which it changes into is much less concentrated and is therefore difficult to re-use.

Key terms

Check that you understand and can explain the following terms:

- ★ energy
- ★ light energy
- ★ heat or thermal energy
- ★ sound energy
- ★ electrical energy
- ★ kinetic energy
- ★ potential energy
- ★ nuclear energy
- ★ concentrated
- ★ chemical energy
- ★ energy transfer

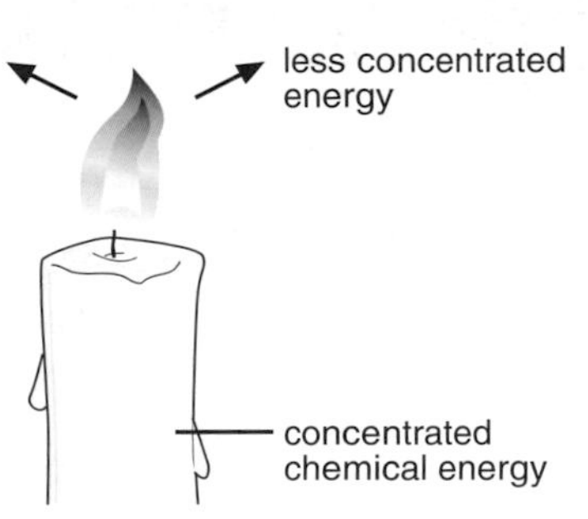

Figure 5 Burning a candle

We can show the amounts of energies in a simple energy transfer in the form of a diagram. Energy is measured in joules (J) or kilojoules (kJ). One kJ is equal to 1000 J.

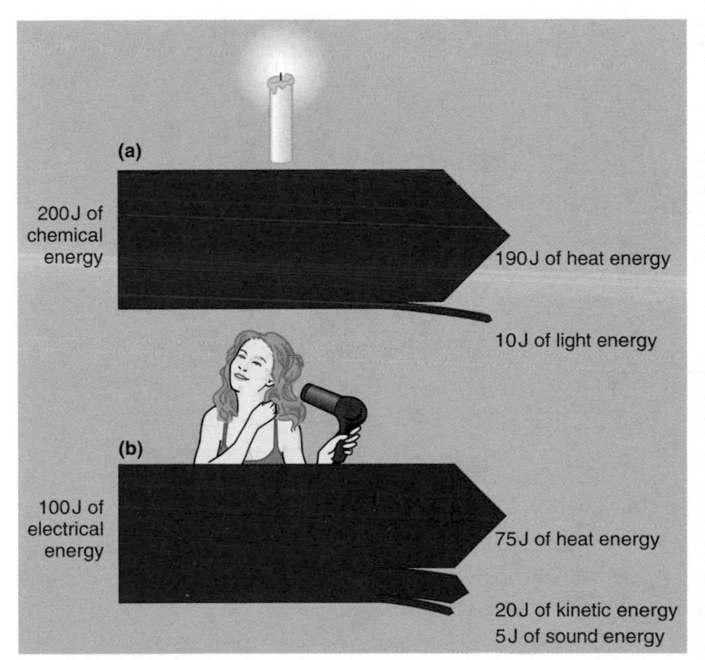

Figure 6 **Energy transfer** diagram for a candle and a hair drier

Questions

1 Name six different types of energy. Give one example of a source for each of these six types of energy.

2 Explain the energy transfer that takes place with **a)** a battery, **b)** a radio, **c)** a television, **d)** a bow and arrow, **e)** a wood fire and **f)** a clockwork car.

3 What device would change
- a) electrical energy into heat or thermal energy
- b) electrical energy into sound energy
- c) chemical energy into kinetic energy
- d) heat or thermal energy into kinetic energy
- e) light energy into chemical energy
- f) light energy into electrical energy?

4 Why are some forms of energy less useful after an energy transfer?

13.2 Energy resources

Energy from food

We obtain the energy we need to live and grow from the **food** we eat. Our age, sex and lifestyle determine the types of food we eat and how much we ought to eat each day. A man who does physical work needs lots of energy. A young child will need much less.

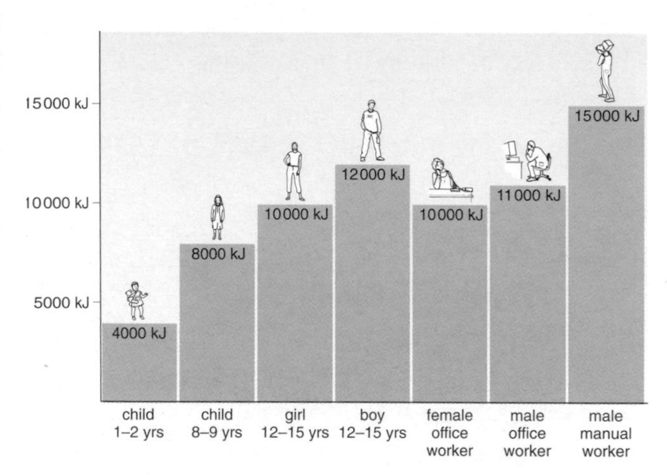

Figure 1 The approximate daily energy needs of different types of people

Different types of food contain different amounts of energy.

The table below shows how much energy is contained in an average portion of different foods.

Food	Energy content in kJ	Food	Energy content in kJ
apple/orange	200	cabbage	80
banana	300	carrots	80
boiled potatoes	400	lettuce	40
chips	1000	bread and butter	400/slice
boiled rice	500	ice cream	500
spaghetti	500	crisps	600
pizza	1200	cake	700
peas	300	chocolate	1500
baked beans	300	lemonade	700

Energy contents of different foods

It is important to remember that to be healthy our diet must be balanced. (See page 20)

Fossil fuels

Fuels are substances which release energy when they are burned. In the UK, the **fossil fuels**, coal, oil and gas provide almost 80% of our energy needs. If we continue to use them at this rate, they will soon be gone. Coal, oil and gas are **non-renewable** sources of energy. This means that once they have been used up, they cannot be replaced.

Millions of years ago, when plants and animals in the sea died, they fell to the sea bed

They were covered by layers of mud and rocks. Over a long period of time, more and more of these layers built up

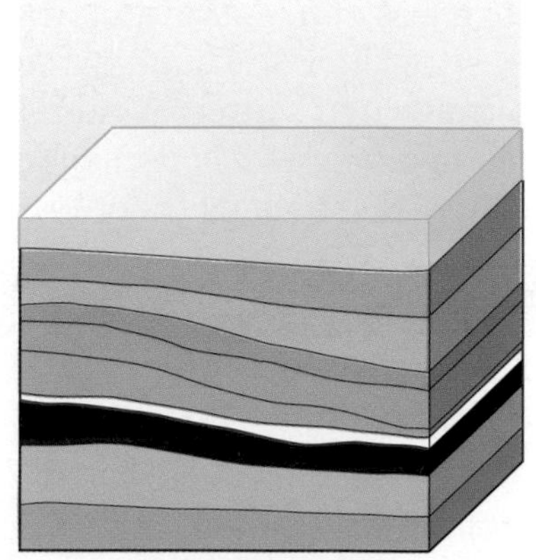

Enormous pressures were created as the layers sank deeper and deeper. These pressures changed the dead plants and animals into coal, oil or gas

Figure 2 The formation of fossil fuels

To avoid running out of fossil fuels we must use them more efficiently and we must look for alternative sources of energy which are **renewable**.

How are fossil fuels formed?

Fossil fuels are formed from dead plants and animals which lived on the Earth millions of years ago (see Figure 2). Those that died and sank to the bottom of lakes and seas became covered with layers of mud. Over a long period of time, more and more layers formed on top, creating enormous pressures and high temperatures which gradually changed the dead plants and animals into coal, oil and gas.

Energy from the Sun

Most of the energy we use originally came from the Sun. The diagram below shows the many different ways in which this energy is captured.

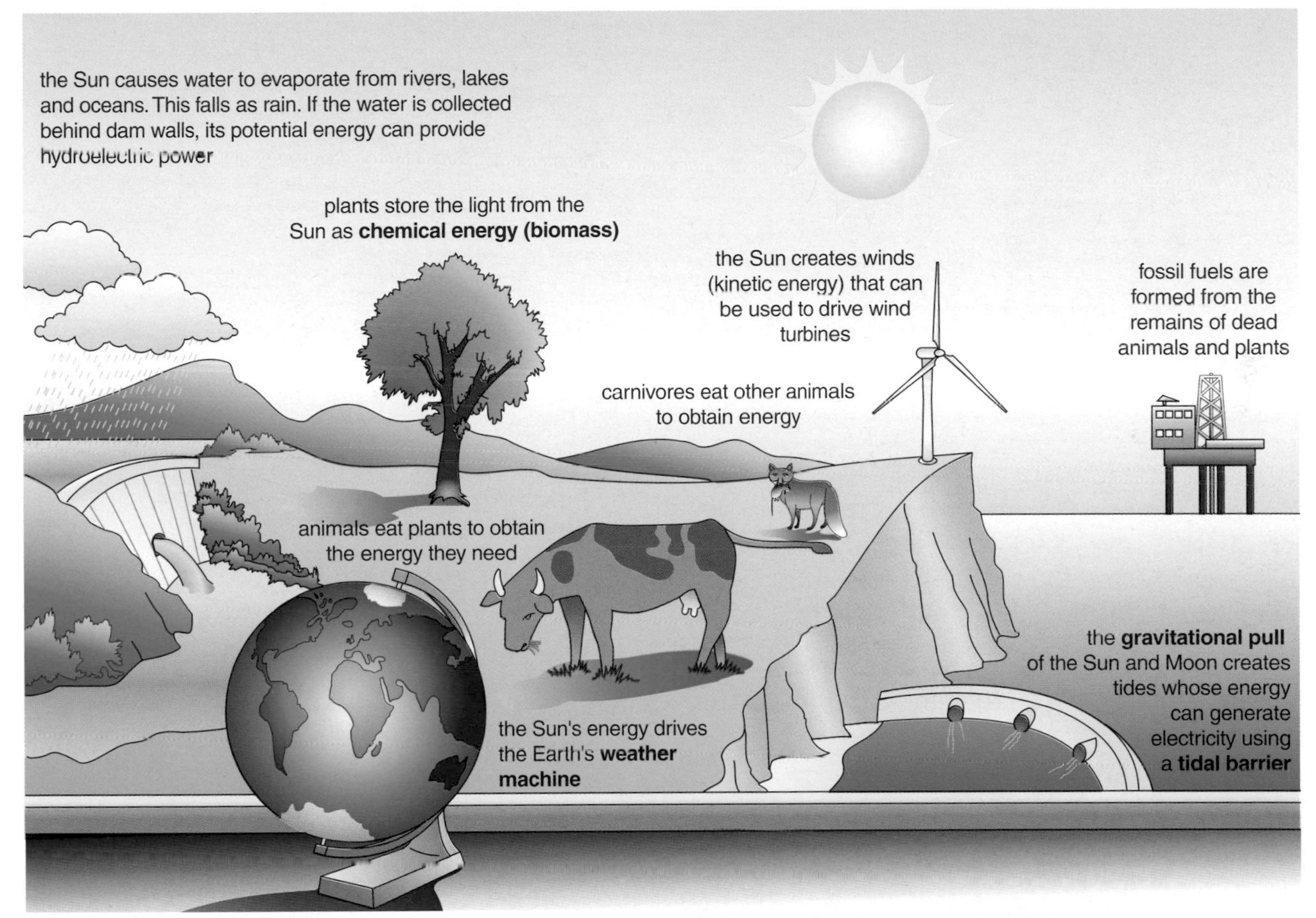

Figure 3 Energy from the Sun creates useful energy resources

Key terms

Check that you understand and can explain the following terms:

- ★ food
- ★ fuel
- ★ fossil fuel
- ★ non-renewable source of energy
- ★ renewable source of energy

Questions

1. Explain why different people may need to eat different amounts of food.
2. Using the information in the table opposite estimate the energy you have gained from the food you have eaten over the last 24 hours.
3. Explain why it is important that we slow down the rate at which we are using fossil fuels.
4. Give some examples which support the statement that 'Nearly all the energy resources on the Earth originate from the Sun'.
5. Explain where the energy in a piece of meat comes from.

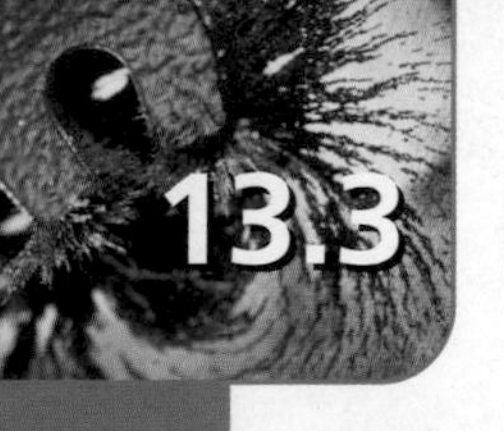

13.3 Renewable sources of energy

Wind energy

Heat energy from the Sun sets up convection currents in the Earth's atmosphere. The kinetic energy of the moving air can be used to turn the blades of wind turbines and so produce electricity. This is an excellent 'low technology' source of energy, but if there is no wind, no electricity is generated.

Figure 1 Wind farms make use of the kinetic energy from moving air

Hydroelectric energy

Rain water stored behind a dam has gravitational potential energy. When it is released, the water's energy can be used to drive turbines and generators to produce electricity (see page 124).

If the upper lake is manmade, i.e. created by flooding, it may have a large impact on the local environment.

Tidal energy

The Moon and the Sun cause waters around the world to rise and fall. If the water is trapped behind a barrier when the tide is high and released when the tide is low, its potential energy can be used like hydroelectric energy.

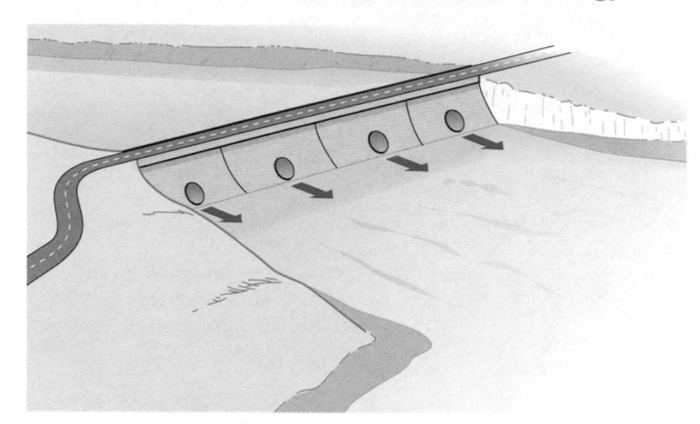

Figure 3 Tidal barriers also use the gravitational potential energy of stored water

Tidal barriers may restrict the passage of vessels up and down a river. They may also create flooding problems.

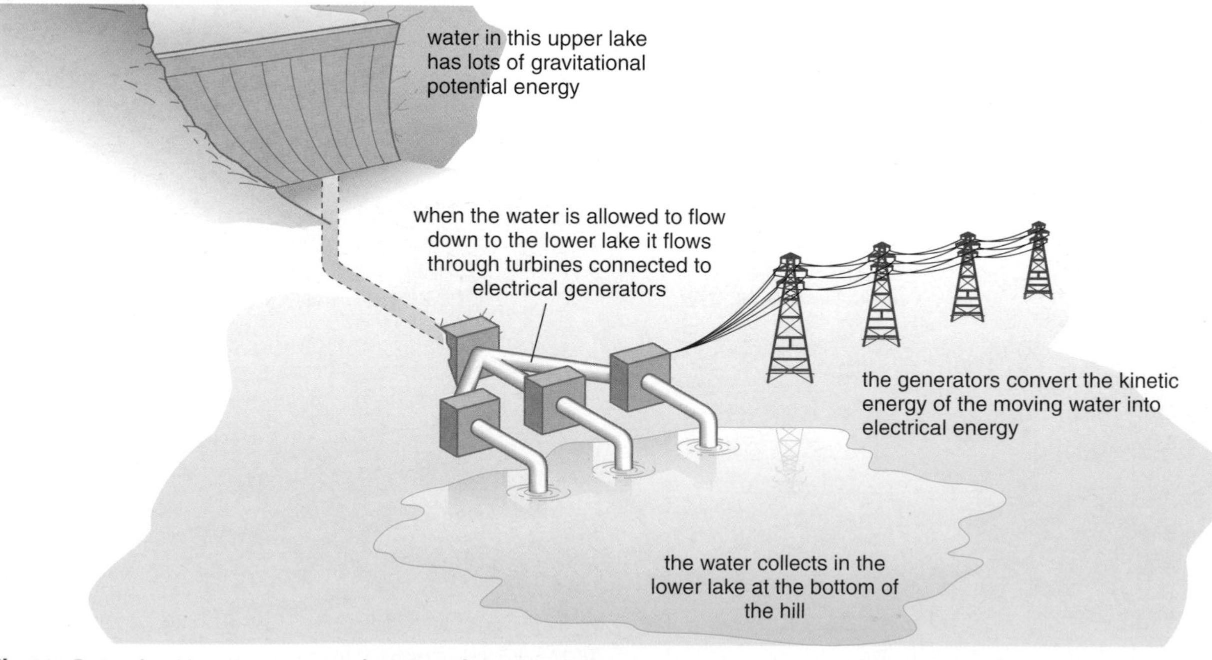

Figure 2 Hydroelectric power plants make use of the gravitational potential energy of stored water

Solar energy

Solar energy can be converted directly into electricity using photocells like those seen in the calculator in Figure 4, or it can be captured by solar panels and used for heating.

As a major source of energy this may be limited to those countries which have many hours of bright sunlight each day.

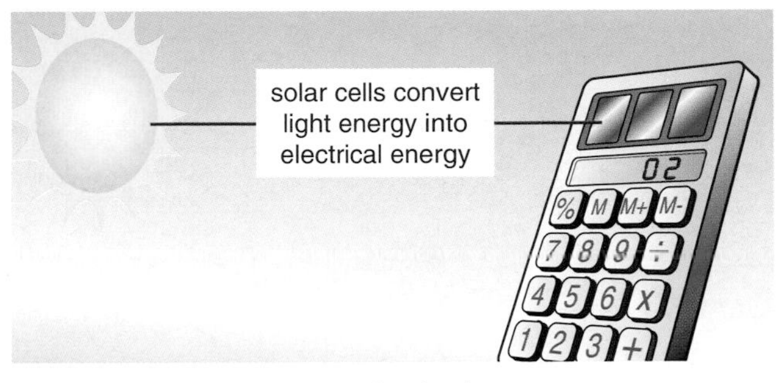

Figure 4 A solar-powered calculator

Biomass

The Sun's energy can be captured by plants and trees and turned into chemical energy which is released on burning. The trees and plants can then be replanted to capture even more of the Sun's energy.

To make a significant contribution to our energy needs, very large areas of land would need to be used to grow sufficient **biomass**.

Wave energy

The constant up and down movement of the surface of the sea can be used to generate electricity.

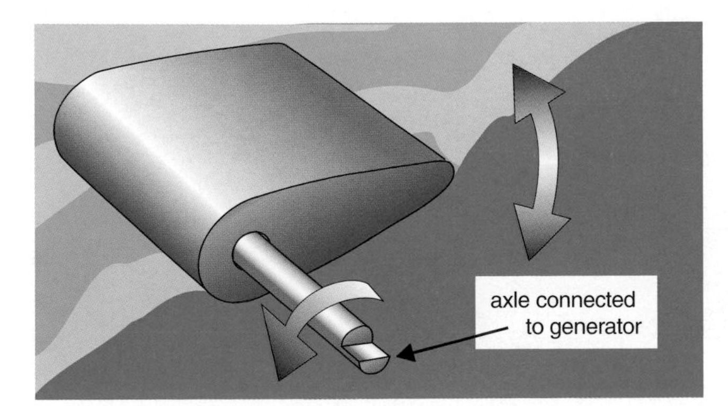

Figure 5 Capturing the energy of water waves

Energy capture by machines like these is poor and very large areas of them would be needed to make a worthwhile contribution to our energy needs.

Geothermal energy

Deep inside the Earth it is very hot because of the radioactive materials which exist there. Cold water which is piped down into the ground returns as steam which can be used to generate electricity.

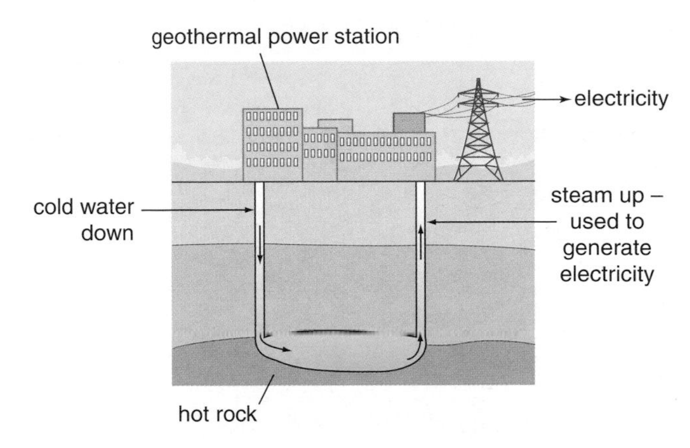

Figure 6 Capturing geothermal energy

The construction cost of geothermal power stations is very high as deep drilling is very expensive. Also there are very few suitable sites, i.e. where the Earth's crust is thin.

Key terms

Check that you understand and can explain the following terms:

- ★ wind energy
- ★ hydroelectric energy
- ★ tidal energy
- ★ solar energy
- ★ biomass
- ★ wave energy
- ★ geothermal energy

Questions

1. Explain what is meant by the phrase 'non-renewable source of energy'. Give two examples of non-renewable sources of energy.
2. Explain what is meant by the phrase 'renewable source of energy'. Give two examples of renewable sources of energy.
3. Which renewable sources of energy depend upon the weather?
4. Which renewable sources of energy might be used in countries which **a)** have lots of mountains and hills, **b)** have lots of sunshine, and **c)** have a coastline?
5. Name one fuel which is a renewable source of energy.
6. Name one renewable source of energy whose construction might have a large impact on the environment.

13.4 Generating electricity

Electrical energy is one of the most convenient forms of energy as it is easily transformed into other types of energy. Most of the electrical energy you use at home is generated at a **fossil fuel power station**.

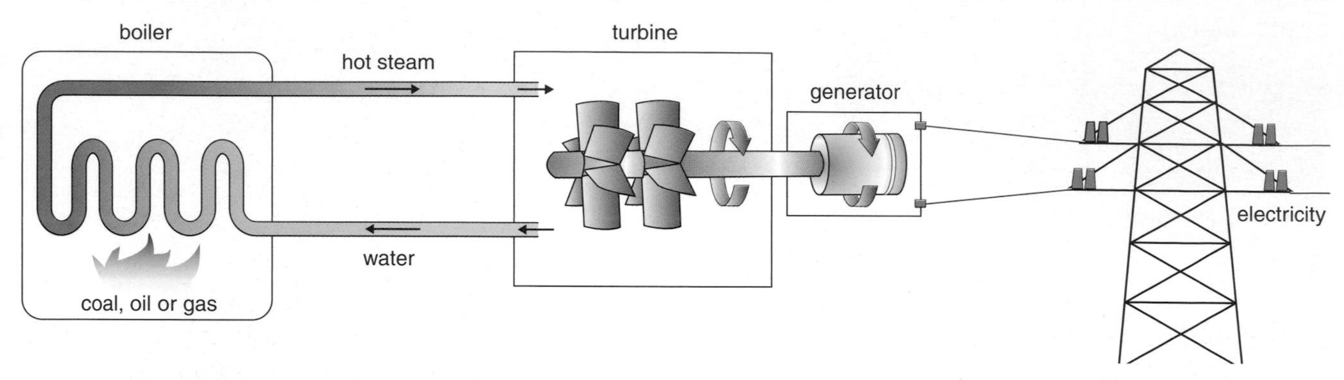

Figure 1 A fossil fuel power station

At a coal, oil or gas power station the fuel is burned to release heat energy. This energy is then used to heat water, changing it into steam.

The energy contained by the steam is then used to turn **turbines**. The turbines turn the **generators** which produce the electricity.

The electrical energy produced travels along a network of wires called the **National Grid** to our homes.

The energy changes involved in this process are:

CHEMICAL ENERGY → HEAT ENERGY → KINETIC ENERGY → ELECTRICAL ENERGY

At a **nuclear power station** the heat energy needed to produce the steam comes from nuclear reactions within radioactive materials such as uranium.

In a **hydroelectric power station** there is no boiler or steam. It is the kinetic energy of the water which is used to drive the turbines.

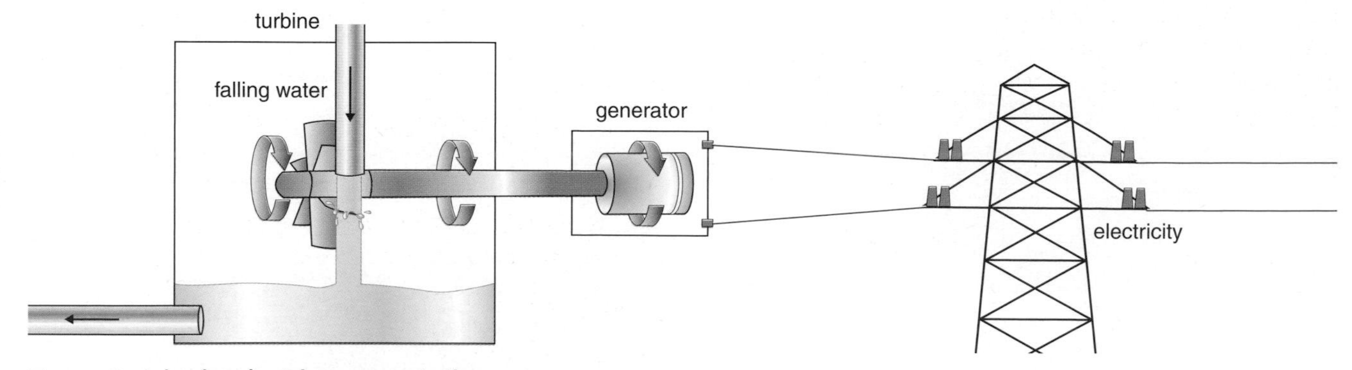

Figure 2 A hydroelectric power station

The energy changes involved are:

GRAVITATIONAL POTENTIAL ENERGY → KINETIC ENERGY → ELECTRICAL ENERGY

Paying for your electricity

As the electrical energy from a power station enters our homes it passes through an electricity meter which measures how much electrical energy we have used. Every three months we receive a bill for the electrical energy we have used.

Figure 3 An electricity meter

Different electrical appliances use or transform electrical energy at different rates. Some like electrical heaters use the energy quickly. Others like light bulbs use the electrical energy much more slowly. The rate at which energy is being used is called the **power** or power rating of the appliance. We measure power in **watts** (W) or kilowatts (kW).

Figure 4 Power rating for a food blender

Key terms

Check that you understand and can explain the following terms:

- ★ fossil fuel power station
- ★ turbine
- ★ generator
- ★ National Grid
- ★ nuclear power station
- ★ hydroelectric power station
- ★ power
- ★ watts

Questions

1 Name three fuels that might be burned at a power station.

2 Draw a block diagram to show the energy changes that take place at a fossil fuel power station.

3 Draw a block diagram to show the energy changes that take place at a hydroelectric power station.

4 In what units do we measure the power of an appliance?

5 Suggest an approximate power rating for
a) a very dim light bulb
b) a very bright light bulb
c) a very powerful electric fire.

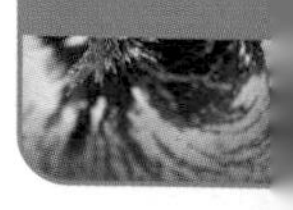

13.5 Temperature and heat energy

Measuring temperature

Temperature is a measure of **hotness**. Human beings are not very good at judging temperatures, so for accurate measurements of temperature we use **thermometers**. There are several different types of thermometer shown below.

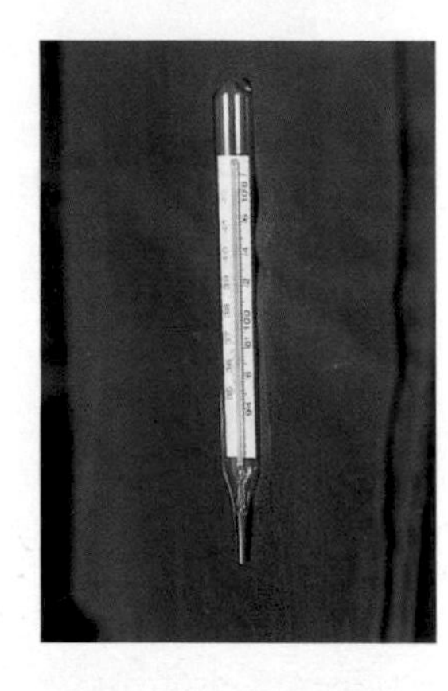

Figure 1 As the liquid mercury becomes warm, it expands and rises. The hotter it is, the further it rises. A scale at the side of the column shows the measured temperature

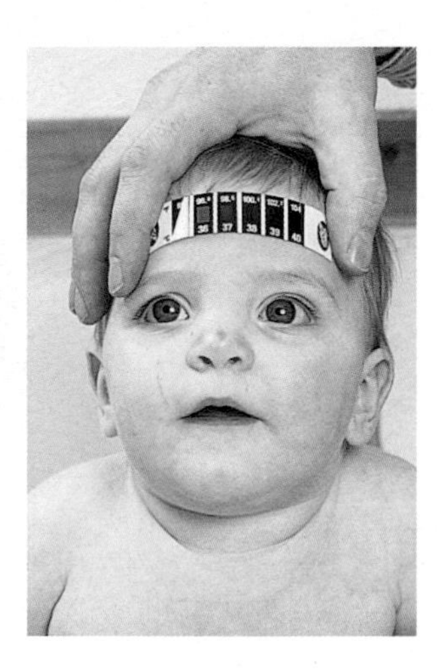

Figure 2 As the temperature of the plastic strip increases or decreases, it's colour changes. The colour of the strip indicates the temperature

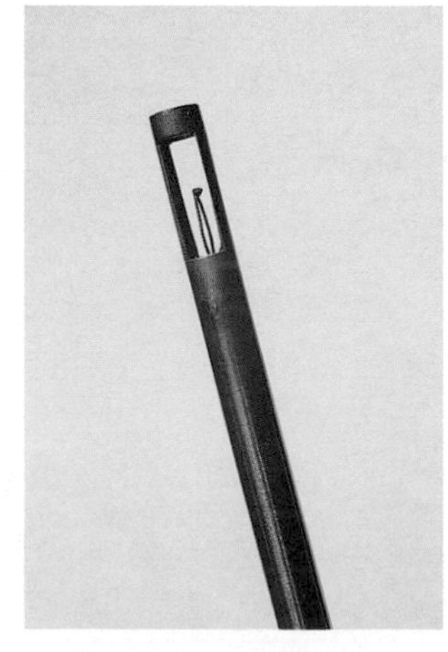

Figure 3 Temperature differences between wires inside a thermocouple create small voltages. The sizes of these voltages indicate the temperature

We measure temperature on the **Celsius scale**, for example the temperature at which pure water boils is 100 °C.

More examples of temperatures on the Celsius scale are shown in the table below.

Where	Temperature	Where	Temperature
absolute zero	−273 °C	boiling water	100 °C
coldest place on Earth	−89 °C	oven	200 °C
water freezes	0 °C	Bunsen flame	1100 °C
average room temperature in UK	20 °C	filament bulb	3500 °C
our body temperature	37 °C	surface of the Sun	6000 °C
maximum temperature on Earth	60 °C	centre of the Sun	15 000 000 °C

Heat energy

Heat energy naturally flows from places of higher temperature to places of lower temperature.

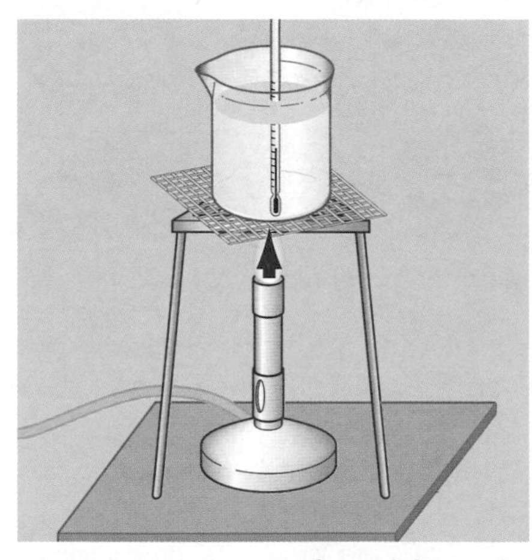

Figure 4 Heat energy is transferred from the Bunsen flame to the water

In Figure 4, heat is flowing from the hot Bunsen flame into the much cooler water. As a result the temperature of the water increases. We can see from this example that although heat and temperature are related, they are not the same. Heat is a form of energy and is therefore measured in joules (J). Temperature is a measure of hotness and is measured in degrees Celsius (°C).

If a solid such as a piece of ice is heated, it takes in heat energy. The graph below shows how the temperature and the **state** of the ice change with time.

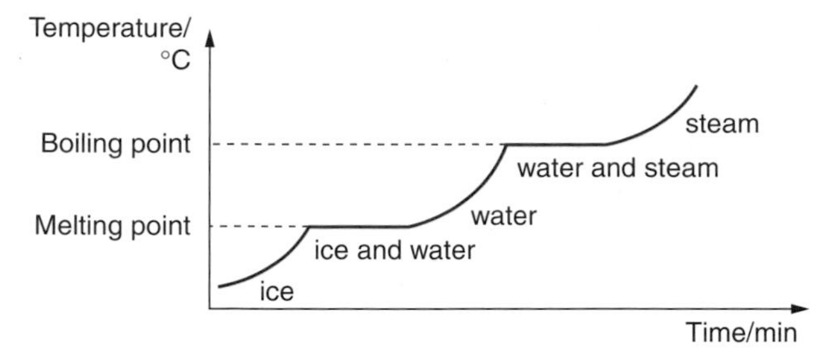

Figure 5 Ice changes to steam as the temperature rises

If the steam is allowed to cool, it loses heat energy. The graph below shows how the temperature and state of the steam change with time.

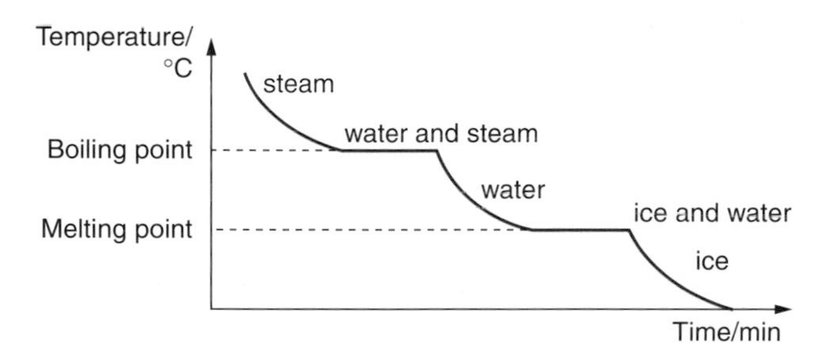

Figure 6 Steam changes back to ice as the temperature is reduced

Thermal expansion and contraction

When an object is heated its **particles vibrate** more vigorously and the object **expands**. When an object cools, the vibrations of its particles become less vigorous and the object **contracts**.

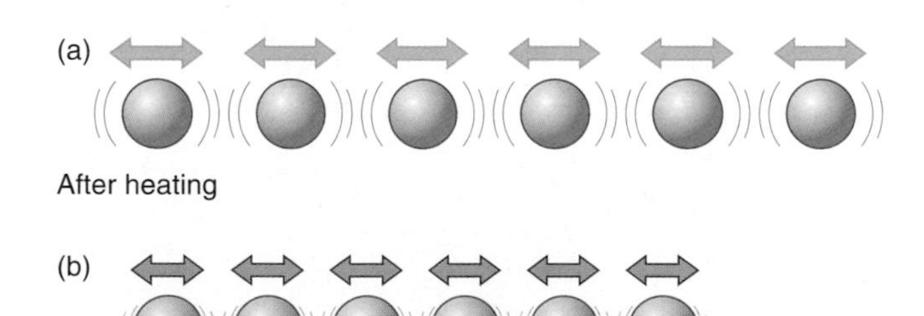

Figure 7 Thermal expansion and thermal contraction

The Firth of Forth railway bridge is approximately one metre longer in the summer than it is in the winter. This change in length had to be taken into account when the bridge was constructed.

When objects expand and contract they can exert very large forces. There are occasions when these large forces can be used to our advantage, as in Figure 8.

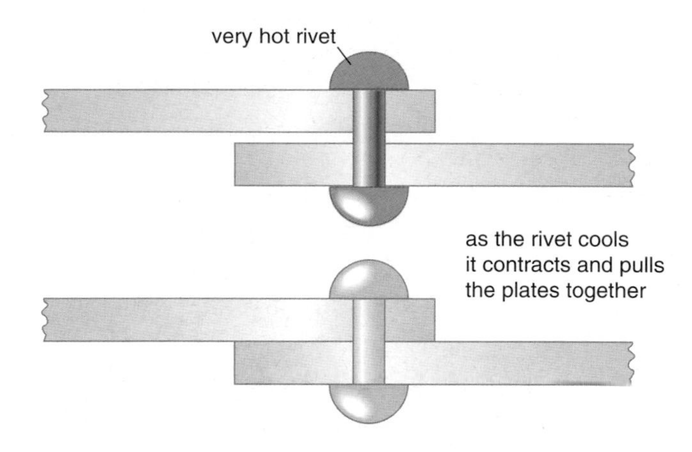

Figure 8

Key terms

Check that you understand and can explain the following terms:

- ★ temperature
- ★ hotness
- ★ thermometer
- ★ Celsius scale
- ★ heat
- ★ change of state
- ★ particles
- ★ vibrate
- ★ thermal expansion
- ★ thermal contraction

Questions

1. Why do we not rely on our sense of touch and feel to measure temperatures?
2. Describe how we would use a mercury thermometer to measure the temperature of a beaker of water.
3. Explain in words what happens to a very cold piece of ice if it is heated until it has all changed into steam.
4. Explain, using the particle model, why objects expand when heated.
5. Give one use of thermal expansion and contraction.
6. Give one disadvantage of thermal expansion and contraction.

13.6 Heat transfer

Conduction

Conduction is the transfer of heat through a material by **vibrating particles**. Materials that do this well are described as good **thermal conductors**. All metals are good conductors of heat. Materials that do not allow good conduction of heat through them are called **insulators**. Materials that are good insulators include plastics, wood and fabrics.

The particles that form the outside of the base of the pan in Figure 1 gain energy from the heater and are made to vibrate more vigorously. Neighbouring particles within the base are then jostled by these excited particles, causing them to vibrate more vigorously as well. Eventually all the particles in the base of the pan will be vibrating vigorously, i.e. heat has been conducted through the pan base. Because we want heat to transfer easily and quickly through the pan, it is made from a metal such as steel or copper.

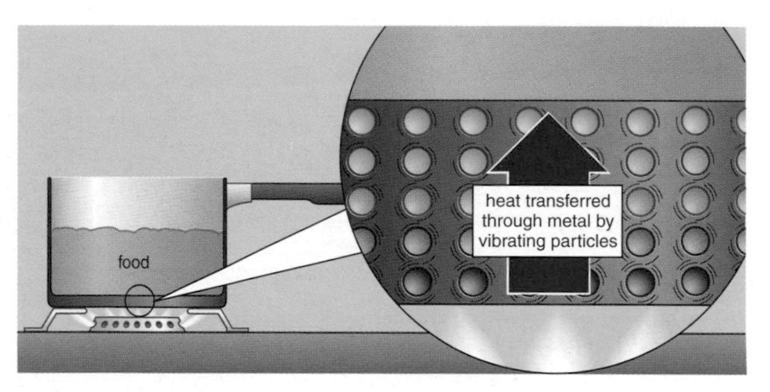

Figure 1 Metals are very good thermal conductors, so are used for pans. The heat from the gas flame is transferred through the metal pan to the food

To prevent the pan handle from becoming too hot to hold, it is made from an insulator, for example wood or plastic.

Comparing the thermal conductivity of different metals

Some metals conduct heat better than others. The apparatus in Figure 2 compares the thermal conductivity of four different metals.

Heat is conducted from the centre of the apparatus out along all the metal bars. Eventually the wax attaching the marbles to the bars will melt and the marbles will fall. The best conductor of the four will transfer heat the fastest and is identified as the bar whose marble falls first.

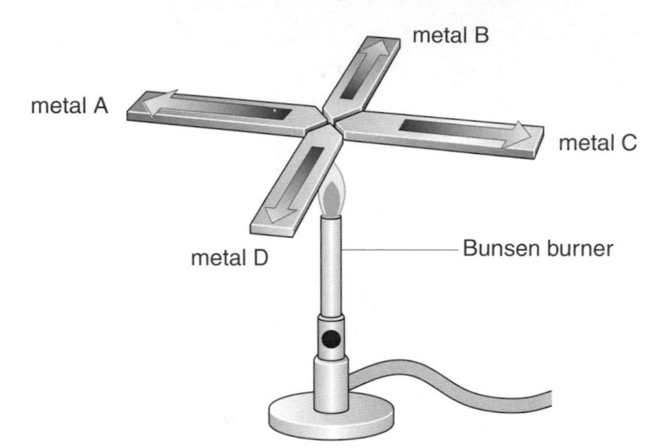

Figure 2 Comparing the conductivities of different metals

Conduction through liquids and gases

Most liquids, including water, are poor conductors of heat.

Gases are extremely poor conductors of heat. In fact they are excellent insulators and are often used where we want to stop heat from escaping, for example double glazing, woollen clothing and fibreglass.

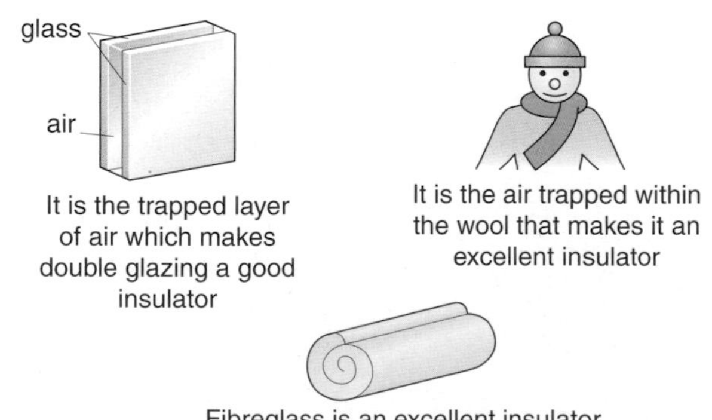

Figure 3 Using the insulating properties of air

Convection

The transfer of heat through a liquid or a gas often occurs by **convection**.

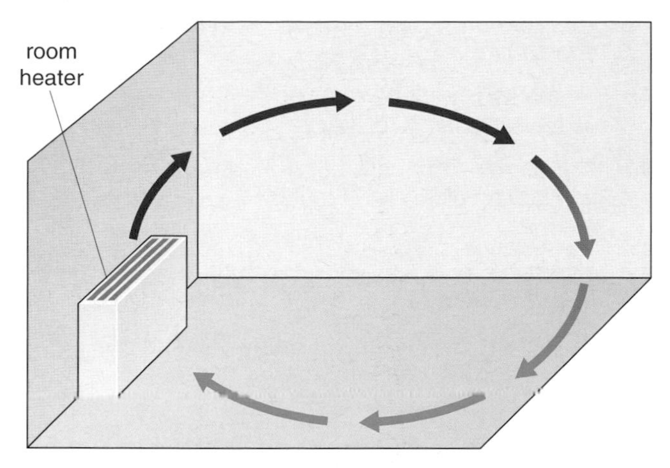

Figure 4 A convection current in a room

This circular movement of air which transfers heat to all parts of the room is called a **convection current**.

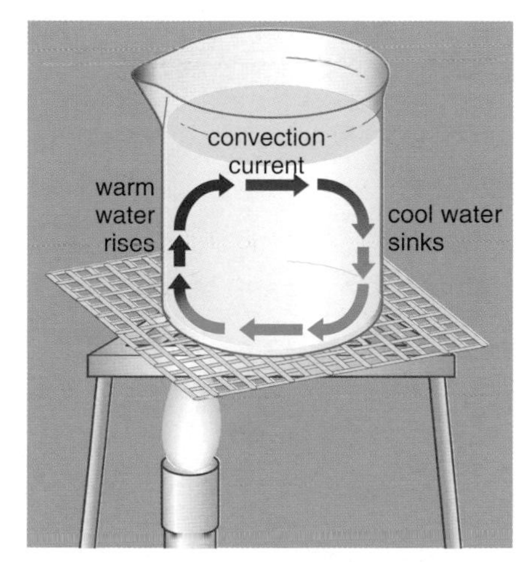

Figure 5 A convection current in a liquid

Radiation

Radiation is the transfer of heat energy by waves (or rays). All the energy which we receive from the Sun travels as radiation. It cannot be transferred by conduction or convection as there are no particles between the Sun and the Earth's atmosphere. Radiation is the only method of heat transfer that can take place in or across a **vacuum**.

When heat radiation strikes an object it may be **absorbed** or **reflected**. If an object absorbs the radiation it becomes warmer.

How much radiation an object absorbs depends upon the nature of its surface. Dark objects with rough surfaces absorb more radiation than objects with shiny, light-coloured surfaces.

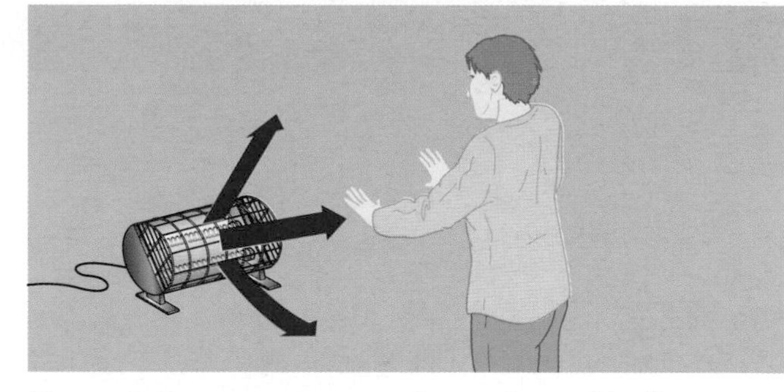

Figure 6 Heat energy travels out from this fire to the boy as waves

Heat loss from the home

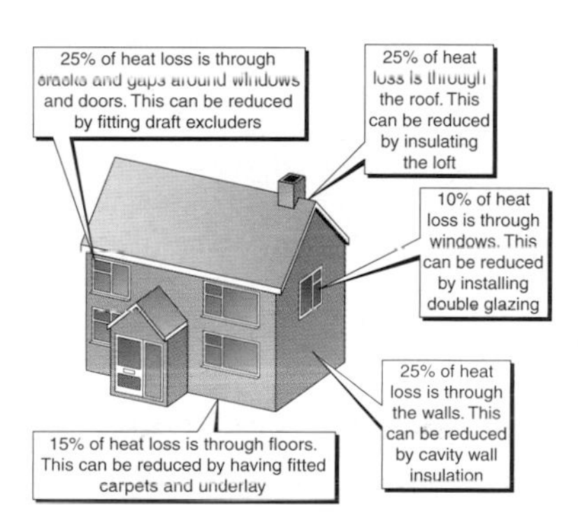

Figure 7 How heat is lost from a house

Key terms

Check that you understand and can explain the following terms:

- ★ conduction
- ★ vibrating particles
- ★ thermal conductor
- ★ thermal insulator
- ★ convection
- ★ convection current
- ★ radiation
- ★ vacuum
- ★ absorbed
- ★ reflected

Questions

1. Give one example of a material which is a good thermal conductor and its use.
2. Give one example of a material which is a good thermal insulator and its use.
3. Where in a beaker of water is the water the warmest? Explain your answer.
4. What is a convection current?
5. Which is the only method by which heat can transfer across a vacuum.
6. Name two things that may happen to heat radiation when it strikes an object.
7. Why are houses in hot countries often painted white?

Chapter 13 Energy and energy resources:

What you need to know

1. We need energy to live, to grow and to work.
2. There are several different forms of energy, including heat (thermal) energy, light energy, sound energy, electrical energy, chemical energy, nuclear energy, potential energy and kinetic energy.
3. Fuels are concentrated sources of energy. When burned they release heat energy.
4. Coal, oil and gas are fossil fuels.
5. Fossil fuels are non-renewable sources of energy.
6. Non-renewable sources of energy cannot be replaced once they have been used, and so must be conserved.
7. Wind, hydroelectric energy, waves, tides, geothermal energy, solar energy and biomass are all renewable sources of energy.
8. Renewable sources of energy can be replaced.
9. The ultimate source of most of the Earth's energy is the Sun.
10. Electricity is generated at power stations using a variety of energy resources.
11. The power or power rating of an electrical appliance tells us how rapidly it is converting energy.
12. Temperature is a measure of hotness. Heat is a form of energy.
13. Heat transfer can take place by conduction, convection and radiation.
14. When objects absorb heat they may increase in temperature, change state and/or expand.

How much do you know?

1 The table below contains some examples of energy resources.

Tick the correct box to show whether an energy resource is renewable or non-renewable.

Energy resource	Renewable	Non-renewable
Coal	☐	☐
Wind	☐	☐
Solar	☐	☐
Oil	☐	☐
Gas	☐	☐
Hydroelectric	☐	☐

6 marks

2 a) A torch battery contains energy. In what form is this energy stored? Tick the correct box.

☐ light energy ☐ chemical energy

☐ kinetic energy ☐ electrical energy

1 mark

b) When the torch is turned on, energy flows from the battery to the bulb. In what form does the energy flow? Tick the correct box.

☐ light energy ☐ electrical energy

☐ chemical energy ☐ heat or thermal energy

1 mark

c) Fill in the missing words to explain what energy change takes place in the bulb.

The bulb changes __________ energy into __________ energy and __________ energy.

3 marks

3 The diagram below shows a ball and ring.

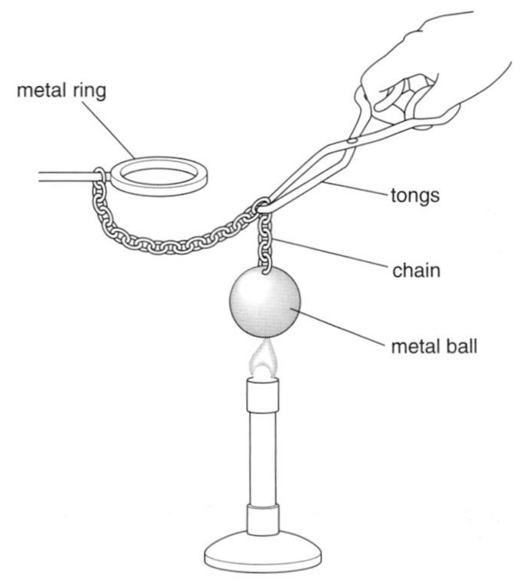

Before being heated the ball will pass through the ring.

a) Explain why the ball will not pass through the ring after it has been heated for several minutes.

____

1 mark

How much do you know? *continued*

b) Explain why, after the ball has been allowed to cool, it is able once again to pass through the ring.

___

1 mark

4 a) Coal, oil and gas are fossil fuels. Explain briefly how they are formed.

___

___

___

3 marks

b) Why are these fuels described as non-renewable energy resources?

___

___

2 marks

c) How is the energy released from a fuel?

___

1 mark

d) Where does the energy stored in fossil fuels originally come from?

___

1 mark

e) Suggest two ways in which we can make fossil fuels last longer.

___

___

2 marks

5 The diagram below shows a black car and a white car parked next to each other.

Explain why on a hot, sunny day the black car is much hotter to touch than the white car.

___

___

3 marks

6 The diagram below is a block diagram of a coal-fired power station. Fill in the gaps in the diagram to show the energy changes that are taking place. The first one has been done for you.

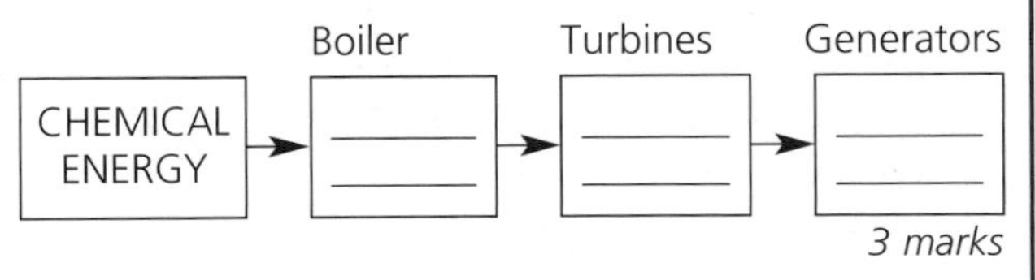

3 marks

7 Explain what energy transfer takes place when a piece of fresh meat is placed

a) in a freezer

___

1 mark

b) in an oven

___

1 mark

8 a) The Sun is the 'ultimate source of the Earth's energy resources'. Explain how the energy from the Sun became stored in the water behind the dam shown below.

___

___

3 marks

b) What kind of energy is stored in the water?

___

1 mark

c) Explain briefly how this energy can be changed into electrical energy.

___

___

4 marks

14.1 Simple circuits

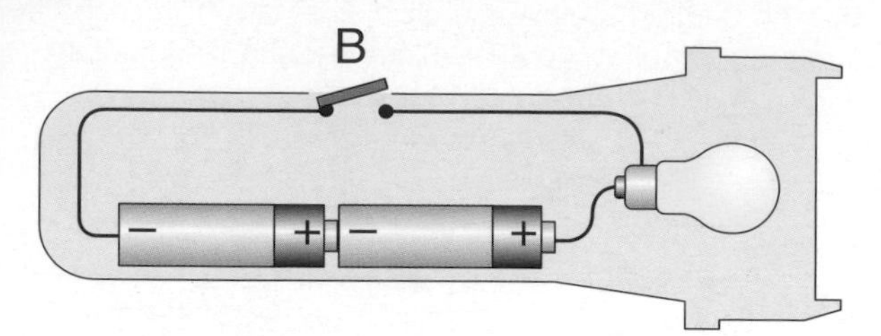

Figure 1 A simple torch circuit

The diagram above shows the workings of a simple torch. When the button B is pressed, electricity flows around the circuit and the torch bulb glows. When the button is released, electricity stops flowing around the circuit and the bulb no longer glows.

Like water flowing through pipes, electricity needs something to travel through. The 'pipes' for electricity are metal wires. The wires, batteries and bulbs form circuits around which currents flow.

Cells and batteries

A **cell** is a kind of pump which makes electricity move. This movement is called an **electric current**. If a larger current is needed several cells can be connected together to form a **battery**.

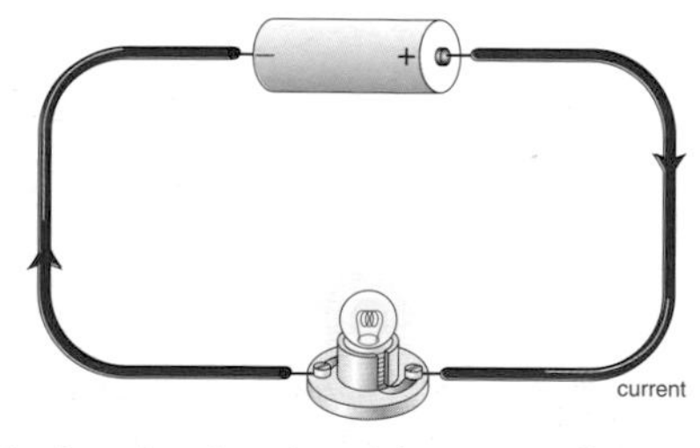

Figure 2 A simple circuit with one cell

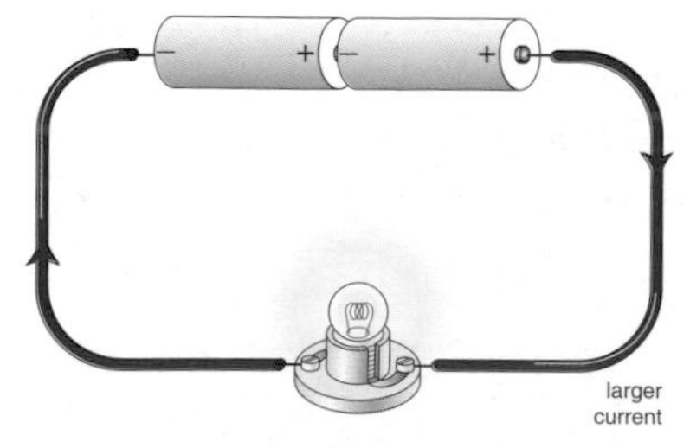

Figure 3 By connecting several cells together to make a battery, a larger current will flow around a circuit

Both circuits shown in Figures 2 and 3 are **complete circuits**. Electricity can flow all the way around them. But if a gap is created in the circuit, electricity will not flow and the bulb will not glow. We describe a circuit like this as being **incomplete**. The position of the gap in the circuit is unimportant. *Current will only flow if the circuit is complete.*

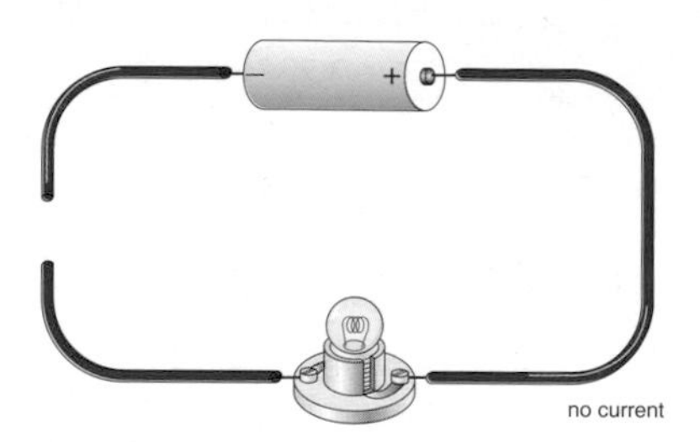

Figure 4 An incomplete circuit

Switches

A switch turns a circuit on and off. It behaves like a drawbridge which can make the circuit complete when closed, or incomplete when open.

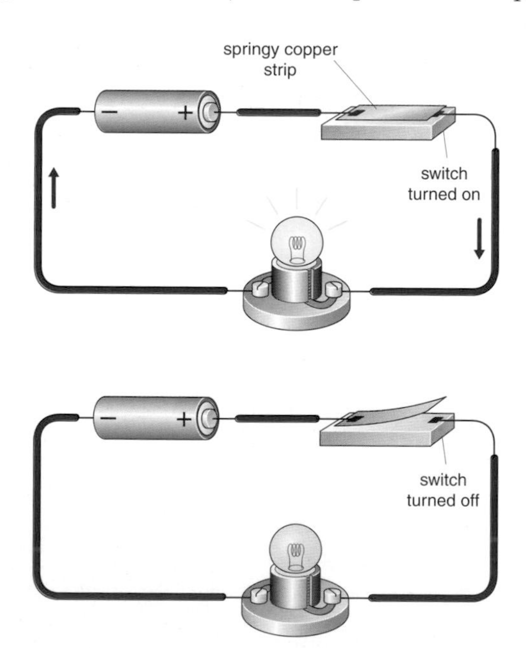

Figure 5 With the switch closed the circuit is complete. With the switch open the circuit is incomplete.

What it is	What it looks like	Symbol	What it does
Cell			Pulls and pushes charges around a circuit
Battery			Provides a larger current than a single cell
Connecting wire			Provides a path through which current can flow
Lamp/bulb			Glows brightly if sufficient current flows through it
Switch			Turns current in a circuit on or off
Resistor			Reduces the current flowing in a circuit
Variable resistor			By altering the value of a variable resistor, the size of the current can be changed

Circuit diagrams

Drawing diagrams such as those shown opposite is not easy. To simplify things scientists and electricians use **circuit diagrams**. These simple diagrams use symbols to represent the various bits and pieces (more properly called components) of the circuit. A list of some of the most common **components** is shown above.

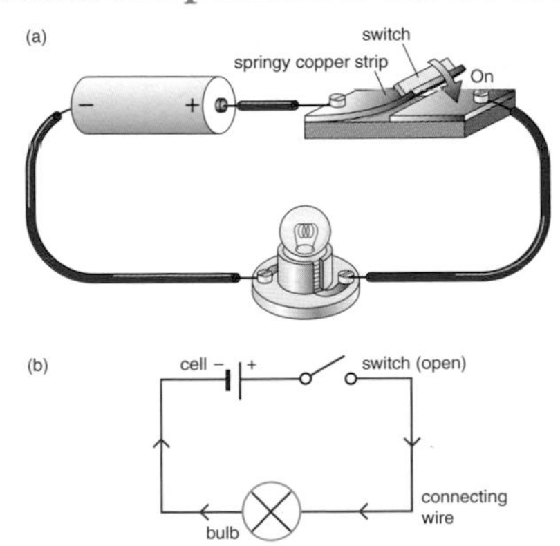

Figure 6 A circuit diagram

Conductors, insulators and test circuits

At present, no current can flow around the circuit in Figure 4 because it is incomplete. If a material which allows electricity to flow through it, i.e. a conductor, is placed across the gap, the circuit is complete and so the bulb glows. If a material which does not allow electricity to flow through it, i.e. an insulator, is placed across the gap, the circuit is still incomplete and so the bulb does not glow. A circuit like this can be used to test a component to discover if it is faulty.

Key terms

Check that you understand and can explain the following terms:

- ★ cell
- ★ electric current
- ★ battery
- ★ complete circuit
- ★ incomplete circuit
- ★ circuit diagram
- ★ components

Questions

1. What do we call two or more cells connected together? Explain why two cells might be connected together.
2. Explain why it is important that the cells in a battery are connected correctly.
3. What is **a)** a complete circuit and **b)** an incomplete circuit?
4. What is the main use for a switch in a circuit?
5. Draw the symbol for **a)** a cell, **b)** a battery, **c)** a bulb, **d)** an open switch, **e)** a closed switch.
6. What is a conductor? Give one example of a material that is a conductor.
7. What is an insulator? Give one example of a material that is an insulator.

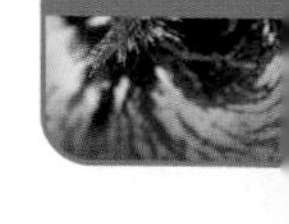

14.2 Series and parallel circuits

There are two kinds of electrical circuit. These are called **series circuits** and **parallel circuits**.

In a series circuit current has only one path to follow, i.e. there are no **branches**.

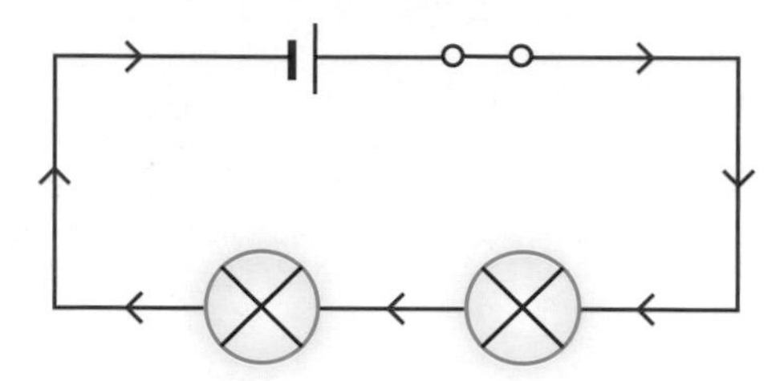

Figure 1 A series circuit

In a series circuit the same current passes through all parts. If one of the bulbs in the circuit below is turned off, they are all turned off.

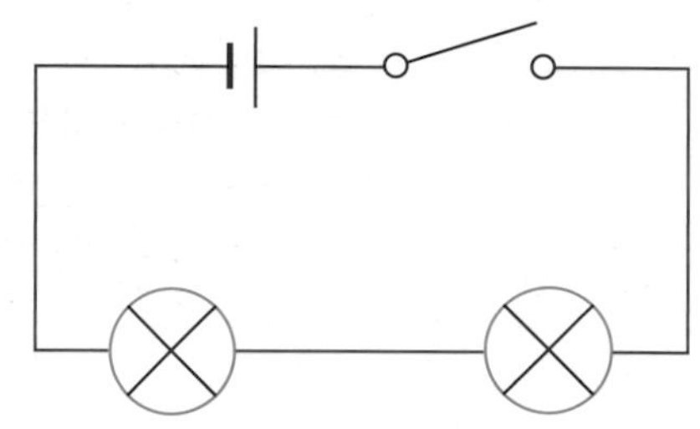

Figure 2 The switch is open, so the circuit is incomplete and the bulbs are off

In a parallel circuit there are several paths electricity can follow, i.e. there are branches.

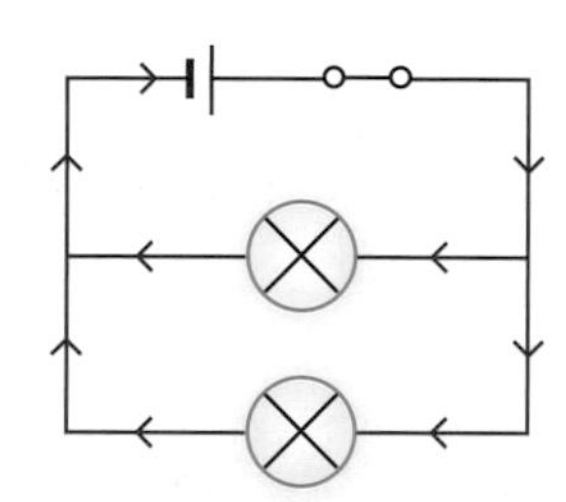

Figure 3 A parallel circuit

In a parallel circuit it is possible to switch off some parts of the circuit and yet leave others on, as shown in Figure 4.

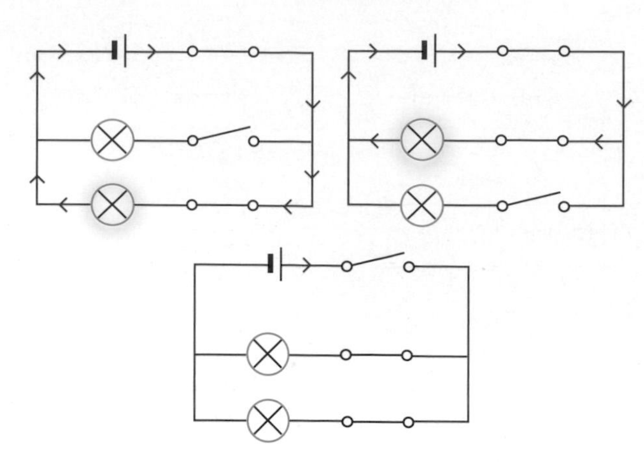

Figure 4 Turning off different parts of a parallel circuit

Measuring current

The size of a current is measured using an instrument called an **ammeter**. Current is measured in units called **amps** (A). The ammeter is placed **in series** with the part of the circuit being investigated.

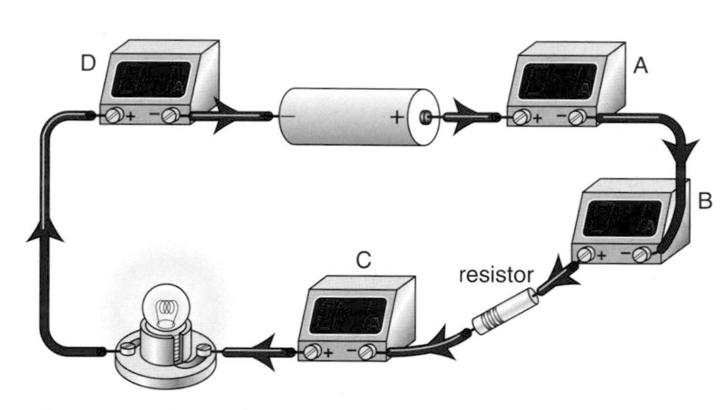

Figure 5 Measuring current in a series circuit

Currents in series circuits

★ Ammeter A is measuring how much current is leaving the cell.

★ Ammeter B is reading how much current is flowing into the resistor.

★ Ammeter C is reading how much current is flowing into the bulb.

★ Ammeter D is reading how much current is flowing back into the cell.

From these readings it is clear that:

1 The current leaving the cell is the same size as the current returning to it. Current is not 'used up' as it flows around a circuit.

2 The size of the current is the same in all parts of a series circuit.

Currents in parallel circuits

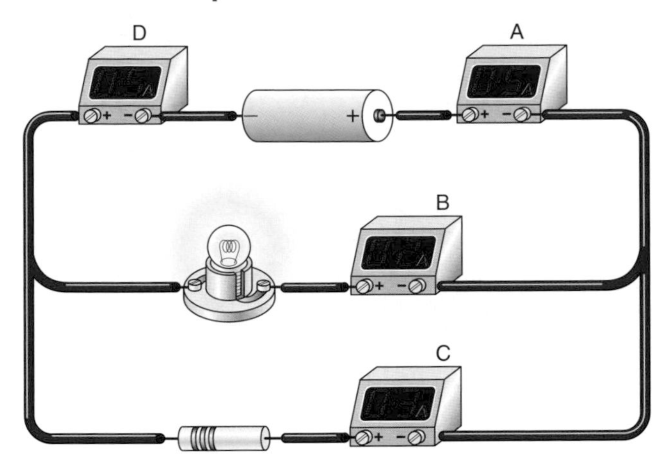

Figure 6 Measuring current in a parallel circuit

- ★ Ammeter A is measuring how much current is leaving the cell.
- ★ Ammeter B is reading how much current is flowing through the bulb.
- ★ Ammeter C is reading how much current is flowing through the resistor.
- ★ Ammeter D is reading how much current is flowing back into the cell.

From these readings it is clear that:

1. The size of the current in the different parts of a parallel circuit is not the same.
2. The current leaving the cell is the same size as the current returning to it.
3. The current entering a junction is equal to the current leaving a junction i.e. 0.5 A = 0.2 A + 0.3 A.

Energy in circuits

As current passes through a cell or a battery it receives electrical energy which it carries around the circuit. This energy is changed into other forms as the current passes through the various components. For example when current passes through a bulb, some of the electrical energy it is carrying is changed into heat and light energy. If the current passes through a buzzer, some of the energy is changed into sound.

We can measure how much energy is given to the electricity as it passes through the cell or battery, and how much electrical energy is transformed in the various components in a circuit using a voltmeter (Figure 7). The **voltmeter** is connected **in parallel** with the part of the circuit we are interested in. Large voltage readings indicate large energy transfers by the components.

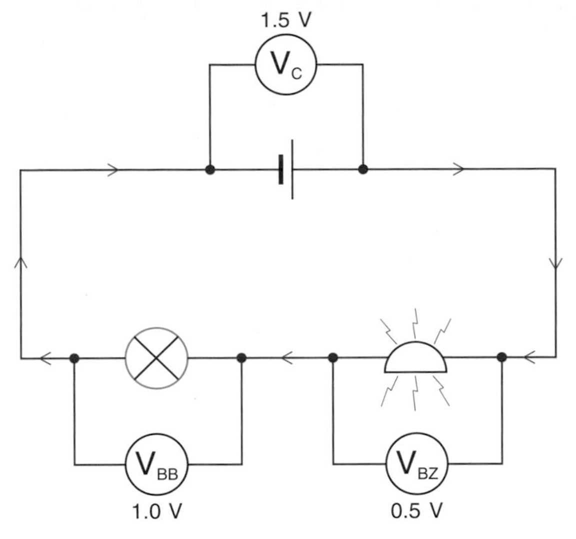

Figure 7

Note that the voltmeter reading across the cell is equal to the sum of the voltmeter readings across the bulb and the buzzer. This indicates that all the electrical energy received by the current as it passes through the cell is converted to other forms of energy by the components in the circuit.

Key terms

Check that you understand and can explain the following terms:

- ★ series circuit
- ★ parallel circuit
- ★ branches/junctions
- ★ ammeter
- ★ in series
- ★ voltmeter
- ★ in parallel

Questions

1. Draw a series circuit which contains a battery, four bulbs and a switch. What happens when the switch is opened and closed?
2. Draw a parallel circuit which contains a cell, four bulbs and two switches. The switches are connected so that one switch turns two bulbs on and off, whilst the second switch turns all four bulbs on and off.
3. Redraw the circuit for Question 2 but include two ammeters to measure the currents leaving and returning to the cell.
4. What do currents receive as they travel through cells and batteries?
5. What do currents carry as they travel around circuits?
6. If a voltmeter is connected across a cell in a circuit, what will the voltmeter reading tell you?
7. If the same voltmeter is connected across a glowing bulb in a circuit, what will the voltmeter reading tell you now?

14.3 Current and resistance

When the switch in this circuit is closed, there is a good flow of electricity through the bulb so it glows brightly.

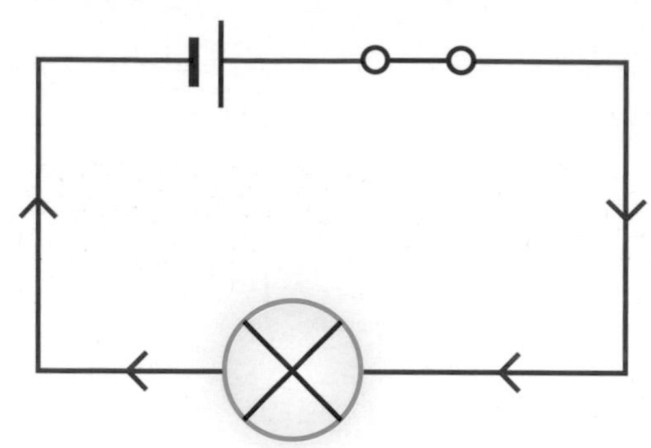

Figure 1 The bulb glows brightly in this circuit showing that the current is large

When a second or third bulb is included in the circuit, they each glow less brightly i.e. the current in the circuit is smaller.

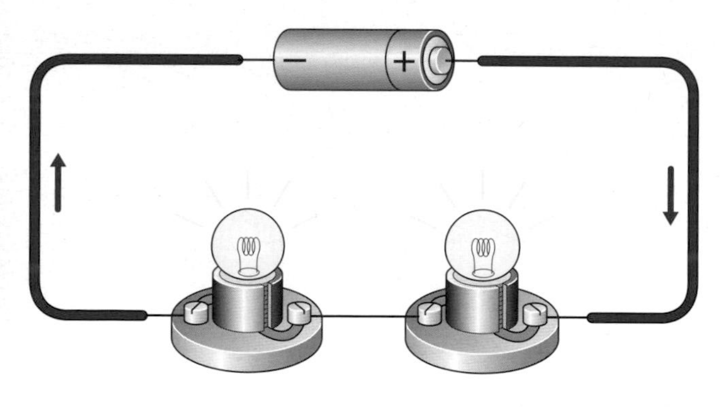

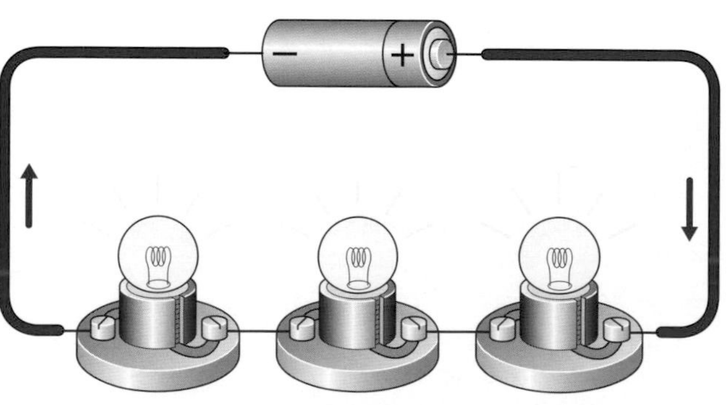

Figure 2 The bulbs glow less brightly when more bulbs are added

The size of the current in any circuit is affected by the number of components in the circuit and what they are. The components **oppose** the flow of electricity. They have **resistance**.

The effect of including a component which has resistance into a circuit can be explained by picturing the current as runners on an athletics track, and the component's resistance as an **obstacle** such as a set of step ladders. Without the component, the electricity flows freely around the track. When the component is included in the circuit, the flow of electricity, i.e. the current, is reduced.

Figure 3 The step ladder is an obstacle for the runners. In the same way, a component's resistance is an obstacle to the flow of electricity in a circuit

Resistors

Components called **resistors** are included in some circuits to control the electricity that flows.

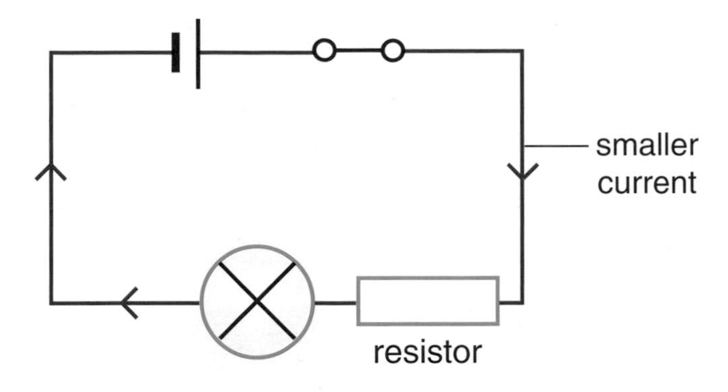

Figure 4 We can imagine the resistance of a component as being an obstacle to the flow of current

Some resistors have a resistance that can be changed or altered. They are called **variable resistors** and are extremely useful for altering the size of current in a circuit.

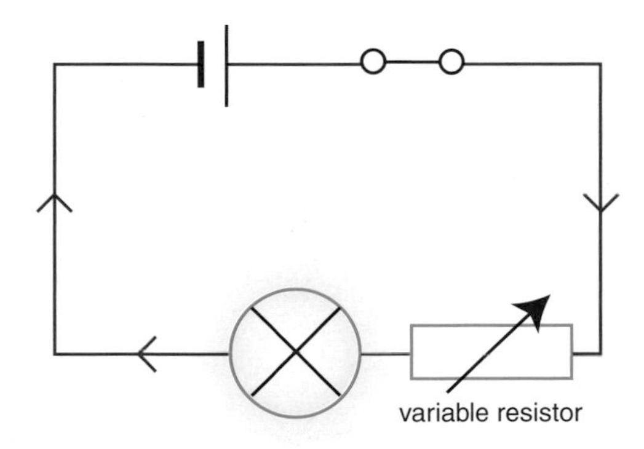

Figure 5

The variable resistor in the circuit in Figure 5 is being used as a **dimmer switch** to control the brightness of the bulb. Variable resistors are also used to alter the loudness of the music from a stereo system, the colour and brightness of the picture on a TV set, and the speeds of electric motors.

We measure the size of a resistor in ohms (Ω). For example a 10Ω resistor will offer half the resistance to the flow of current of a 20Ω resistor.

Key terms

Check that you understand and can explain the following terms:

- ★ oppose
- ★ resistance
- ★ obstacle
- ★ resistor
- ★ variable resistor
- ★ dimmer switch

Questions

1 Explain in your own words why the bulbs in circuit A will glow more brightly than those in circuit B.

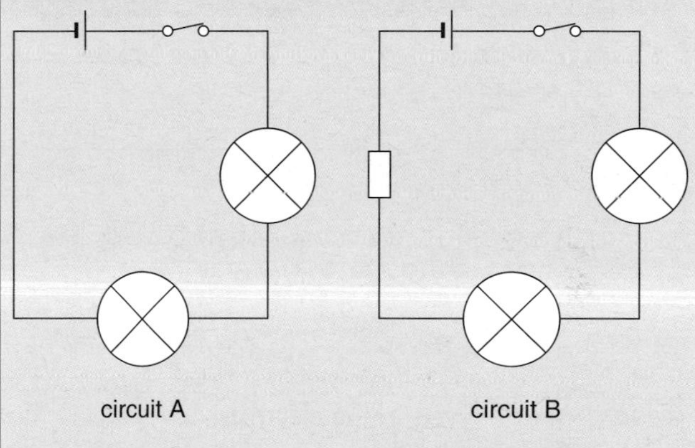

2 Why are resistors often included in electrical circuits?

3 Give four examples of circuits that contain variable resistors.

Chapter 14 Electricity:

What you need to know

1. Conductors allow currents to pass through them, insulators do not.
2. Electricity will flow around complete circuits, but not around incomplete circuits.
3. Series circuits have no junctions. Current at all points in a series circuit is the same.
4. Parallel circuits have junctions. Current may be different in different parts of a circuit.
5. Currents carry energy around circuits. This energy is changed into other forms by the components in the circuit. We can measure how much energy is being transferred using a voltmeter.
6. The components in a circuit may offer opposition to the flow of electricity. This opposition is called resistance. We can use the resistance of a component to control the size of current in a circuit.

How much do you know?

1 a) What do we call several cells connected together?

____

1 mark

b) Explain why it is important that the cells are connected the right way around.

____

1 mark

c) What is given to the electricity as it flows through the cells?

____

1 mark

2 a) Draw a circuit diagram for a series circuit containing a cell, a switch and four bulbs.

1 mark

b) Draw a circuit diagram for a parallel circuit containing a cell and four bulbs.

1 mark

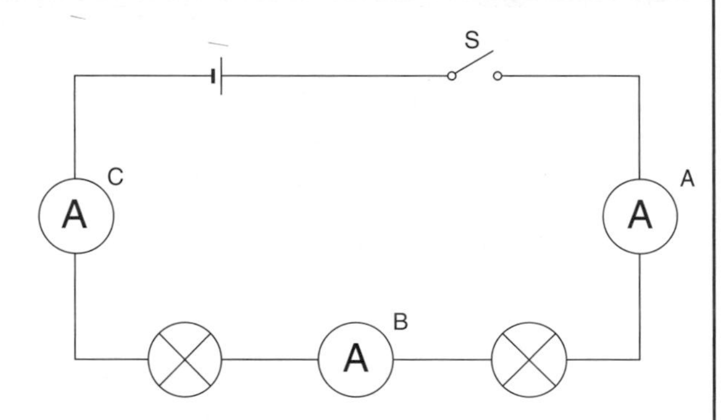

c) Explain why no electricity flows in the above circuit when the switch S is open.

____

1 mark

When switch S is closed a current of 0.2 A flows through ammeter A.

d) What current flows through ammeter B?

____

1 mark

e) What current flows through ammeter C?

____

1 mark

3 The diagram below shows two Christmas trees decorated with lights. Both trees have one bulb which is broken. None of the lights on tree B are glowing. All of the lights on tree A are glowing apart from the one that is broken.

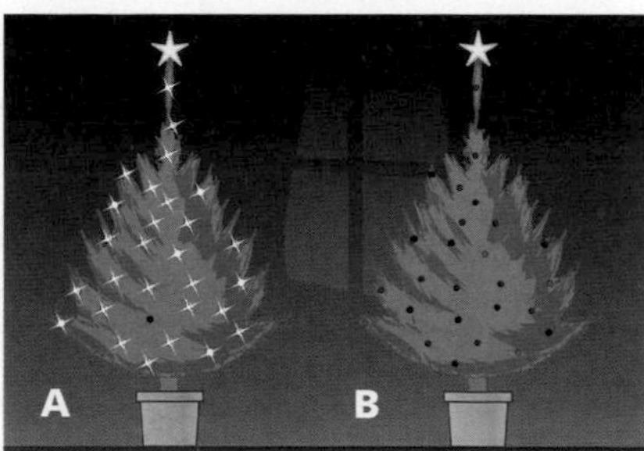

How much do you know? *continued*

a) What kind of lighting circuit does tree B have? Explain your answer.

____

____

____

3 marks

b) What kind of lighting circuit does tree A have? Explain your answer.

____

____

____

3 marks

4 Katy built the circuit shown below.

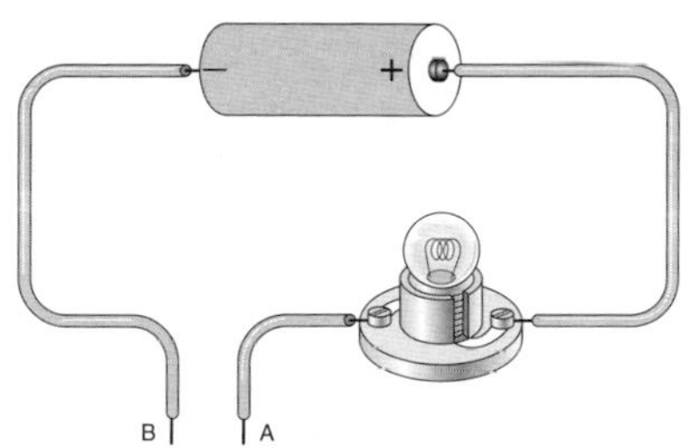

a) Why did the bulb not glow?

____

1 mark

b) Katy placed each of the objects below in turn, across the gap AB.

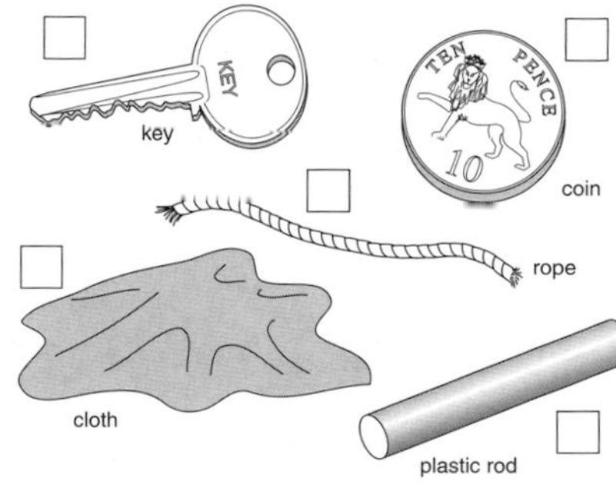

Which two objects made the bulb glow? Tick the appropriate boxes.

2 marks

c) Why did the bulb glow when either of these two objects was placed across the gap AB?

____

____

2 marks

5 In the circuits shown here all the cells and all the bulbs are identical.

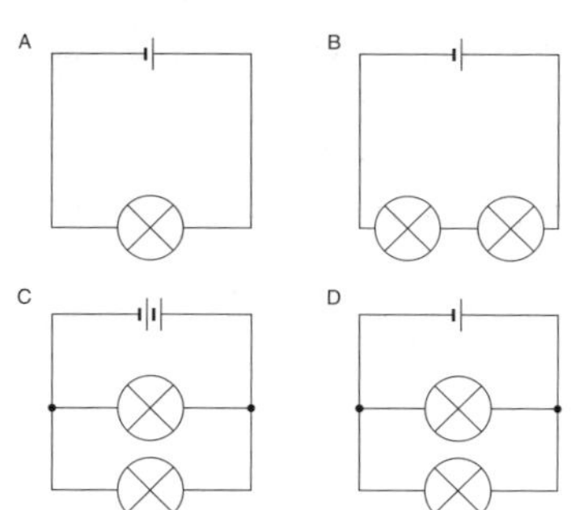

a) In which circuit will the bulb(s) glow the brightest?

____

1 mark

b) In which circuit will the bulb(s) glow least?

____

1 mark

c) In which two circuits will the bulbs have the same brightness?

____

____

2 marks

d) What energy changes are taking place when electricity flows through each of the bulbs?

____

1 mark

6 The diagram below shows two similar circuits.

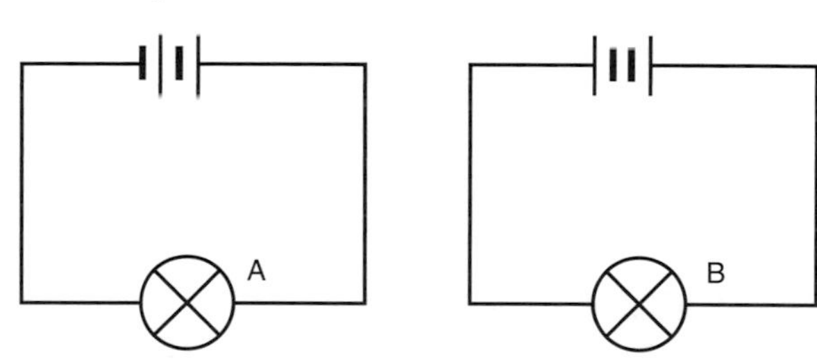

Explain why bulb A glows but bulb B does not glow.

____

____

2 marks

15.1 Magnets and magnetism

Magnetic materials

Magnets are able to attract objects which are made from certain materials, e.g. iron, steel, nickel and cobalt. These are called **magnetic materials**.

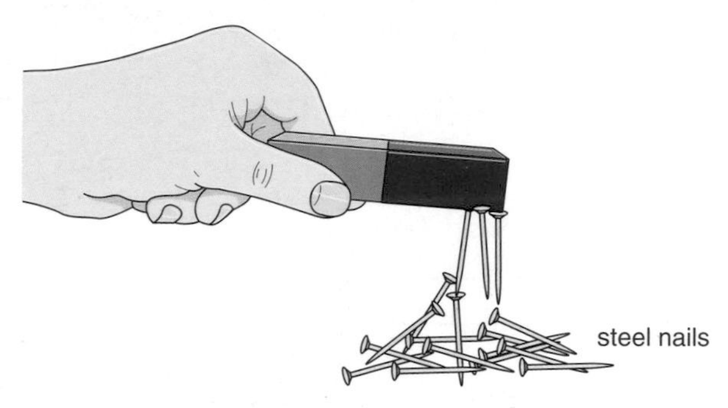

Figure 1 Steel is a magnetic material

Non-magnetic materials

Magnets are unable to attract objects which are made from materials such as paper, plastic and copper. These are called **non-magnetic materials**.

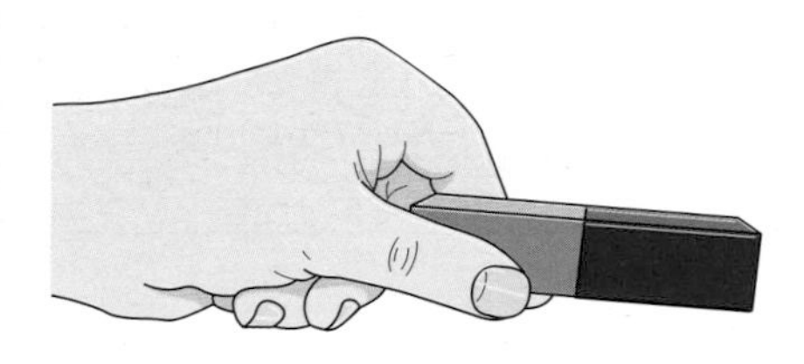
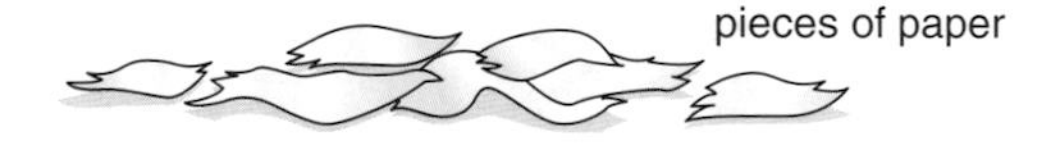

Figure 2 Paper is a non-magnetic material

Poles of a magnet

If iron filings are sprinkled over a bar magnet or a horseshoe magnet, most of them will stick to the two ends. These are the strongest parts of the magnets and are called the **poles**. Magnets have two poles, a **north pole** and a **south pole**.

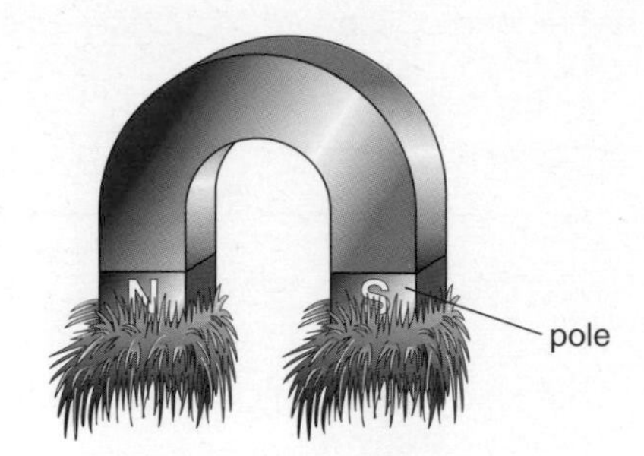

Figure 3 Magnetism is strongest at the poles of a magnet

Attraction and repulsion between poles

Figure 4 If two opposite poles are placed close together they **attract**

Figure 5 If two similar poles are placed close together they **repel**

The domain theory of magnetism

We believe that magnetic materials such as iron and steel have inside them small **molecular magnets**. These small magnets are contained in tiny cells called **domains**. Within each domain all the molecular magnets point in the same direction.

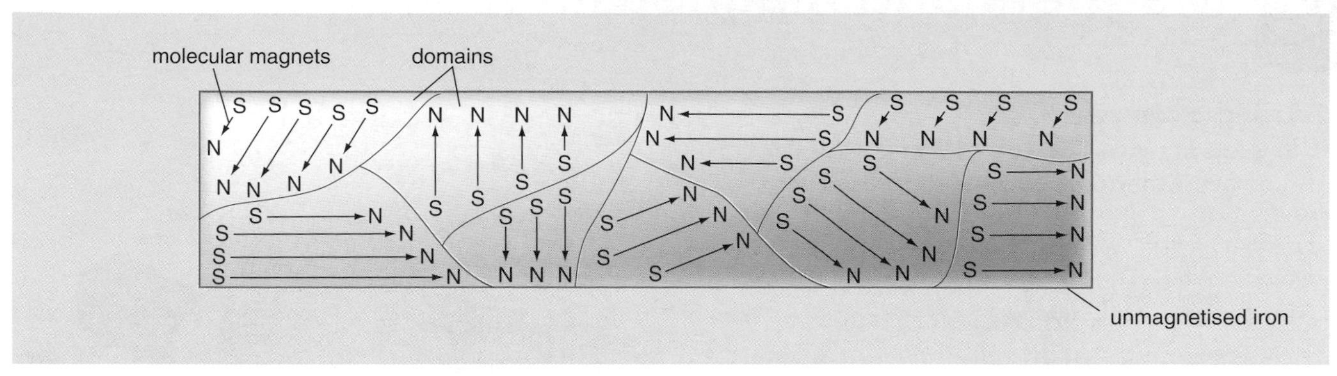

Figure 6 In an **unmagnetised** piece of iron the molecular magnets in each of the domains point in different directions and there is no overall magnetic effect

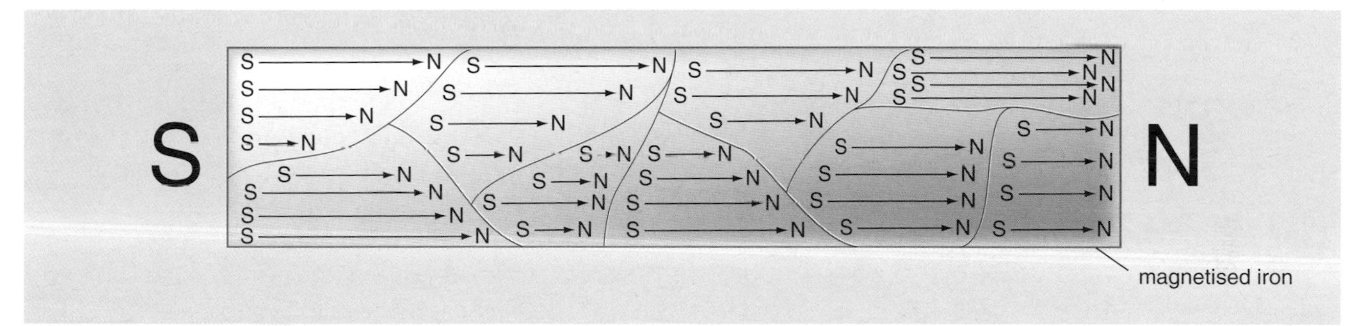

Figure 7 In a **magnetised** piece of iron all the molecular magnets in all the domains point in the same direction, producing a magnet

Making a magnet

If a steel rod is stroked 15–20 times with one end of a bar magnet, the domains within the rod can be made to line up so that they are all pointing in the same direction. The rod is then magnetised.

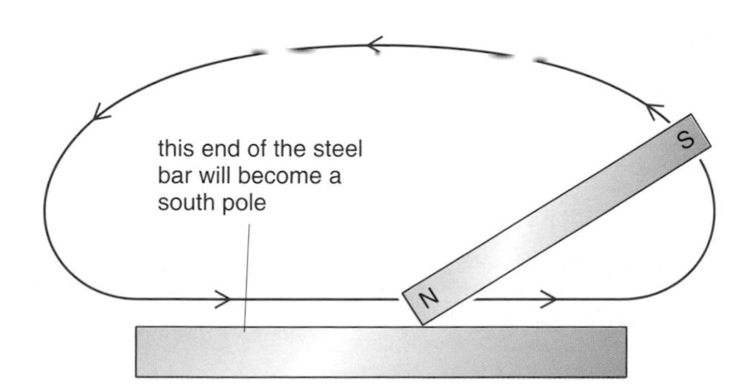

Figure 8 The single touch method of making a magnet

Magnetic fields

When an object made of iron or steel is placed close to a magnet, it is attracted towards it. This happens because the object is inside the magnet's **magnetic field**.

Discovering the shape of a magnetic field

1 Using iron filings

A piece of paper is placed over the magnet. Iron filings are gently sprinkled over the paper. The pattern formed by the filings shows the shape of the magnetic field.

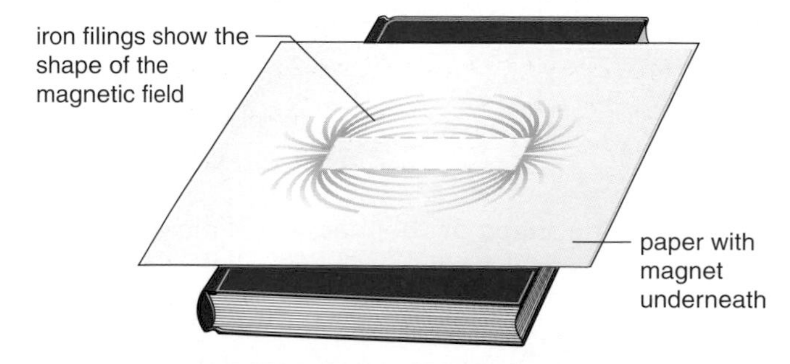

Figure 9 Using iron filings to discover the shape of the magnetic field around a bar magnet

Magnets and magnetism (*continued*)

2 Using a compass

A magnet is placed on top of a piece of paper and drawn around. A **compass** is placed next to the magnet, a circle drawn around it and the direction of the compass needle marked inside the circle. The compass is then moved so that its tail is next to the compass point just drawn. A circle is again drawn around the compass and its direction recorded. This process is repeated all over the paper. The compass needles on the paper will show the shape and direction of the magnetic field.

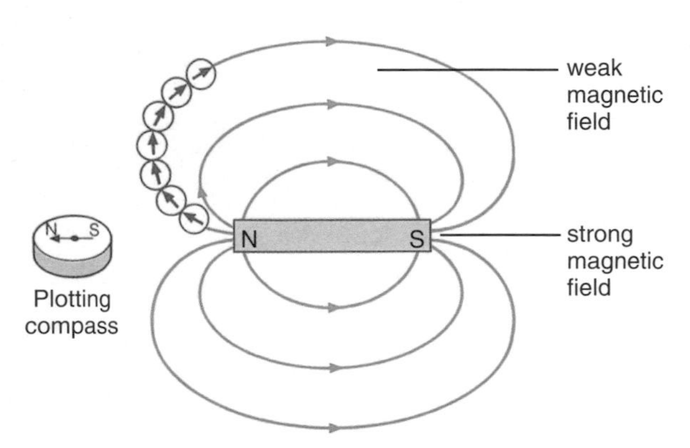

Figure 10 Using a compass to discover the shape and direction of the magnetic field around a bar magnet

Magnetic field pattern around a bar magnet

Both methods show that the field pattern around a bar magnet is as given in Figure 11.

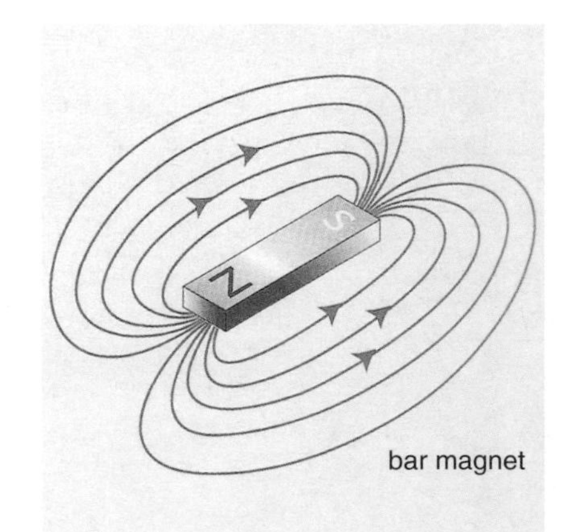

Figure 11 The shape of the magnetic field around a bar magnet

Where the field is strong, the lines are drawn close together. Where the field is weak, the lines are drawn well apart.

The Earth's magnetic field

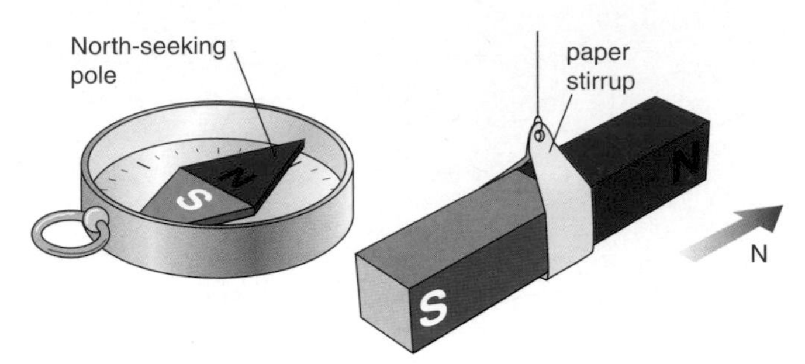

Figure 12 Magnet and compass inside the Earth's magnetic field

If a bar magnet is suspended so that it is free to rotate, it will eventually come to rest with its north pole pointing northwards and its south pole pointing southwards. The magnet is therefore behaving like a simple compass. Magnets and compasses do this because they are inside the Earth's magnetic field.

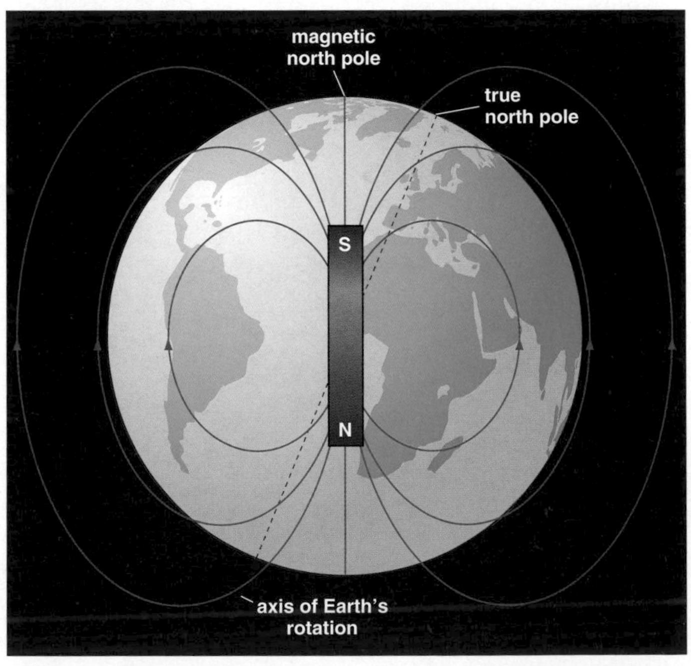

Figure 13 The shape of the Earth's magnetic field is as if there is a giant magnet buried within the Earth, lying almost parallel to the axis of rotation

It is this magnetic field which has allowed travellers to navigate using a compass.

Magnetic shielding

When the magnet below the wooden table in Figure 14 is moved, the footballer above is made to move. If you tried to play this game on a table with a steel or iron top, moving the magnet would have no effect on the footballer. Magnetism is able to pass through non-magnetic materials such as wood, but cannot pass through magnetic materials such as iron or steel. Magnetic materials are often used to shield objects such as sensitive electrical circuits from stray magnetic fields.

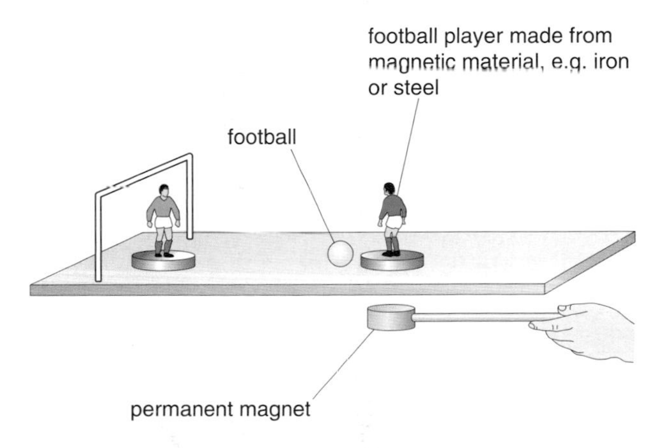

Figure 14 Playing table football using magnets

Key terms

Check that you understand and can explain the following terms:

- ★ magnetic material
- ★ non-magnetic material
- ★ pole
- ★ north pole
- ★ south pole
- ★ repel
- ★ attract
- ★ molecular magnet
- ★ domain
- ★ unmagnetised
- ★ magnetised
- ★ magnetic field
- ★ compass
- ★ magnetic shielding

Questions

1. Explain why a magnet can attract a piece of iron, but not a piece of paper.
2. Draw a diagram showing the field pattern around a bar magnet. Mark with an A a place on your diagram where the field is strong and with a B a place where the field is very weak.
3. What happens when **a)** two similar poles are placed next to each other, **b)** two dissimilar poles are placed next to each other?
4. Draw a diagram of the Earth and its magnetic field. Give one use for the Earth's magnetic field.
5. Which of the following materials could be used to screen electrical equipment from stray magnetic fields – copper, plastic, iron or wood?
6. Explain how the North pole of a bar magnet got its name.
7. Find out two situations where it is important to shield an object from stray magnetic fields.

15.2 Electromagnetism

When a current flows through a wire, a circular magnetic field is created around it. This field can be seen using either iron filings or compasses.

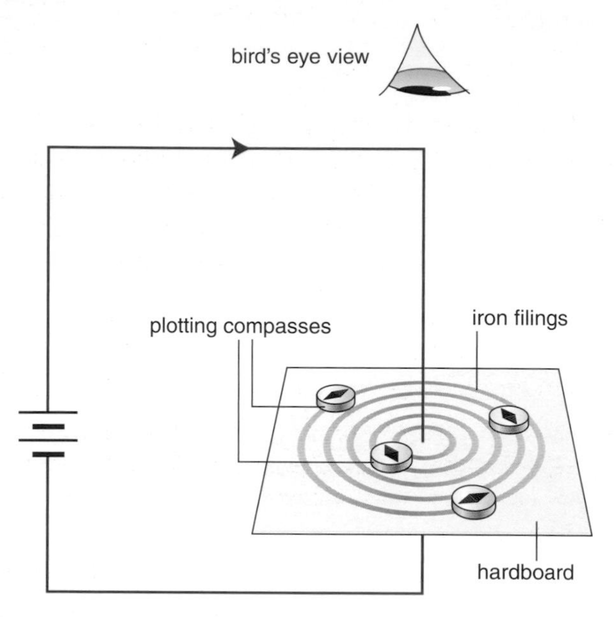

Figure 1 The magnetic field around a current-carrying wire

The field only exists when the current is flowing. If it is turned off, the field disappears. If the current flows in the opposite direction, the field changes direction.

Field strength

The field created when a current flows in a single piece of wire can be quite weak, but its **strength** can be increased by

★ increasing the current through the wire,

★ increasing the number of pieces of wire, i.e. making the wire into a **coil**.

Field shape

The magnetic field around a long coil of current-carrying wire is the same shape as that around a bar magnet.

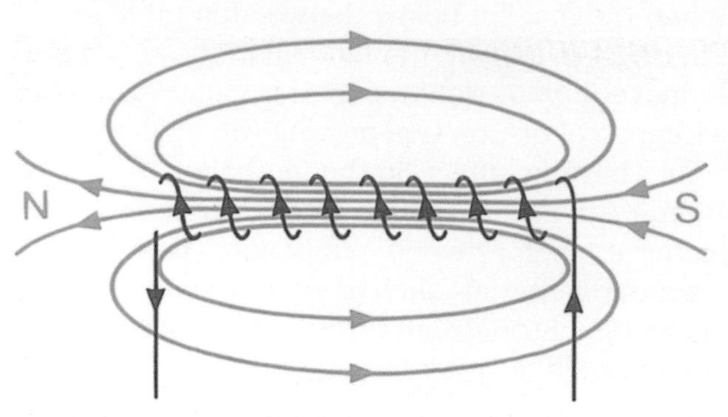

Figure 2 The shape of the magnetic field around a long coil

Electromagnets

If a piece of iron is inserted down the centre of a long coil of wire, this increases the strength of its magnetic field. This combination of **iron core** and coil is known as an **electromagnet**. When current flows around the coil, the iron becomes magnetised. When current stops flowing, the iron loses its magnetism.

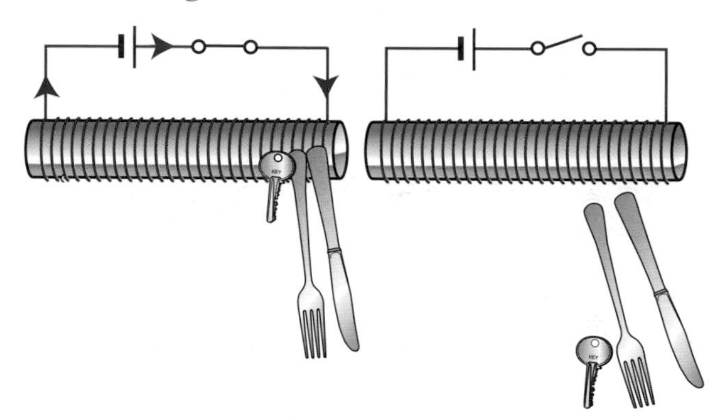

Figure 3 The iron core becomes magnetised when the current flows, but loses its magnetism when the switch is open and current can't flow

Electromagnets are very useful because

★ they can be turned on and off,

★ they can be made stronger and weaker.

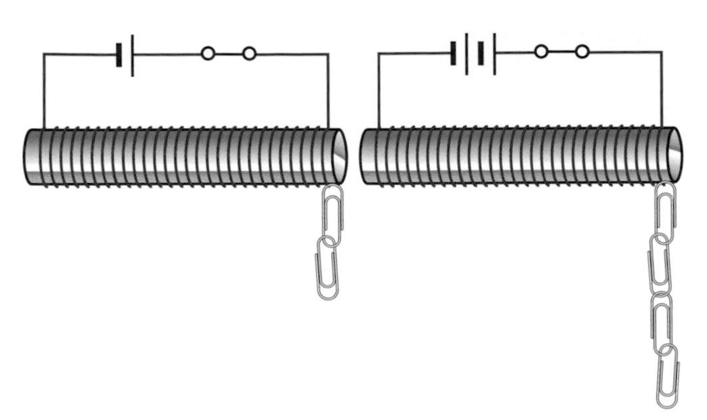

Figure 4 The strength of an electromagnet can be increased by increasing the current flowing through the coils, or by increasing the number of turns on the coil

The electric bell

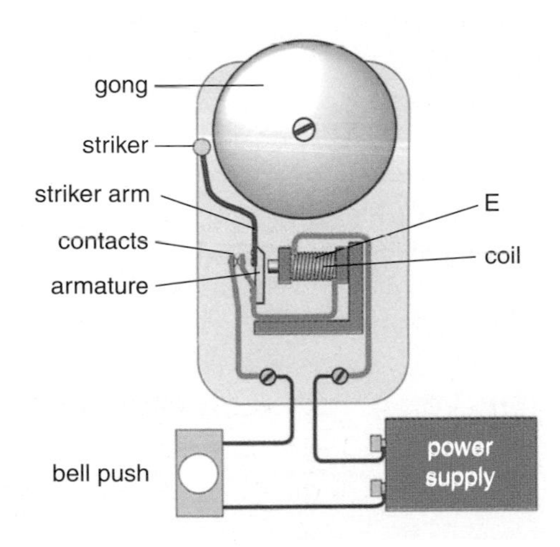

Figure 5 The electric bell

When the button is pressed, current flows around the circuit and the electromagnet at E becomes magnetised. The striker arm is attracted towards it and the striker hits the bell. While this is happening, a gap appears at the contacts. The circuit is now incomplete, the electromagnet loses its magnetism and the striker arm moves back to its original position.

The whole process then begins again. The bell continues to ring as long as the button is pressed.

The relay switch

Sometimes it is useful to be able to control the current flowing in one circuit by using a second circuit. This is especially true if the current flowing in the first circuit is large. This can be done by using a **relay switch**.

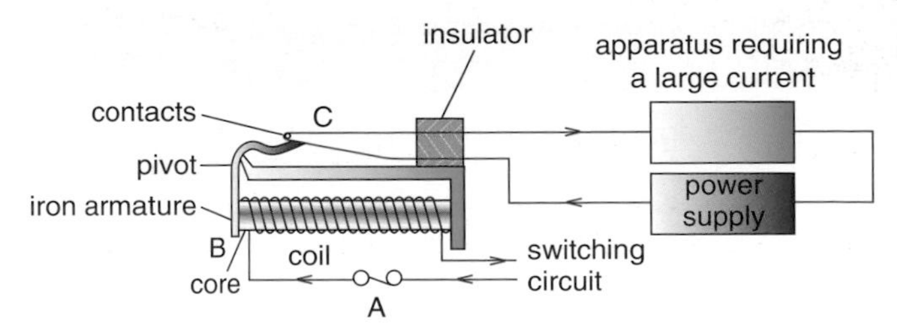

Figure 6 The relay switch

In Figure 6, when switch A is closed, the iron core of the coil becomes magnetised and attracts the **iron armature**. As the end of the armature (B) is attracted, its other end pushes the wires at C together, turning the second circuit on.

If switch A is opened, the electromagnet loses its magnetism, the armature returns to its original position and the second circuit is turned off.

Key terms

Check that you understand and can explain the following terms:

- ★ electromagnetism
- ★ electromagnet
- ★ field strength
- ★ relay switch
- ★ coil
- ★ iron armature
- ★ iron core

Questions

1. Draw a diagram of the field pattern around a piece of wire through which a current is flowing.
2. Draw a diagram of the field pattern around a coil through which a current is flowing.
3. Suggest three ways in which the field around a coil can be made stronger?
4. Explain why an electromagnet would be very useful in a scrapyard, but a permanent magnet would not be.
5. Draw a diagram of an electric bell and in your own words, explain how it works.

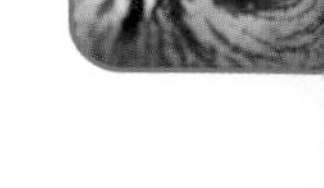

Chapter 15 Magnets and electromagnets:

What you need to know

1. A magnetic material can be attracted by a magnet. Iron, steel, nickel and cobalt are magnetic materials. Magnetic materials can be used to shield apparatus from magnetic fields.
2. A non-magnetic material can not be attracted by a magnet. Plastic, wood, paper, copper and aluminium are examples of non-magnetic materials.
3. Magnetic materials contain domains and molecular magnets. When all the molecular magnets in an object are pointing in the same direction, the object is magnetised.
4. The strongest parts of a magnet are its poles.
5. Similar poles repel. Opposite poles attract.
6. There is a magnetic field around a permanent magnet, for example the Earth and an electromagnet which is turned on.
7. A current flowing through a wire produces a magnetic field.
8. An electromagnet can be made by wrapping a length of wire around an iron core.
9. The strength of the magnetic field around an electromagnet can be altered by changing the current flowing through the wire, or by altering the number of turns of wire there are on the coil.
10. Electromagnets are used in sorting metals, in electric bells, in circuit breakers and in relay switches.

How much do you know?

1 The diagram below shows two bar magnets.

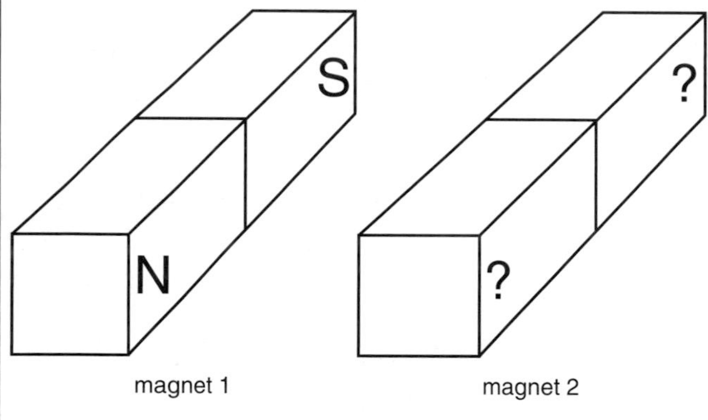

a) Explain how you would use magnet 1 to discover which end of magnet 2 is a north pole.

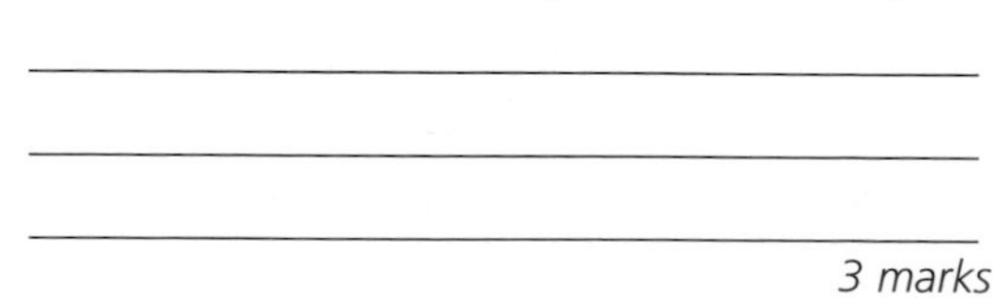

3 marks

b) In the space below draw the magnetic field that exists around a permanent bar magnet

3 marks

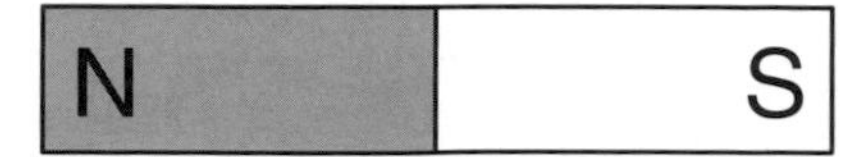

c) The diagram below shows the domain of the bar magnet. Draw in the molecular magnets showing how they are arranged in a magnetised material.

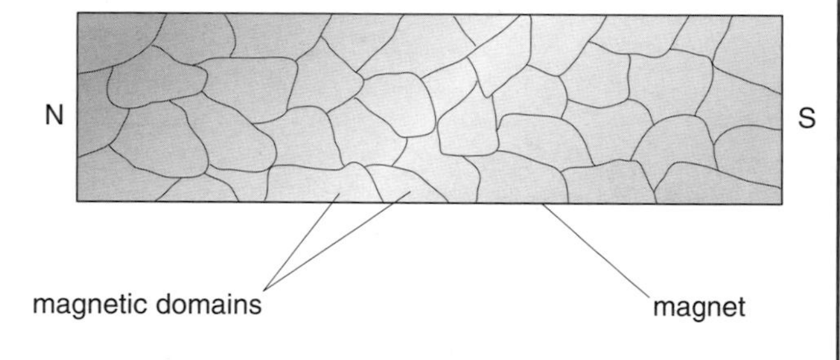

3 marks

2 The sentences below are about magnetic and non-magnetic materials.

Tick the box if you think the statement is true.

☐ A magnet will attract an iron rod.

☐ A magnet will not attract a steel nail.

☐ A magnet will attract a copper pipe.

☐ A magnet will not attract a piece of wood.

2 marks

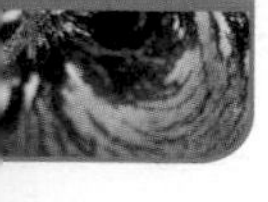

How much do you know? *continued*

3 The diagram below shows a simple electromagnet.

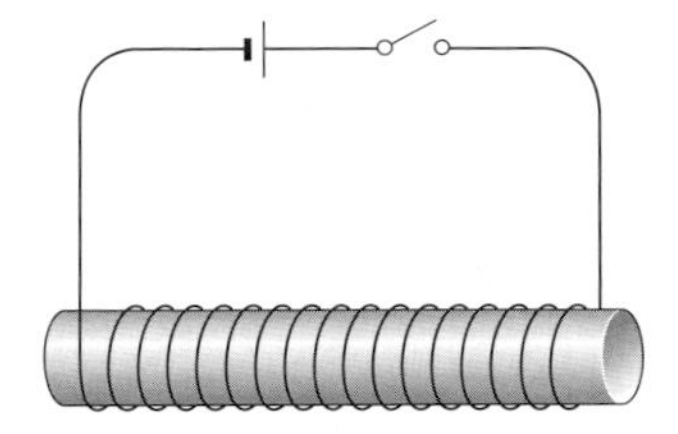

a) Draw on the diagram above the magnetic field that is produced when current passes through the coil.

3 marks

b) Suggest two ways in which you could increase the strength of the magnetic field around this electromagnet.

___

___

2 marks

4 The diagram below shows an electric bell.

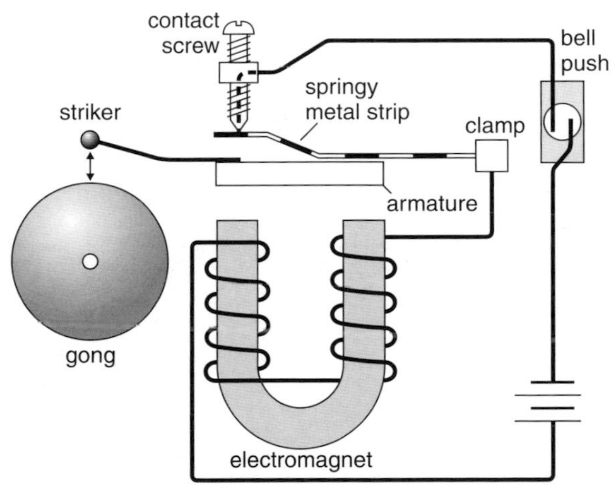

Explain why the striker hits the gong when the button is pressed

___

___

___

2 marks

5 A circuit breaker is a device which is used to turn off a circuit if the current is too high.

The diagram below shows a circuit breaker.

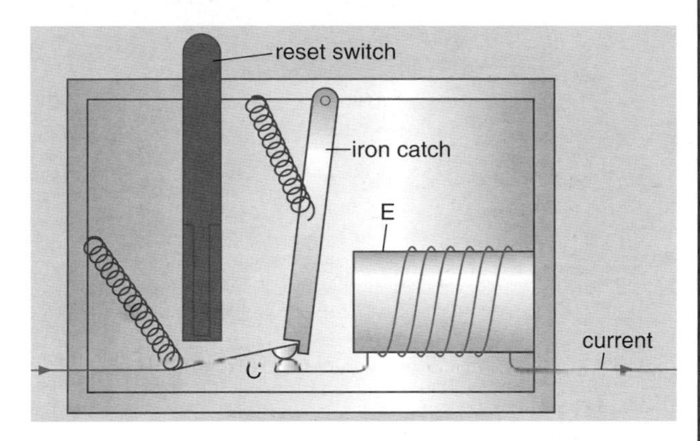

a) Explain in your own words why this circuit breaker does not turn the circuit off when the current is small.

___

___

1 mark

b) Explain why the circuit is turned off if the current is too large.

___

___

2 marks

16.1 Speed

Figure 1

The athlete in Figure 1 can travel 100 m in 10 s. On average he will travel 10 m each second. In other words his **average speed** is 10 m/s.

You can calculate the average speed of an object using the equation

$$\text{speed} = \frac{\text{distance travelled}}{\text{time taken}}$$

In this example, speed $= \frac{100\,\text{m}}{10\,\text{s}}$

Average speed of sprinter = 10 m/s

Figure 2 Concorde

Concorde travels 3000 km in just 1½ hours.

$$\text{speed} = \frac{\text{distance travelled}}{\text{time taken}}$$

$$\text{speed} = \frac{3000\,\text{km}}{1.5\,\text{h}}$$

Average speed of Concorde = 2000 km/h

Figure 3

This cyclist took 5 s to travel 100 m.

$$\text{speed} = \frac{\text{distance travelled}}{\text{time taken}}$$

$$\text{speed} = \frac{100\,\text{m}}{5\,\text{s}}$$

His average speed = 20 m/s

The table below gives some typical speeds.

Light	300 000 000 m/s
Earth orbiting the Sun	2 200 000 m/s
Rifle bullet	approx. 700 m/s
Sound	340 m/s
Jumbo jet 747 (maximum speed)	270 m/s
Cheetah (maximum speed)	28 m/s
Olympic sprinter	12 m/s
Man walking briskly	1.3 m/s
Snail	approx. 0.001 m/s

Using graphs to show motion

It is often useful to show the movement of an object as a graph.

Distance–time graphs

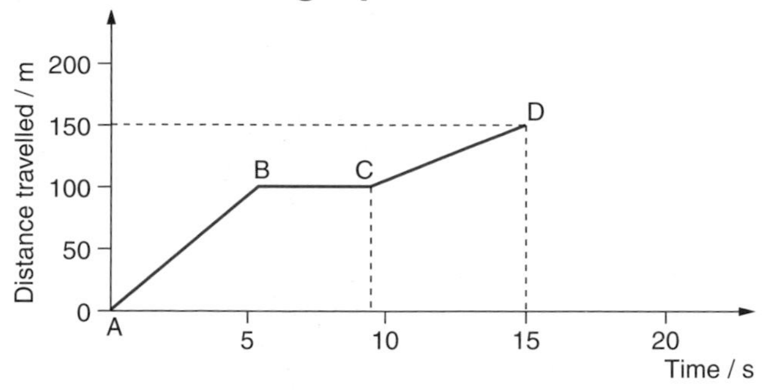

Figure 4 Distance–time graph for a cyclist

Figure 4 shows that the cyclist

- ★ travels 100 m in 5 s between A and B,
- ★ is stationary for 4 s between B and C,
- ★ travels 50 m in 6 s between C and D.

Speed–time graphs

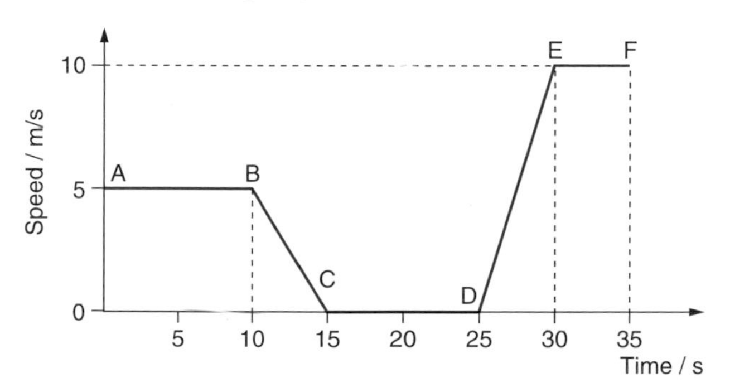

Figure 5 Speed–time graph for a runner

Figure 5 shows that the runner

- ★ travels at a constant speed of 5 m/s for 10 s between A and B,
- ★ slows down between B and C,
- ★ is stationary for 10 s between C and D,
- ★ increases speed between D and E,
- ★ travels at a constant speed of 10 m/s for 5 s between E and F.

Figure 6 This girl is accelerating

An object which increases its speed is **accelerating**.

Figure 7 This drag car is decelerating

An object which is slowing down is **decelerating**.

You can calculate the acceleration of an object using the equation:

$$\text{acceleration} = \frac{\text{change in speed}}{\text{time taken}}$$

Example: If the girl on the sledge in Figure 6 increases her speed from 0 m/s to 20 m/s in 5 s

$$\text{acceleration} = \frac{20\text{ m/s} - 0\text{ m/s}}{5\text{ s}} = 4\text{ m/s/s}$$

If the girl has an acceleration of 4 m/s/s this means she increases her speed by 4 m/s each second.

Key terms

Check that you understand and can explain the following terms:

- ★ average speed
- ★ distance–time graph
- ★ speed–time graph
- ★ accelerating
- ★ decelerating

Questions

1 Fill in the gaps in the table below.

	Distance travelled	Time taken	Average speed
1	100 m	5 s	
2	400 m	8 s	
3	250 km	5 h	
4	750 km	3 h	
5	10 km	300 min	
6	18 km	1.5 h	

2 Calculate the speed of the cyclist in Figure 4 between points
a) A and B, b) C and D.

3 Between which points in Figure 5 is the runner a) accelerating and b) decelerating?

4 A train travels at an average speed of 60 km/h for 3 h. Calculate the total distance travelled.

5 A car travels a total distance of 250 km at an average speed of 100 km/h. How long did the journey take?

6 A car increases its speed from 10 km/h to 90 km/h in 4 s. Calculate the acceleration of the car.

16.2 Effects of forces

There are many different types of **force**. These include pushes, pulls, twists and stretches.

If you apply a force to an object it may:

★ make it start to move
★ make it move faster
★ slow it down
★ make it stop
★ change the direction in which it is moving
★ change its shape.

These effects are illustrated below:

Figure 1 Different types of forces

Sometimes it is not necessary to be **in contact** with an object in order to apply a force to it.

The steel nails in Figure 2 are lifted by a force. This **attractive force** exists between magnets and magnetic materials such as iron and steel (see page 140).

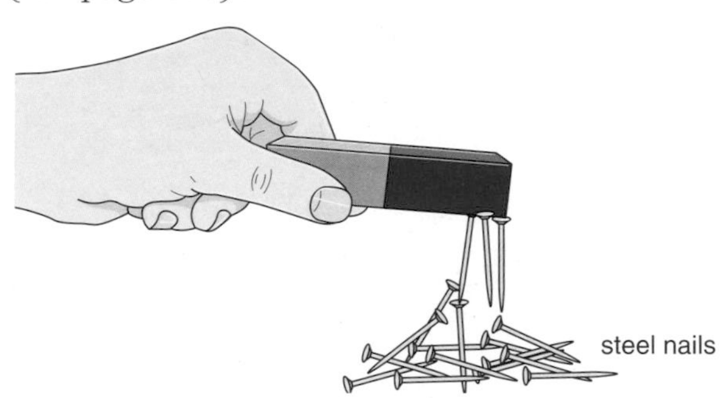

Figure 2 The nails are attracted to the magnet by a force

The bungee jumper in Figure 3 has just jumped out of the basket and is feeling the force of **gravity** pulling him downwards. The common name given to this force is **weight**.

Figure 3 The force of gravity pulls the bungee jumper downwards

There are **gravitational** forces between all objects, but they are often very weak. They are only noticeable when one or both of them is massive, such as a planet, moon or star. It is gravitational forces of attraction that hold the Moon in orbit around the Earth and hold the planets in orbit around the Sun (see page 184).

Mass and weight

The **mass** of an object is the amount of matter it contains and is measured in **kilograms** (kg). The weight of an object is the size of gravitational attraction between the object and the planet (or moon) it is on. We measure weight in **newtons** (N).

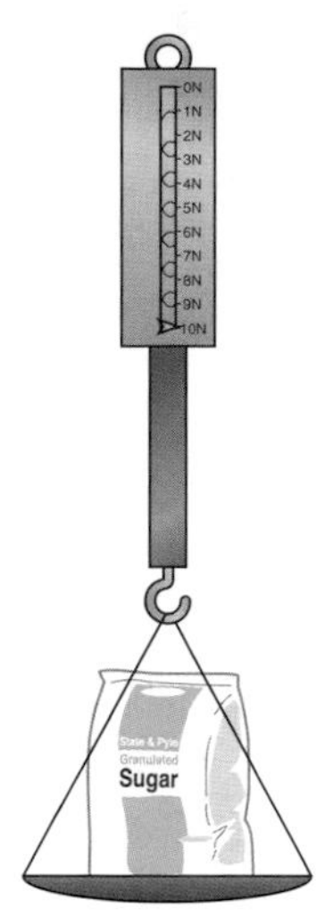

Figure 4 The sugar in this bag has a mass of 1 kg. On Earth it weighs 10 N

The table below gives some examples of the masses and weights of different objects on the Earth and on the Moon.

Object	Mass on Earth	Weight on Earth	Mass on Moon	Weight on Moon
bag of sugar	1 kg	10 N	1 kg	1.6 N
apple	0.1 kg	1 N	0.1 kg	0.2 N
sack of potatoes	25 kg	250 N	25 kg	42 N
small boy	40 kg	400 N	40 kg	67 N
large man	100 kg	1000 N	100 kg	160 N

We can see from this table that gravity on the Moon's surface is only ⅙th of that on the Earth. So objects here only weigh ⅙th of what they would weigh on Earth. The masses of all the objects do not change. They are the same here on the Earth as they are on the Moon.

Figure 5 This girl weighs six times more on the Earth than on the Moon

Newtonmeter

We can measure the size of a force using a **newtonmeter**. The larger the force, the more the spring stretches.

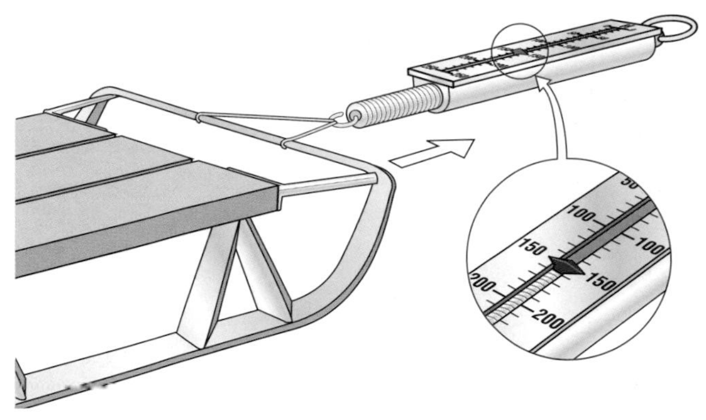

Figure 6 Measuring a force using a newtonmeter

Key terms

Check that you understand and can explain the following terms:

★ force
★ in contact
★ attractive force
★ gravity
★ weight
★ gravitational forces
★ mass
★ kilogram
★ newton
★ newtonmeter

Questions

1 Name five things that may happen to an object if a force is applied to it. Give one example for each.

2 Describe two situations in which a force is applied to an object without being in contact with it.

3 Find out how much force you apply to the floor when you are standing up. (*Hint*: 1 kg provides a force of approximately 10 N.)

4 A lady has a mass of 60 kg and weighs 600 N on the Earth. What will be the mass and weight of the lady on the surface of the planet Saturn if gravity here is just ⅘th that on the surface of the Earth?

5 Draw a diagram of a newtonmeter and explain how it can be used to find the force necessary to open your classroom door.

16.3 Balanced and unbalanced forces

In everyday life it is rare for an object to be acted upon by no forces or a single force. It is much more likely that it will experience several forces. These forces may be **balanced** or **unbalanced**.

Figure 1 Balanced force

If the two tug of war teams in Figure 1 pull with the same force, the forces are balanced and there is no movement.

Figure 2 Unbalanced force

If one of the teams pulls with a force which is greater than that of the opposition, the forces are unbalanced and there is movement (Figure 2).

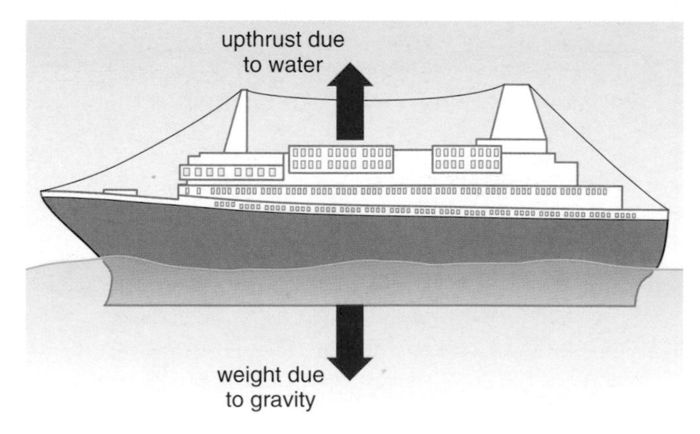

Figure 3 The upthrust exerted by the water on this ship is equal to the ship's weight

The ship above is floating in water and is stationary. There are several forces acting upon it, so these forces must be balanced. **Gravitational forces**, i.e. the **weight** of the ship, are pulling it downwards but a second force called the **upthrust** from the water is pushing upwards. If too much cargo is loaded onto the ship, the weight may become larger than the upthrust and the ship may sink.

An object placed on a table also experiences balanced forces.

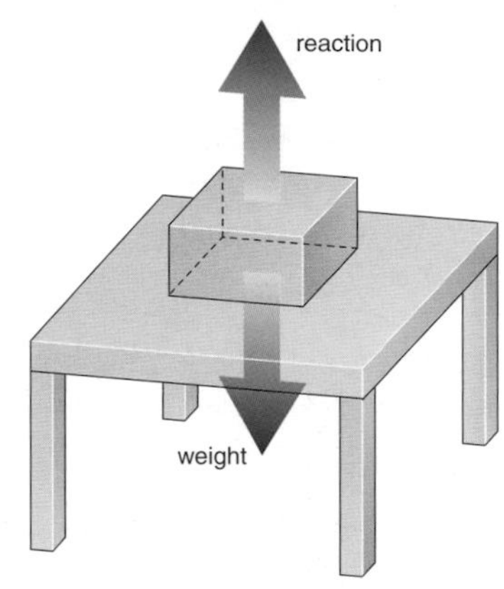

Figure 4 The reaction applied to the box by the table is equal to the weight of the box

If an object is moving and balanced forces are applied to it, the object will continue to move in the same direction and at the same speed. The aeroplane below is experiencing four forces: its weight, a lifting force from its wings, thrust from its engines driving it forwards and drag from the air trying to resist the forward motion.

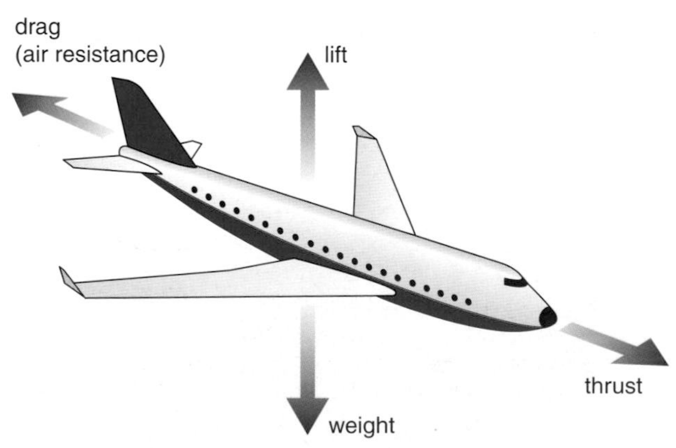

Figure 5 The balanced forces acting on an aeroplane in flight

If the lift and weight forces are balanced, there is no vertical motion. The aeroplane stays at the same height.

If the thrust and drag forces are balanced, the aeroplane will travel at a constant speed in a straight line.

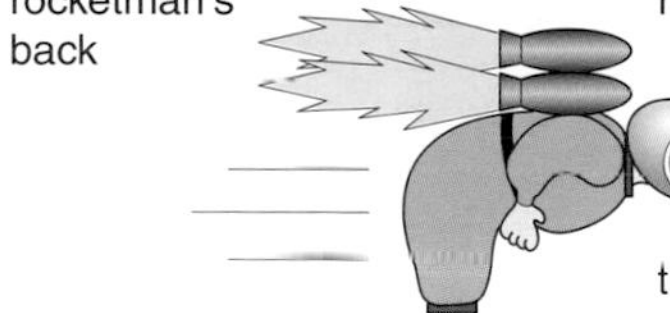

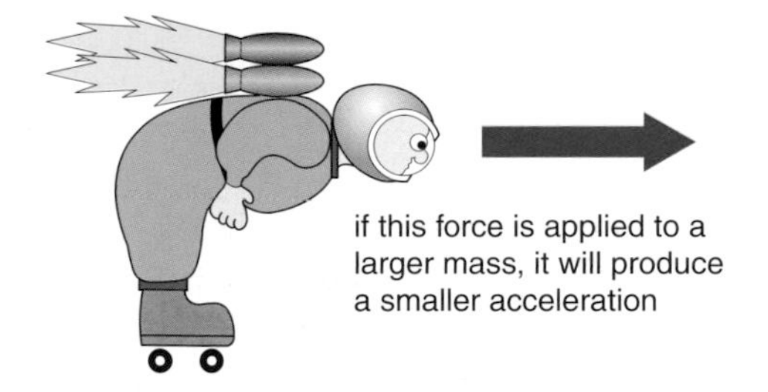

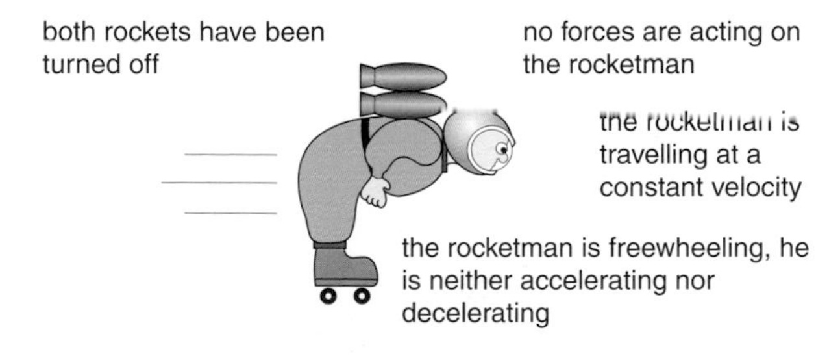

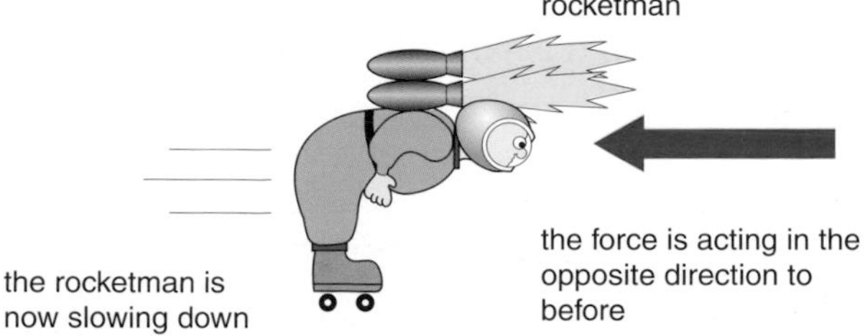

Figure 6 Forces causing acceleration and deceleration

Accelerating and decelerating forces

If unbalanced forces are applied to an object, they may cause it to **accelerate** or **decelerate**. The size of the acceleration depends on:

★ the size of the applied force

★ the mass of the object.

We can summarise this by saying that unbalanced forces produce a change in the motion of an object, while balanced forces do not.

Key terms

Check that you understand and can explain the following terms:

★ balanced forces

★ unbalanced forces

★ gravitational forces

★ weight

★ upthrust

★ acceleration

★ deceleration

Questions

1 Give one example of a situation where an object is being acted upon by balanced forces. Draw a diagram to show the sizes and the directions of the forces.

2 Give one example of a situation where an object is being acted upon by unbalanced forces. Draw a diagram to show the sizes and directions of the forces and state what the overall effect of these forces is on the object.

3 Describe the motion of an object which is **a)** stationary and **b)** moving at a constant speed, if several forces which are balanced are applied to it.

4 Describe the motion of an object which is **a)** stationary and **b)** moving at a constant speed, if forces which are unbalanced are applied to it.

5 Suggest two ways in which engineers could alter the design of a racing car so that it would have a greater acceleration.

16.4 Friction

One of the most common forces which can act upon an object is **friction**. Whenever an object moves or tries to move, friction is present. Friction is a force which opposes motion

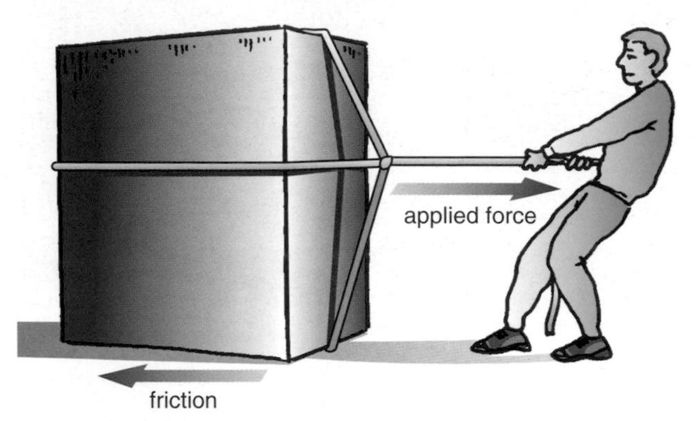

Figure 1 The friction between the box and the floor makes it hard to pull

On some occasions friction can prove very useful. For example, when you walk or run, you push yourself forward by pushing backwards on the ground. Friction between your foot and the floor helps you to do this. If there was no friction, i.e. like on a slippery ice rink, your feet would slip!

Figure 2 The importance of friction

Smooth surfaces reduce the friction between objects, while rough surfaces increase the **frictional forces**. Trainers and football boots are designed to prevent your foot from slipping by increasing the frictional forces between you and the ground. In contrast, skates and skis are designed with smooth surfaces which keep friction to a minimum.

Where there is contact between surfaces, friction can be reduced using a **lubricant** such as oil, water or ice.

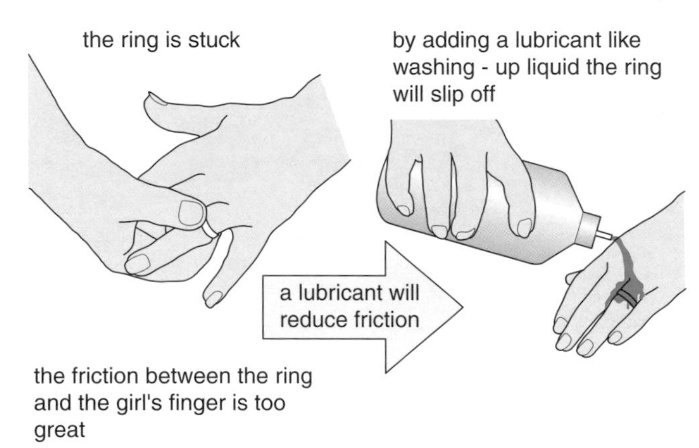

Figure 3 Friction is preventing this girl from removing the ring from her finger. By adding a lubricant she is able to reduce these forces

Friction between two surfaces can cause the surfaces to wear away and become hot.

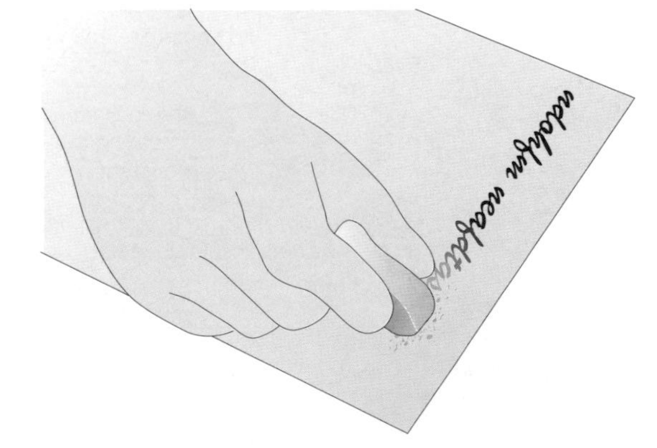

Figure 4 Friction is wearing away this eraser

Figure 5 Friction is being used here to start a fire

Drag

Whenever an object moves through the air it experiences frictional forces or **drag**, which try to prevent its motion. The faster the object moves, the greater the drag. To reduce these forces objects such as cars, trains and aircraft are shaped so that they cut through the air. They are **streamlined** to reduce **air resistance**.

Figure 6 This train has been streamlined to reduce air resistance

Animals such as dolphins, sharks and penguins have a streamlined shape so that when they move through the water, frictional forces are as small as possible.

Figure 7 Dolphins are very streamlined

Terminal velocity

As a car accelerates, the frictional forces it experiences increase until eventually the driving force from the engine is balanced by the drag. Under these conditions the car will travel at a constant speed known as its **terminal velocity**.

Figure 8 When the driving force is greater than the drag, the car will accelerate

Figure 9 When the driving force and the drag are balanced, the car will travel at its terminal velocity

16.4 Friction (*continued*)

Speed–time graph for a skydiver

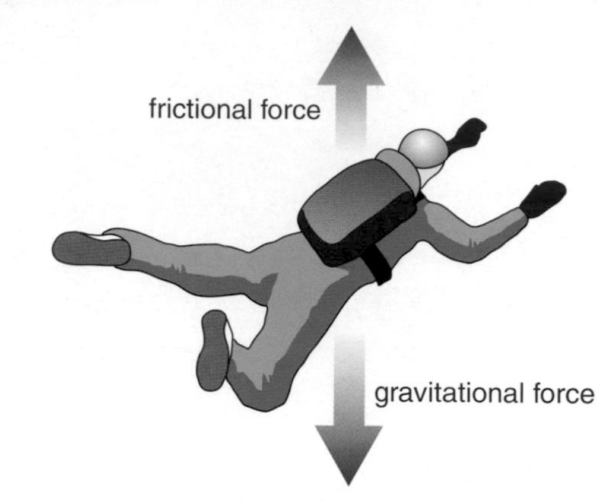

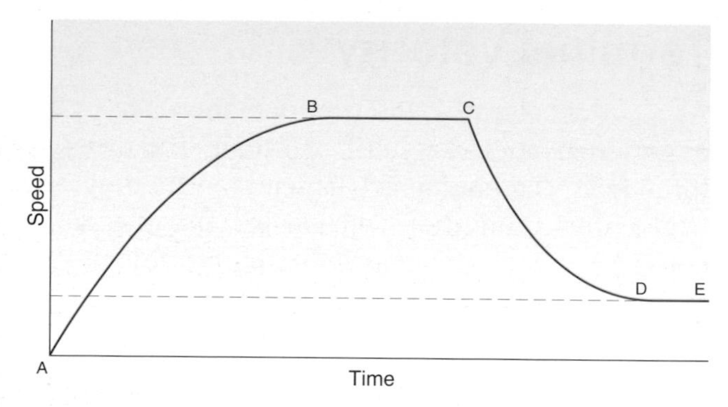

Figure 10 Speed–time graph for a skydiver

AB: As the skydiver jumps from her aircraft or balloon she accelerates as she falls. As her speed increases, the frictional forces of the air increase, causing her acceleration to decrease, i.e. she does not increase her speed so quickly.

BC: Eventually the frictional forces of the air and the gravitational forces pulling her downwards balance. The skydiver now falls at her terminal velocity.

CD: When the skydiver opens her parachute, she increases the frictional forces of the air. Because these are acting upwards she decelerates, i.e. she slows down.

DE: As she slows, the frictional forces become less. Once again the upward forces and the downward forces become balanced and she falls at a much lower terminal velocity.

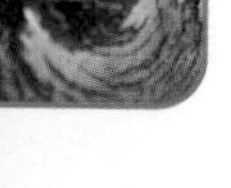

Friction and the motor car

In order that a car can change its speed or direction, there must be friction between its tyres and the road surface. A new tyre with lots of tread has a rough surface and will provide lots of grip. A worn tyre with a smoother surface will provide less grip and the car will be much more difficult to control and stop.

Having tyres which are in good condition is important if you need to stop quickly, but there are several other important factors which will affect how quickly a car can stop. These are:

★ the reaction time of the driver,

★ the efficiency of the braking system of the car,

★ the speed of the car,

★ the weather/road conditions.

Figure 11 illustrates how the speed of a car affects the total distance it will travel before it stops. The total stopping distance is equal to the thinking distance + the braking distance.

at 13 m/s (30mph)

thinking distance 9m braking distance 14m total stopping distance 23m

at 22 m/s (50mph)

thinking distance 15m braking distance 38m total stopping distance 53m

at 30 m/s (70mph)

thinking distance 21m braking distance 75m total stopping distance 96m

Figure 11 Stopping distances of a car in ideal conditions

Key terms

Check that you understand and can explain the following terms:

- ★ friction
- ★ frictional forces
- ★ lubricant
- ★ drag
- ★ streamlined
- ★ air resistance
- ★ terminal velocity
- ★ thinking distance
- ★ braking distance
- ★ stopping distance

Questions

1 What is friction? Give two examples of situations in which friction is an advantage and two examples where it is a disadvantage.

2 How can the friction between two objects be **a)** increased and **b)** decreased?

3 What is drag? How can drag be decreased?

4 Give two effects of applying friction to an object, other than altering its motion.

5 Explain why rockets have a streamlined shape, but satellites that orbit the Earth are not streamlined.

6 Under what conditions will a parachutist fall at his terminal velocity?

Chapter 16 Forces and motion:

What you need to know

1. The speed of an object can be found using the equation: speed = distance travelled/time taken.
2. Pulls, pushes and twists are all different kinds of forces.
3. The size of a force is measured in newtons (N).
4. Weight is the force exerted on an object due to the pull of gravity.
5. The mass of an object is a measure of how much matter it contains and is measured in kilograms.
6. If several forces act upon an object, they may be balanced or unbalanced.
7. Unbalanced forces cause an object to change its motion i.e. change speed or direction.
8. Friction is a force which opposes motion.
9. In some situations friction is needed, in some situations it is unwanted.
10. As objects move through liquids or gases, they experience frictional forces or drag.
11. Having a streamlined shape can reduce drag.
12. The stopping distance of a moving car depends on how fast the car is travelling, the friction between the car tyres and the road surface and the reactions of the driver.

How much do you know?

1 The rocketman in the diagram below is travelling forwards at a constant speed.

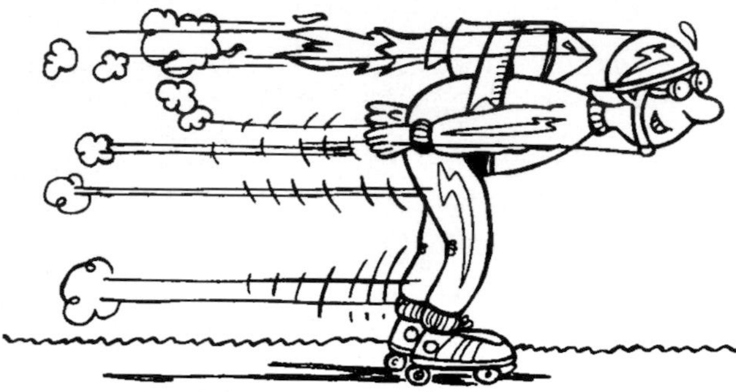

a) Tick the box which describes what happens to his speed when the rocket motor is turned on.

☐ He continues at the same speed.

☐ He slows down.

☐ He goes faster.

1 mark

b) The rocketman travels 40 m in just 8 s. Calculate his average speed.

____

____

2 marks

c) Calculate the acceleration of the rocketman if he increases his speed from the value calculated in part b) of this question to 20 m/s in 3 s.

____

____

____

3 marks

2 The graph below shows the journey of a cyclist.

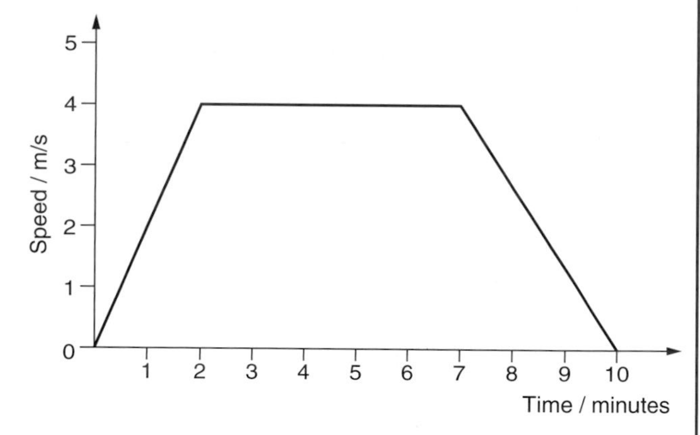

a) How long did the whole journey take?

____

1 mark

b) What was the greatest speed the cyclist reached?

____

1 mark

c) For how long was the cyclist travelling at a constant speed?

____

1 mark

How much do you know? *continued*

3 The diagram below shows a wagon moving at a constant speed and the forces which are being applied to it.

a) What happens to the wagon if the frictional forces become greater than the pull of the engine?

1 mark

b) What happens if the frictional forces are less than the pull of the engine?

1 mark

4 The photo below shows a golf ball being hit by a golf club.

What can you see in the photo which shows that the club must be applying a force to the ball?

1 mark

5 A man of mass 84 kg weighs 840 N on the Earth. Gravity on the surface of the Moon is ⅙th of that on the Earth's surface.

a) What will be the mass of the man on the surface of the Moon?

1 mark

b) What will be the weight of the man on the surface of the Moon?

1 mark

6

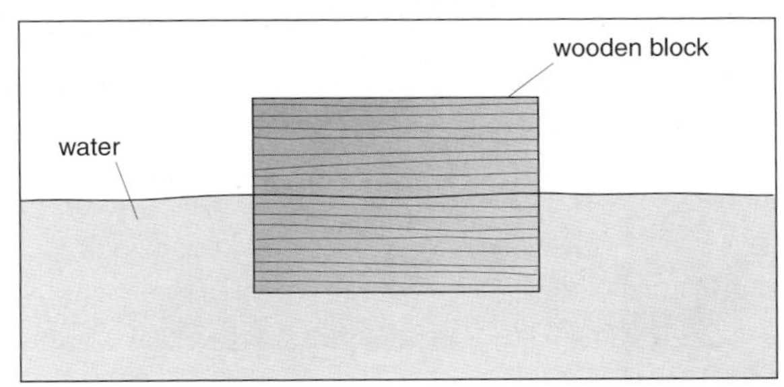

a) Add labelled arrows showing two forces that are being applied to the block in the diagram above.

4 marks

b) Explain why the block is floating.

1 mark

c) Why might a ship sink if it has too much cargo on board?

2 marks

7 a) Explain, in your own words, the sentence: 'A dolphin has a streamlined shape'.

2 marks

b) Give two examples of man-made objects that have streamlined shapes.

2 marks

8 a) Explain why icy roads and bald tyres will greatly increase the stopping distance needed for a car.

2 marks

b) How will the speed at which a car is travelling affect its stopping distance?

1 mark

17.1 Rays of light and reflection

Figure 1 Seeing luminous and non-luminous objects

You see **luminous objects** such as the Sun, light bulbs and fires because some of the rays of light they emit enter your eyes. You see **non-luminous objects** because of the light they are reflecting into your eyes. If there is no light entering your eyes, for example because you are wearing a blind fold or there is no light source, you are unable to see.

How does light travel?

On a bright day when the Sun shines through the clouds, you might notice that rays of light travel in straight lines. It is possible to shine light through a straight tube, but not along one which is bent. This again suggests that light travels in straight lines.

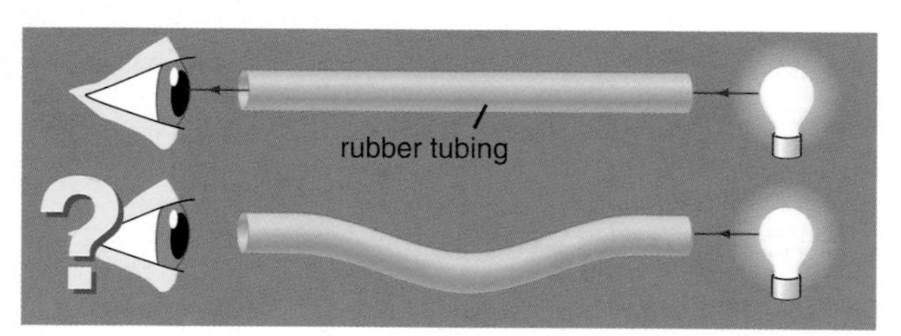

Figure 2 Light will travel in straight lines but not around a bend

Light travels very, very quickly – at 300 000 km/s. It takes a ray of light just 8.5 minutes to travel from the Sun to the Earth, and just 1.3 seconds to travel from the Moon.

What can light travel through?

Light can travel through many different materials, such as clear solids (glass), clear liquids (water) and clear gases (air). It can even travel through a vacuum. Objects which allow the light to pass through and which you can see through are described as being **transparent**. Some objects such as tracing paper or frosted glass allow light to travel through, but we are unable to see through them. These objects are described as being **translucent**. Objects which do not allow light to travel through them, such as a piece of card or wood, are described as being **opaque**.

Figure 3 Transparent, translucent and opaque objects

If an opaque object is placed in front of a source of light, an area of darkness called a **shadow** is created. The shape of the shadow is the same as the shape of the object. This supports the idea that light travels in straight lines.

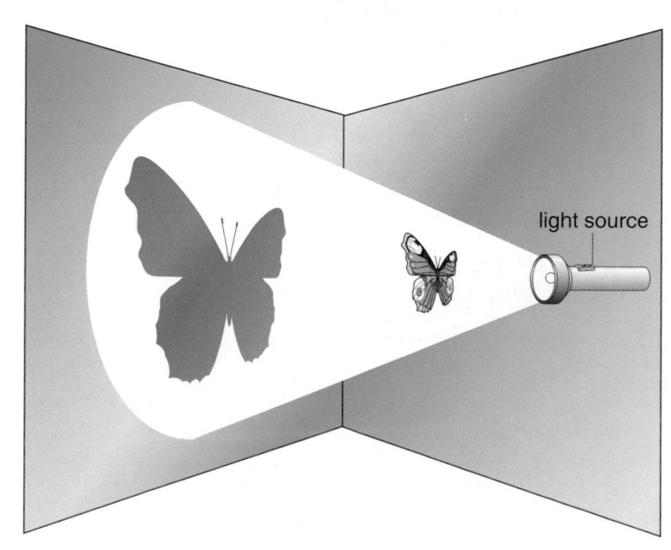

Figure 4 How a shadow is created

Reflection of light

When a ray of light strikes a flat or plane surface, it is always **reflected** at the same angle. In other words, the **angle of incidence** is equal to the **angle of reflection**.

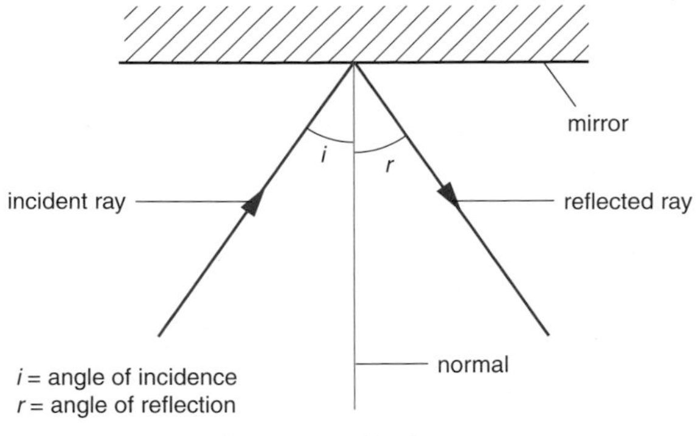

Figure 5 The reflection of light

A simple periscope, like the one in Figure 6, makes use of this idea so that you can see over high objects or round corners.

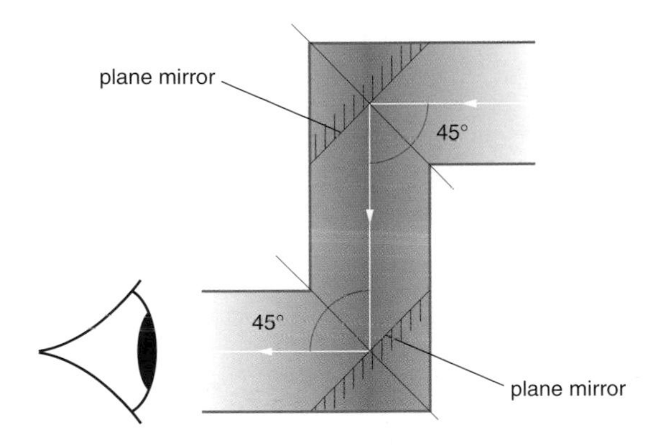

Figure 6 A simple periscope

Images in a plane mirror

If you stand in front of a plane mirror you will see an **image** of yourself. This image is:

★ upright
★ the same height as you
★ the same distance behind the mirror as you are in front
★ **laterally inverted**, i.e. your left-hand side appears on the right of the image and your right-hand side appears on the left of the image.

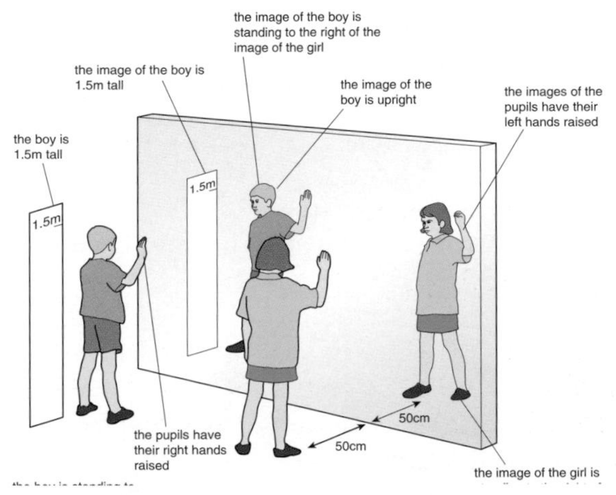

Figure 7 A plane mirror creates images of objects placed in front of it

The diagram below shows how an image is created in a plane mirror.

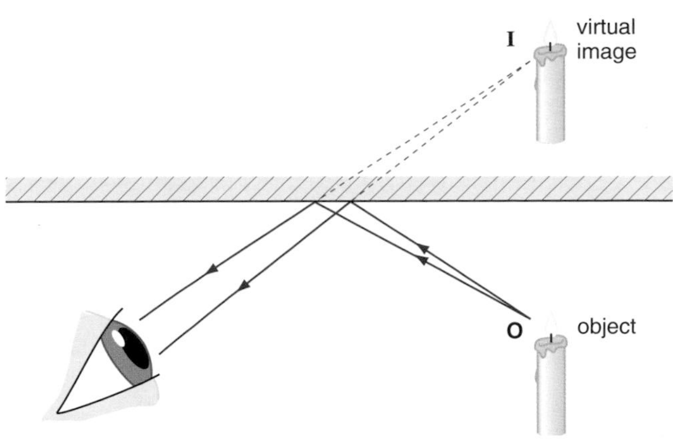

Figure 8 A ray diagram showing how a plane mirror creates an image of an object

Light from the object O is reflected by the mirror into the eyes of the observer. The observer believes that light travels in straight lines and so sees the object at I. This is the image seen in the mirror.

Key terms

Check that you understand and can explain the following terms:

★ luminous object
★ non-luminous object
★ transparent
★ translucent
★ opaque
★ shadow
★ reflect
★ angle of incidence
★ angle of reflection
★ image
★ laterally inverted

Questions

1 Explain the difference between a luminous object and a non-luminous object. Give three examples of each.

2 Why are you unable to see when you close your eyes?

3 Explain the difference between a transparent object and a translucent object. Give one example of the use of **a)** a transparent object and **b)** a translucent object.

4 A ray of light strikes a plane mirror at an angle of incidence of 55°. What is the angle of reflection of this ray? Draw an accurate and labelled diagram to show the reflection of this ray.

5 Draw a diagram of a periscope and explain how you might use it to see over a wall.

6 Describe the properties of the image created when you stand in front of a plane mirror.

17.2 Refraction and colour

Light is able to travel through many different materials such as air, water, glass and perspex. These materials are called **media**. When a ray travels between media, it often changes direction as it crosses the boundary. This changing of direction is called **refraction**. It happens because the light changes speed as it crosses the boundary.

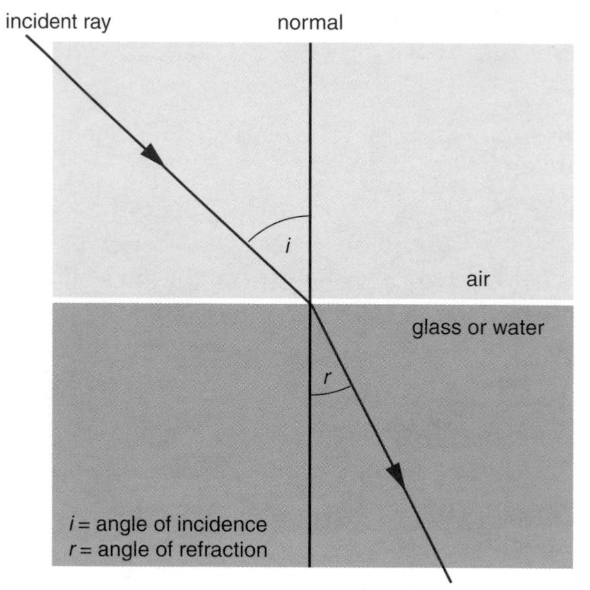

Figure 1 A ray of light is refracted towards the normal as it enters the glass or water

A ray of light travelling from air into glass or water slows down and bends *towards* the **normal**. The normal is an imaginary line drawn at right angles to the surface at the point where the ray changes between media.

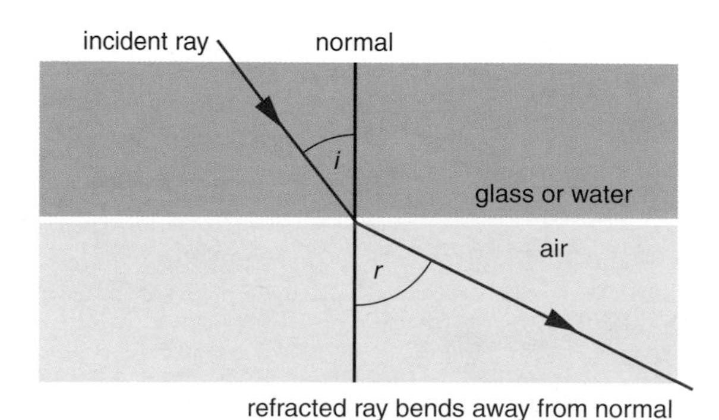

Figure 2 A ray of light is refracted away from the normal as it emerges from the glass or water

A ray travelling from glass or water into air speeds up and bends *away* from the normal.

But if a ray of light strikes the boundary at 90° to the surface, it passes through undeviated, i.e. it is not refracted.

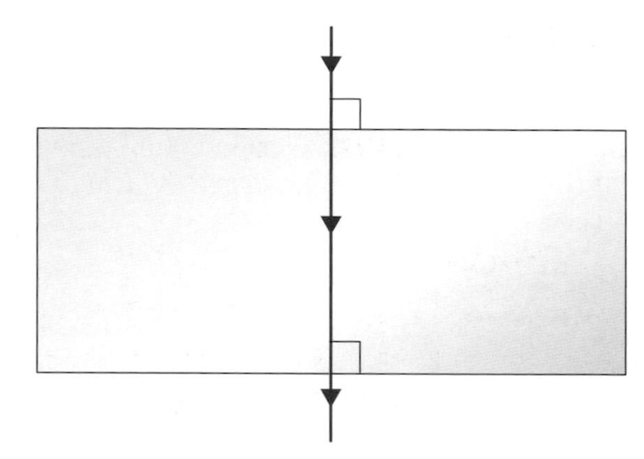

Figure 3 Striking the boundary at 90° this ray is undeviated

Effects of refraction

Because we usually expect light to travel in straight lines, refraction can cause some strange optical illusions.

If you place a pencil in a beaker of water it appears to be bent. This happens because the rays of light from the part of the pencil under the water change direction as they cross the water/air boundary.

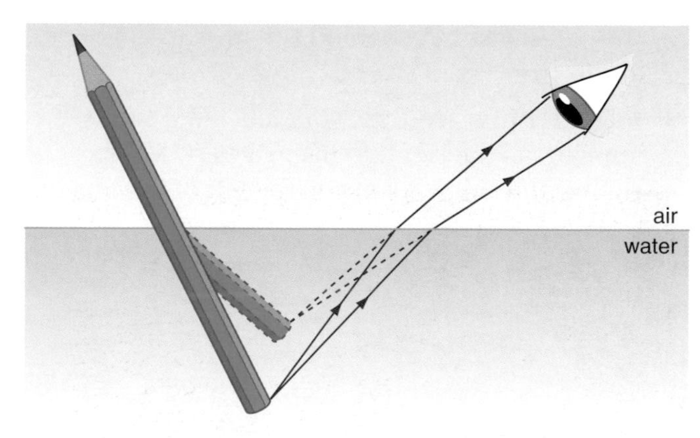

Figure 4 The pencil appears to be bent because of the refraction of light

Refraction may cause this hunter to miss his prey.

Figure 5 Because of refraction the hunter sees his fish at B and not at A

Dispersion of white light

If you shine a ray of white light into a glass prism you will see it refract as it enters and as it leaves. These refractions cause the white light to split into a band of colours called a **spectrum**. This effect is called **dispersion**.

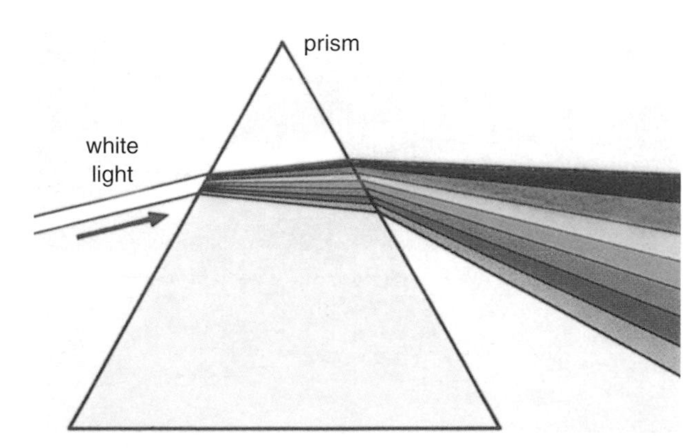

Figure 6 The spectrum

If a second identical prism is inverted and placed behind the first, the colours can be made to recombine producing a beam of white light.

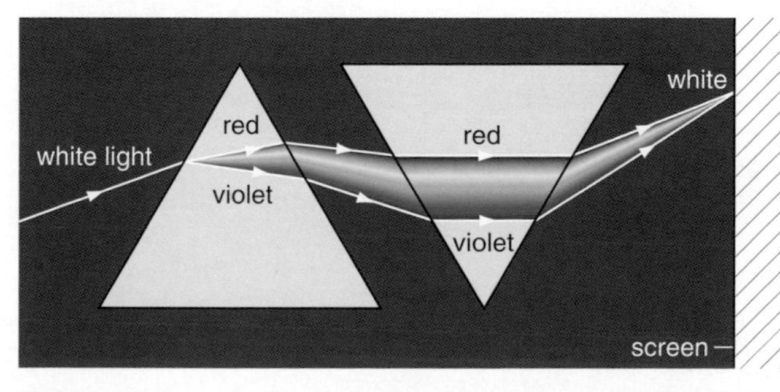

Figure 7 A second prism combines the colours of the spectrum to produce white light again

This experiment shows that white light is a mixture of coloured lights – the colours of the rainbow. The colours in order are red, orange, yellow, green, blue, indigo and violet. The order is easily remembered by the saying, **R**ichard **O**f **Y**ork **G**ave **B**attle **I**n **V**ain.

Coloured objects

Most objects have some colour, for example grass is green, ink is blue. They have these colours because they contain a chemical called a **dye**. The dye **absorbs** all the colours of light which strike it except for its own colour which is reflected. For example, when white light from the Sun hits the red car in Figure 8, all the colours are absorbed except for red. This is reflected into your eyes, so the car looks red.

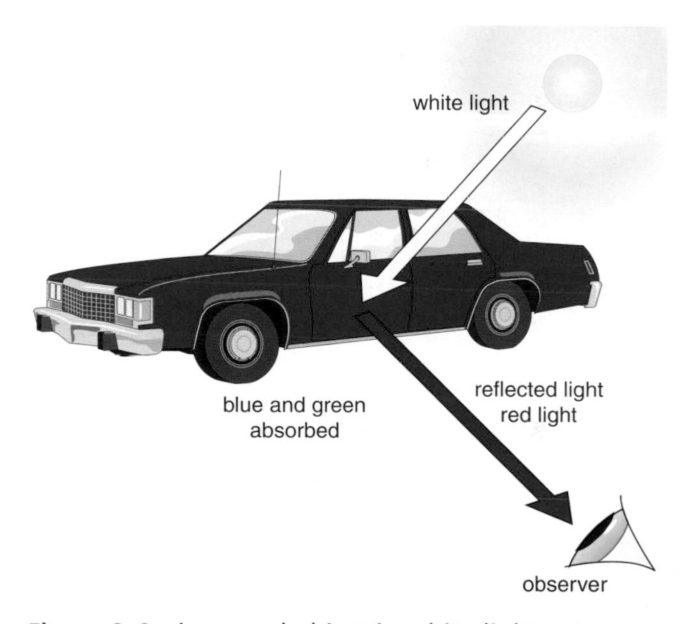

Figure 8 Seeing a red object in white light

Figure 9 shows that when white light from the Sun hits the white feather, none of the colours are absorbed. As they are all reflected into your eyes, the feather looks white.

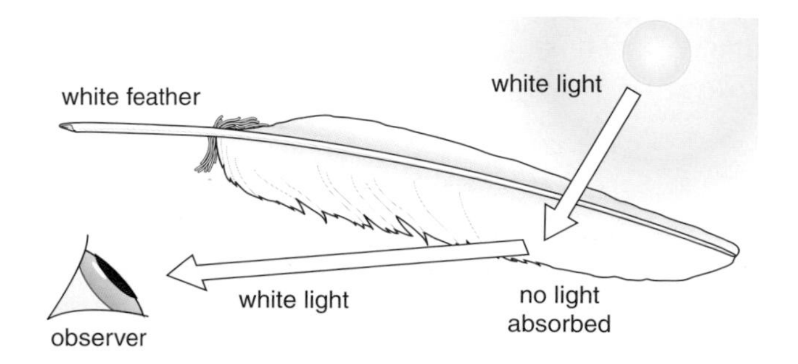

Figure 9 Seeing a white object in white light

17.2 Refraction and colour (*continued*)

Mixing coloured lights

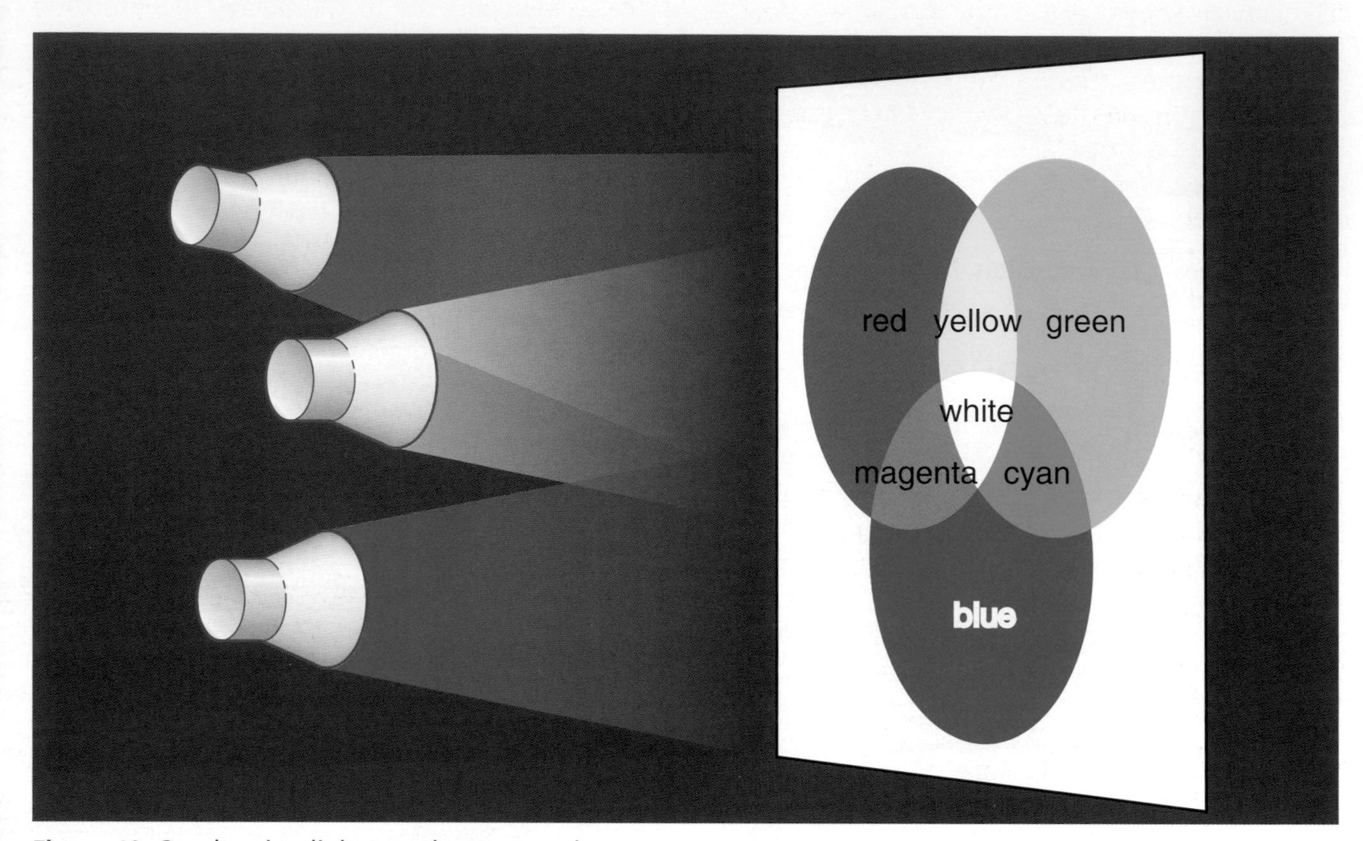

Figure 10 Overlapping lights produce new colours

If you shine coloured lights onto a white screen you can seen that where the lights overlap, new colours are produced. Three colours that cannot be produced by mixing other colours are red, green and blue. These are called the **primary colours**. Any other colour can be made by mixing two or more of the primary colours in different amounts.

If all three colours are mixed together in equal amounts they produce white light. If equal amounts of any two of the primary colours are added together, they produce yellow, cyan or magenta. These are known as the **secondary colours**.

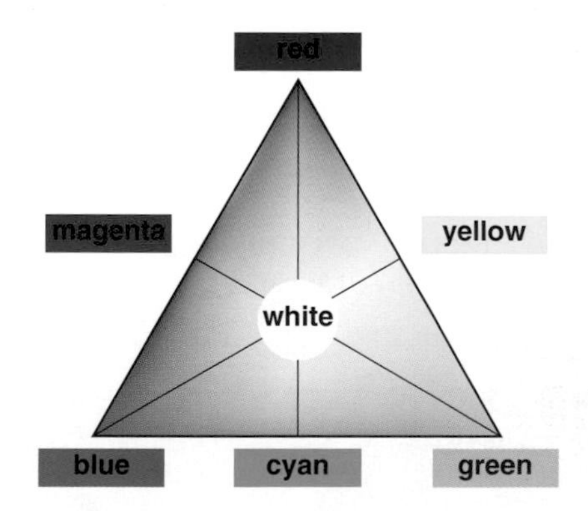

Figure 11 The colour triangle shows us which colours are produced when coloured lights mix

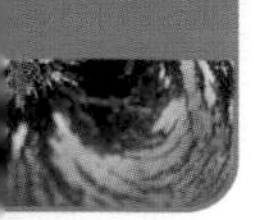

Colour television

A colour television contains three electron guns. One gun produces a red spot on the screen, the second a blue spot and the third a green spot. The guns are controlled by the signals received through the television aerial. When a red colour signal is received, the red gun is aimed at a particular point on the screen. Electrons released by the gun cause this spot to glow red. The image we see on the screen is created by these glowing spots. Different combinations of the glowing coloured spots produce images of different colours. For example, equal numbers of red and green glowing dots will produce a region of yellow on the screen.

Coloured filters

A **filter** is a clear piece of plastic which has been coloured by a dye. They are often used to produce a beam of light of a particular colour, for example a spotlight in a theatre, or they are used to remove certain colours of light from a beam. Light which is the same colour as the filter can pass through it but other colours cannot. These colours are absorbed.

In Figure 12, red light passes through a red filter but the orange, yellow, green, blue, indigo and violet are absorbed. When the red light strikes the green filter it is unable to pass through it. It is absorbed.

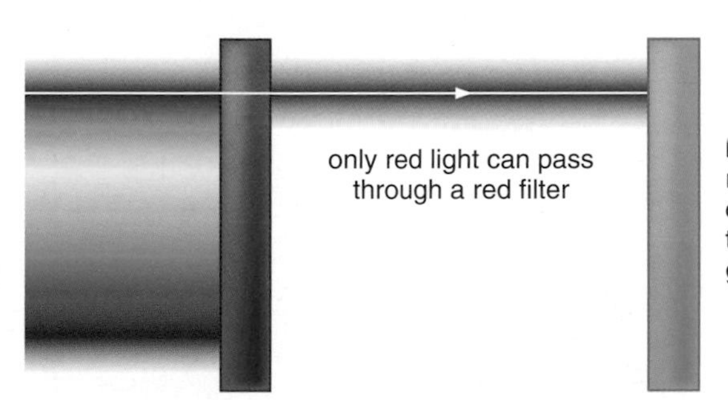

Figure 12 A filter will only allow certain colours of light to pass through it. Other colours are absorbed

Key terms

Check that you understand and can explain the following terms:

- ★ media
- ★ refraction
- ★ normal
- ★ spectrum
- ★ dispersion
- ★ dye
- ★ absorb
- ★ primary colour
- ★ secondary colour
- ★ filter

Questions

1. Draw a labelled diagram to show what happens to a ray of light when it travels through a glass block, striking the top surface at an angle of 45 °.
2. Draw a labelled diagram to show what happens to a ray of light which hits a rectangular glass block at an angle of 90 ° to one of its surfaces.
3. Explain why a pencil half immersed in water may appear to be bent.
4. Explain what happens to a ray of white light as it enters and passes through a glass prism. Include a diagram with your answer.
5. Explain with a diagram why a blue book appears to be blue when white light shines on it.
6. What are the primary colours and why are they given this name?
7. Explain with a diagram what happens when white light tries to pass through a blue filter.

Chapter 17 Light:

What you need to know

1. The Sun, stars and fires are all sources of light. They are luminous objects.
2. We see non-luminous objects because light is reflected from them into our eyes.
3. Light travels in straight lines.
4. Light travels at 300 000 km/s.
5. Opaque objects do not allow light to pass through them.
6. Translucent objects allow light to pass through them, but do not allow us to see through them.
7. Transparent objects allow light to pass through them and we can see through them.
8. Shadows are areas of darkness created when opaque objects are placed in front of a source of light.
9. Flat surfaces, such as plane mirrors, reflect light in a predictable direction.
10. The image of an object created in a plane mirror is upright, the same size as the object, the same distance behind the mirror as the object and laterally inverted.
11. Rays of light may refract as they cross the boundary between two different media.
12. White light is a mixture of coloured lights.
13. Coloured objects absorb all the colours of the spectrum except their own, which is reflected.
14. Coloured filters absorb all the colours of the spectrum except their own, which they allow to pass through.

How much do you know?

1 Why is it that the boy wearing the blindfold in the diagram below is unable to see any of the other people in the room?

____

____

2 marks

2 The diagram below shows a ray of light striking a plane (flat) mirror. Which letter, A, B, C or D, shows the direction in which the ray will be reflected by the mirror.

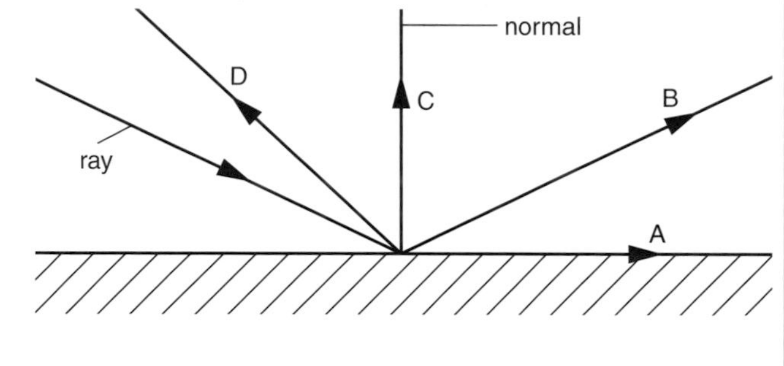

____

1 mark

3 Complete the diagram below showing how the rays of light striking this bicycle reflector are reflected.

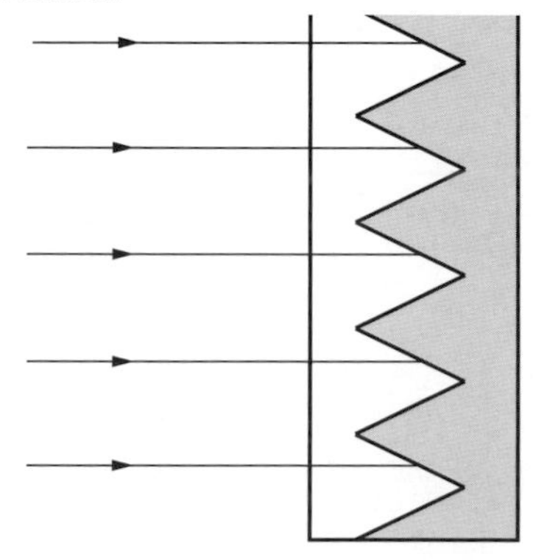

2 marks

How much do you know? *continued*

4 The diagram below shows a ray of white light striking a glass prism.

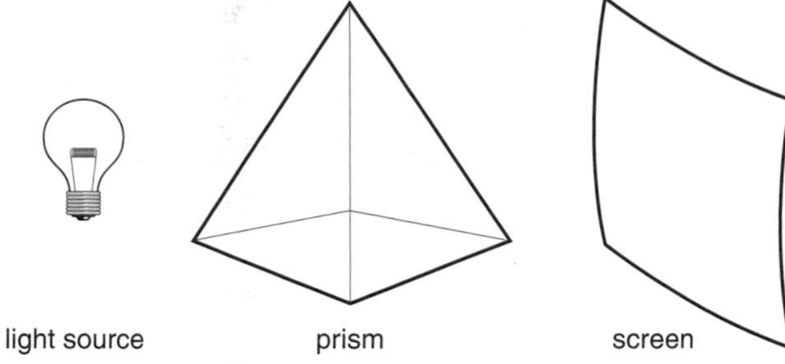

a) What is seen on the white screen?

1 mark

b) A red filter is placed between the prism and the screen. What is now seen on the white screen?

1 mark

5 A pupil places a book in front of a plane mirror. The book is 20 cm tall and stands 30 cm from the mirror.

a) How tall is the image of the book which is seen in the mirror?

1 mark

b) How far behind the mirror is the image of the book?

1 mark

c) Explain why it is not possible to read the title of the book by looking at the image in the mirror

1 mark

6 A white board is illuminated by the three circular, coloured lights as shown below.

a) What colour is seen at point X?

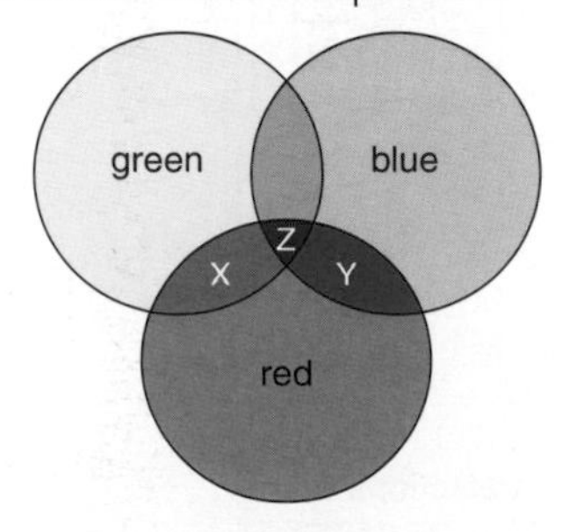

1 mark

b) What colour is seen at point Y?

1 mark

c) What colour is seen at point Z?

1 mark

d) Why are red, green and blue known as the primary colours?

1 mark

e) Name one secondary colour

1 mark

7 The diagram below shows the construction of a simple periscope

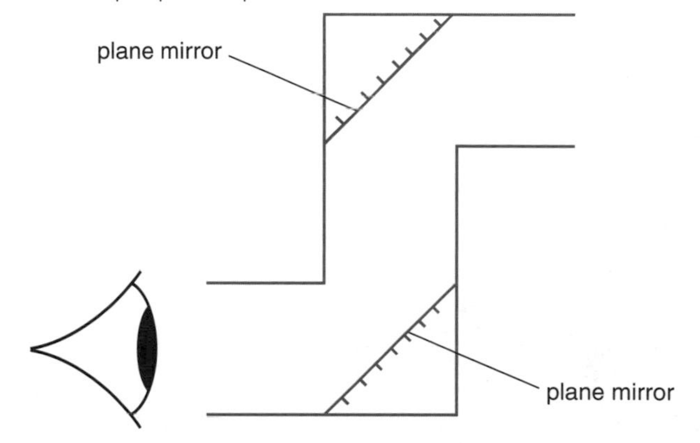

a) Using a pencil and a ruler draw in the path of a ray of light from the object which passes through the periscope.

4 marks

b) Give one use of a periscope

1 mark

8 Explain why a red car looks a) red in white light, b) red in yellow light and c) black in blue light.

3 marks

18.1 Hearing sounds

Producing sounds

All sounds are produced by objects that are **vibrating**. These sounds travel outwards from the source as **sound waves**.

For example:

★ The skin on a drum vibrates to produce the sound of the drum beat.

★ When the clanger hits the inside of a bell, the vibrations are heard as a ringing sound.

★ The buzzing sound you hear from a bee is created by the vibration of its wings.

★ In a loudspeaker, electrical energy is changed into vibrations which we hear as sounds or music.

Creating and hearing sound waves

Sound waves travel from vibrating objects to our ears by means of sound waves. Figure 1 shows how a sound wave is created in air.

As the object vibrates to the right, it pushes the air particles close together, creating a **compression**. As the object vibrates to the left, a region of *more spread out particles* is created. This is called a **rarefaction**. When the object has vibrated several times, a series of compressions and rarefactions have been created that are moving away from the object. This is a sound wave. A model of a moving sound wave can be created by pushing and pulling one end of a slinky.

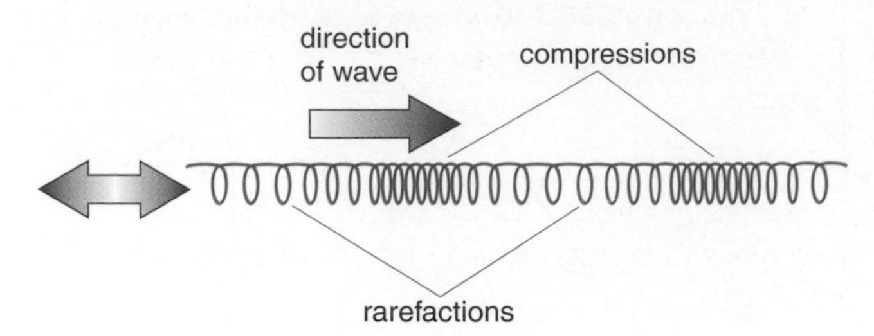

Figure 2 Model of a sound wave

The ear

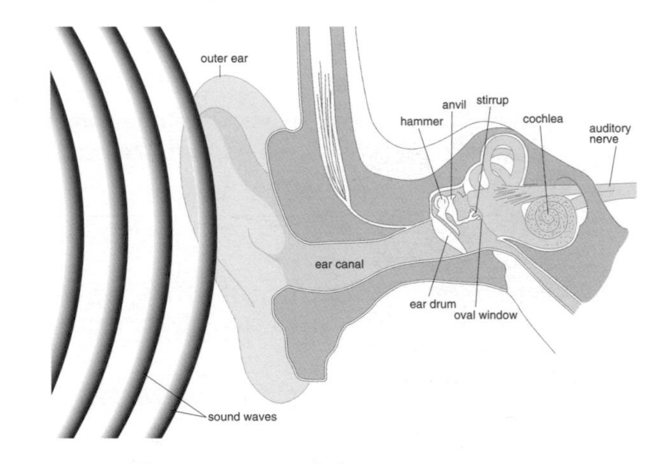

Figure 3 The structure of the ear

Sound waves are collected and directed into the ear canal by the outer ear. When the sound waves strike the **ear drum**, they make the eardrum vibrate. Three tiny bones called the hammer, the anvil and the stirrup **amplify** these vibrations and then transmit them through a liquid inside a long coiled structure called the cochlea.

Figure 1 A sound wave in air

Hairs along the inside of the cochlea are stimulated by the vibrations and impulses are sent from the hairs to the brain through the **auditory nerve**. The listener then hears the sound.

Wax in the ears, damage to the eardrum or amplifying bones, and infection in the cochlea can all lead to impaired hearing and possible deafness.

Sound waves must have something to travel through

Figure 4 Sound waves can travel through solids, liquids and gases

Sound waves can travel through solids, liquids and gases. But they cannot travel though a space where there are no particles, i.e. they cannot travel through a **vacuum**.

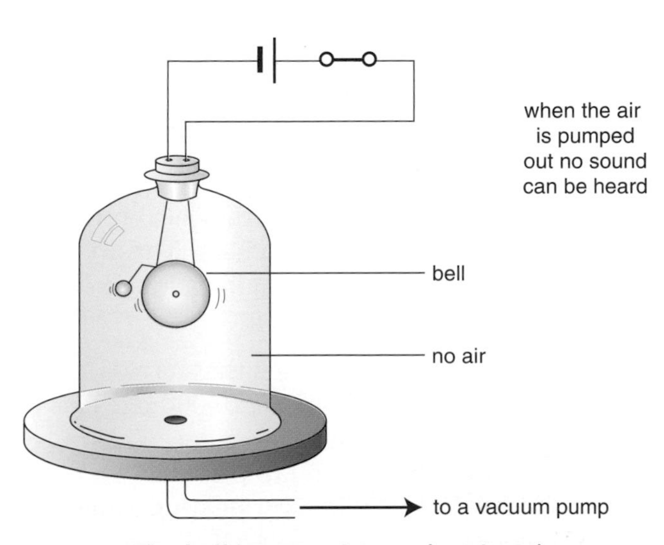

Figure 5 The bell jar experiment showing that sound cannot travel through a vacuum

When there is air in this bell jar we are able to hear the bell ring when the switch is closed. If the air is pumped out of the jar and the switch closed, we can see the bell ringing, but we cannot hear it. This experiment proves that light waves can travel through a vacuum but sound waves cannot.

Speed of sound

Sound waves travel much more slowly than light waves. This is why sometimes we see an event before we hear the sound, e.g. thunder and lightning or the flash of an exploding rocket followed seconds later by a loud bang.

Sound waves travel at different speeds in different materials. In air the speed of sound is approximately 340 m/s, in water it is approximately 1500 m/s and in steel it is 6000 m/s. Sounds travel most quickly through solids and least quickly through gases.

Key terms

Check that you understand and can explain the following terms:

★ vibrating
★ sound wave
★ compression
★ rarefaction
★ ear drum
★ amplify
★ auditory nerve
★ vacuum

Questions

1 All sounds begin with an object which is ________.

2 Explain why it is not possible for sound waves to travel through a vacuum.

3 Suggest three ways in which someone's hearing may be impaired.

4 Explain why we always see the flash of the lightning before we hear the sound of thunder.

5 Through what group of materials do sound waves travel **a)** fastest and **b)** slowest?

18.2 Different kinds of sounds

Frequency of vibration and pitch

Large objects, like one of the strings of a double bass, vibrate slowly when plucked. Its frequency of vibration is low and only a small number of sound waves are produced each second. The note produced by the string will have a **low pitch**.

Figure 1 Low-pitched sounds produced by the double bass as seen on an oscilloscope

It is possible for you to *see* a picture that represents these waves using a piece of apparatus called a **cathode ray oscilloscope** (CRO).

Small objects, like a violin string, vibrate more quickly when plucked. Its frequency of vibration is higher so more sound waves are produced each second. The note produced by this string will have a **higher pitch**.

Figure 2 Higher-pitched sounds produced by the violin as seen on an oscilloscope

Small objects vibrate quickly producing high-pitched sounds.

Large objects vibrate slowly producing low-pitched sounds.

The frequency of an object or wave is the number of complete vibrations it performs each second. It is measured in hertz (Hz) where 1 Hz is one vibration per second

Musical instruments

Like all musical instruments, the double bass and the violin can produce notes with a wide range of frequencies and pitch. The changes in frequency produced by the vibrating strings are achieved by altering

★ the length of the string by changing the positions of the fingers on the fret board. The longer the vibrating string, the lower the pitch of the note produced

★ the thickness of the string. The thicker the string, the lower the pitch of the note produced

★ the tension of the string. The greater the tension, the higher the pitch of the note produced by the string.

Changes in the frequencies of the notes produced by wind instruments are achieved by altering the lengths of vibrating air columns.

Figure 3 Altering the length of a trombone alters the pitch of the note it produces

Hearing range

Some objects vibrate so quickly you may be unable to hear the sounds they produce. These sounds have a very **high frequency** and are called **ultrasounds**. Some animals such as dogs, bats and dolphins are able to hear ultrasounds.

Some objects vibrate so slowly they also produce sounds you may not be able to hear. These sounds have very **low frequencies** and are called infrasounds.

Human beings can hear sounds that have frequencies between 20 Hz and 20 000 Hz. This is called their **hearing** or **audible range**. This range can vary a little from person to person. Older people usually have a narrower hearing range than the young.

Loudness and amplitude of vibration

If an object vibrates with a large **amplitude** it will produce a **loud** sound. This is seen on the oscilloscope as a tall wave.

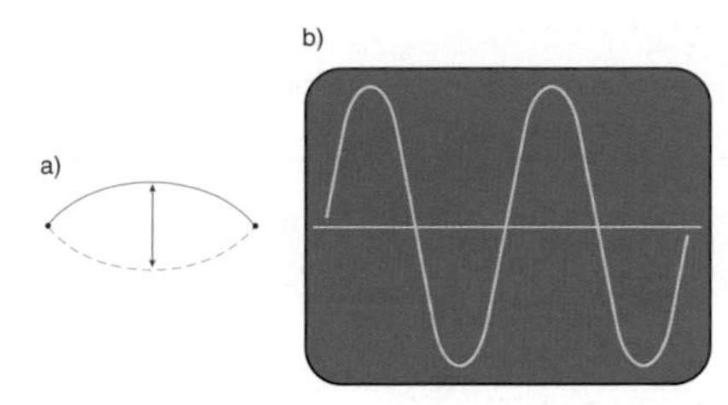

Figure 4 A loud sound as seen on an oscilloscope screen

A wave like this carries a lot of sound energy. It could be produced by plucking a string strongly or hitting a drum skin hard.

If you pluck the string or hit the skin of a drum gently it will vibrate with a small amplitude, producing a **soft or quiet** sound. This is seen on the oscilloscope as a wave which is small in height

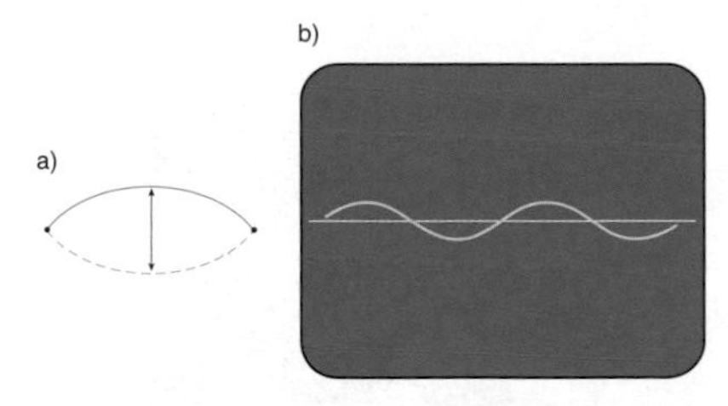

Figure 5 A quiet sound as seen on an oscilloscope screen

Loudness and the decibel scale

Very loud noises, such as those made by aircraft, machinery in factories, heavy goods lorries and personal stereos with the volume turned up high, can permanently damage your hearing. Ear protectors should be worn by people operating noisy machinery.

The loudness of a sound is measured on the **decibel scale**.

Type of sound	Loudness in decibels
A bird singing quietly	20
People talking reasonably quietly	60
A dog howling close by	80
A noisy factory	90
A noisy disco	110
A jet aircraft taking off	130
A bomb exploding	more than 150

Noise pollution

Unwanted sound is called **noise**. Noise pollution can be a real problem. It can cause stress and make it more difficult to concentrate. There are several ways in which planners, builders, architects and engineers try to reduce this problem:

★ place barriers such as walls or rows of trees between the source of the noise and nearby buildings

★ fit double glazing in buildings close to the source of the sound e.g. homes close to airports

★ develop machinery that is quieter e.g. better insulated car engines.

Key terms

Check that you understand and can explain the following terms:

★ low pitch
★ cathode ray oscilloscope
★ high pitch
★ high frequency
★ ultrasound
★ low frequency
★ hearing (audible) range
★ amplitude
★ loudness
★ soft or quiet
★ decibel scale
★ noise

Questions

1 Write down the names of four different types of musical instruments you would find in an orchestra. Which of these instruments would produce high-pitched notes and which would produce low-pitched notes? Explain your answers.

2 Draw a diagram to show the kind of picture you would see on a CRO if the note being played is **a)** low pitched and **b)** high pitched.

3 Name two animals that can hear sounds which are outside the human hearing range.

4 Draw a diagram of the picture you would expect to see on a CRO for **a)** a loud note and **b)** a quiet note.

5 What number on the decibel scale would you record for **a)** a jet aircraft taking off close by and **b)** two people having a normal conversation?

6 What might happen to your hearing if you are exposed to very loud sounds for a long period of time? Give two examples of places where you might find these very loud sounds.

Chapter 18 Sound and hearing:

What you need to know

1 Sounds are made when objects vibrate.
2 The greater the amplitude of vibration, the louder the sound.
3 The higher the frequency of vibration, the higher the pitch of the sound.
4 Sound energy travels in waves.
5 Sounds can travel through solids, liquids and gases but not through a vacuum.
6 Sound waves cause the ear drum to vibrate.
7 Loud sounds can damage the ear drum and cause deafness.
8 Different people have different hearing ranges.

How much do you know?

1 The diagram below shows a man playing a guitar.

a) Suggest two ways in which the guitarist could alter the note being produced by one of the strings.

____

____

2 marks

b) Draw a picture of what you would expect to see on an oscilloscope screen if the sound produced by the guitarist is

(i) high pitched but quiet

(ii) low pitched but loud

4 marks

2 The diagram below shows a young girl watching a firework display.

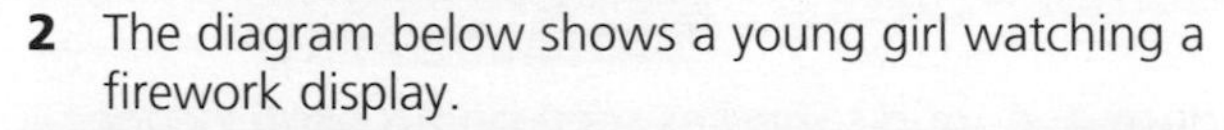

Why does she see the flash of an exploding rocket before she hears the bang?

____

____

2 marks

3 The gong in the diagram below is vibrating gently and producing a quiet sound.

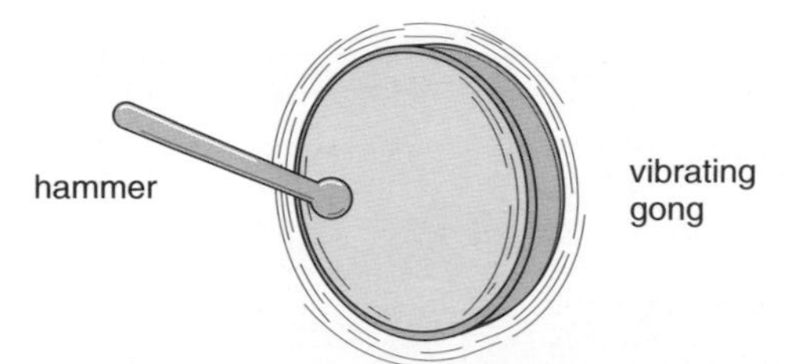

a) What would you do to the gong to make it produce a loud sound?

____

1 mark

How much do you know? *continued*

b) Which of these two bells would you shake if you wanted to produce a lower-pitched sound?

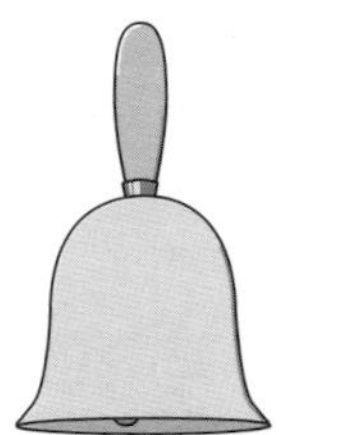
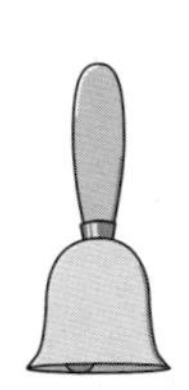

____

1 mark

4 The table below contains several sources of sounds. In the second column enter an approximate decibel level for these sounds.

Source of sound	Loudness on decibel scale
A normal conversation between two people	
Someone shouting as loud as they possibly can	
A firework rocket exploding	
A small alarm clock ringing	

5 Richard likes to listen to loud music.

a) What may happen to Richard's hearing if he does this too often?

____

1 mark

b) What two things could Richard do to avoid this problem?

____

____

2 marks

6 a) What is noise? What effect might noise have on a person?

____

____

2 marks

b) Suggest two ways in which noise levels can be reduced.

____

____

2 marks

c) Give one example of someone who should wear ear protectors.

____

1 mark

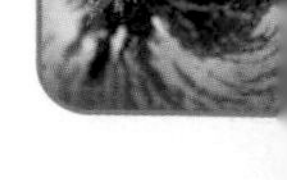

19.1 Forces and pressure

If you were to stand on someone's foot while wearing stiletto heels, you would cause them considerable pain, as all your weight would be concentrated in a small area. If, however, you were to stand on someone's foot while wearing hiking boots, you would cause them considerably less pain as your weight would be spread over a larger area.

Figure 1

If a force is **concentrated** into a **small area** it creates a **large pressure**. If a force is spread over a large area it creates a **small pressure**.

If you wear snowshoes whilst walking in snow, your weight will be spread over a large area and you won't sink. Without the snowshoes the force is more concentrated and you are likely to sink.

Figure 2 Snowshoes spread out your weight, so stop you from sinking

You should NEVER walk across a frozen pond or lake. The pressure your weight creates may be sufficient to crack the ice. Rescuers can avoid this problem by using a long ladder which spreads their weight and so reduces the pressure they exert on the ice.

Figure 3 The ladder reduces the pressure the rescuer exerts on the ice

The pressure created at the point of this nail is large enough for the point to pierce the wood.

Figure 4

If the point is blunt, the pressure is less and the point will not pierce the wood.

Figure 5

Because there are a large number of nails on this fakir's bed, the pressure on each nail point is too small to pierce his skin.

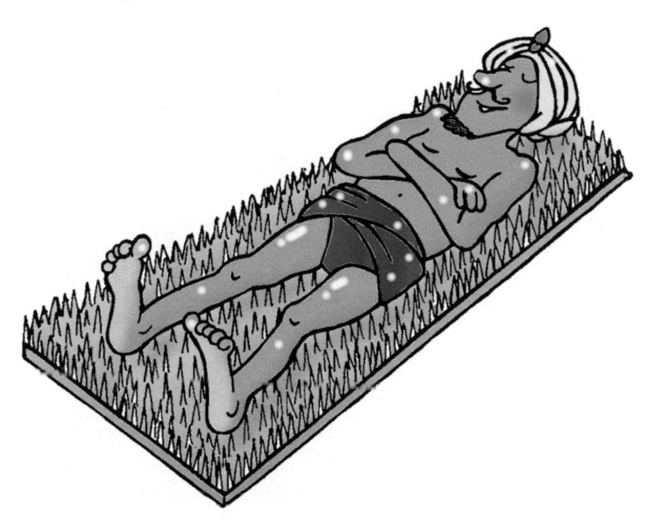

Figure 6

Calculating pressure

The pressure created by a force can be calculated using the equation:

$$\text{pressure} = \frac{\text{force}}{\text{area}}$$

Pressure is measured in **pascals** (Pa). 1 Pa is the same as 1 N/m^2.

Examples

Figure 7 shows three crates of equal weight, with different areas in contact with the ground. The dimensions are labelled in each case. An accompanying calculation shows how the pressure created by each crate is calculated.

Key terms

Check that you understand and can explain the following terms:

- ★ concentrated
- ★ small area
- ★ large pressure
- ★ small pressure
- ★ pascal

Questions

1. Explain why it is easier to cut a piece of hard cheese with a sharp knife rather than a blunt one.
2. Why do sprinters wear shoes that have spikes in their soles?
3. Suggest two situations where you would want to try to avoid creating high pressures. How can you achieve this?
4. Why does a drawing pin have a large head?
5. A girl weighing 500 N is wearing shoes whose soles have a total surface area of 0.1 m^2. Calculate the pressure she creates beneath her feet.
6. A camel weighs 2000 N. If each of its feet has a surface area of 0.01 m^2 in contact with the ground, calculate the pressure created beneath each foot. What assumption have you made?
7. Why do caravan owners often put pieces of wood under the legs of the caravan when they wind them down?
8. When is the most dangerous time for a fakir who likes to lie on a bed of nails? Explain your answer.

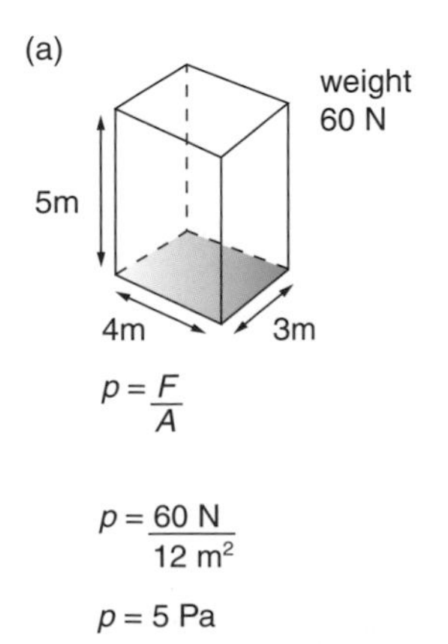

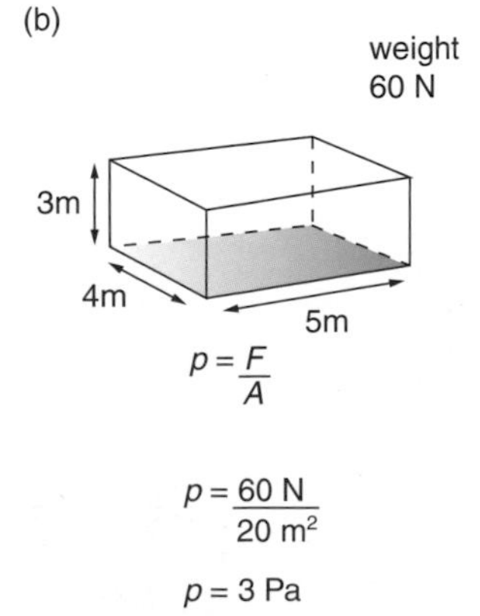

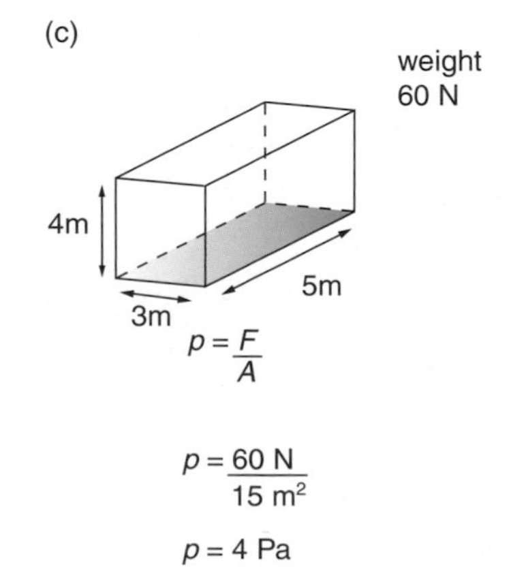

Figure 7

19.2 Pressure in liquids and gases

Pressure in a liquid

The diver in the photograph below feels water pressure all around her. Her diving suit will help her survive the high pressures she will feel deep under the surface of the water. The deeper she goes, the greater the water pressure around her.

Figure 1 Water exerts pressure on a diver deep below the surface

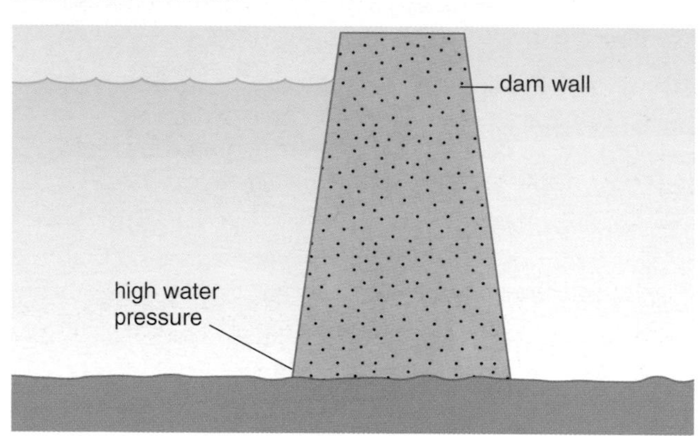

Figure 2 Dams are always thicker at their base because this is where the pressure due to the water is greatest

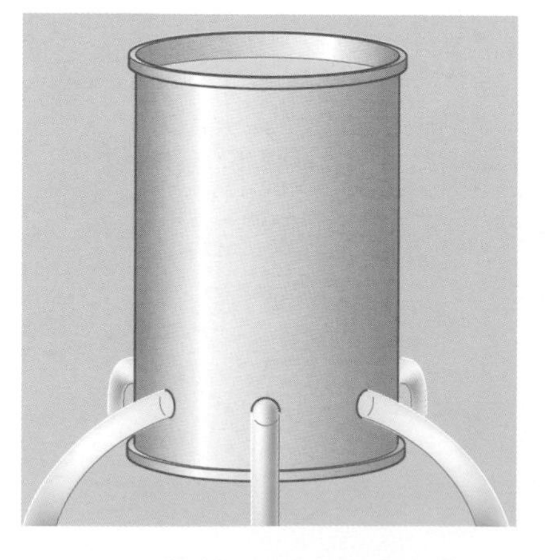

Figure 3 Water flows out of this can at the same rate in all directions, showing that the pressure in a liquid is the same in all directions

Transmitting pressure and forces through liquids

Liquids are **incompressible**. Their particles cannot easily be pushed together.

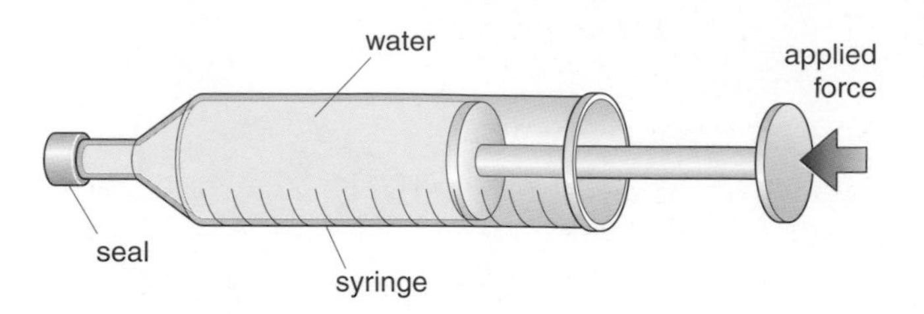

Figure 4 It is impossible to push the piston into the syringe. The syringe is blocked and the water is incompressible

If a force is applied to a liquid, it is **transmitted** or passed through the liquid. The study of transmitting forces and pressures through liquids is called **hydraulics**. It has some very useful practical applications.

The hydraulic jack

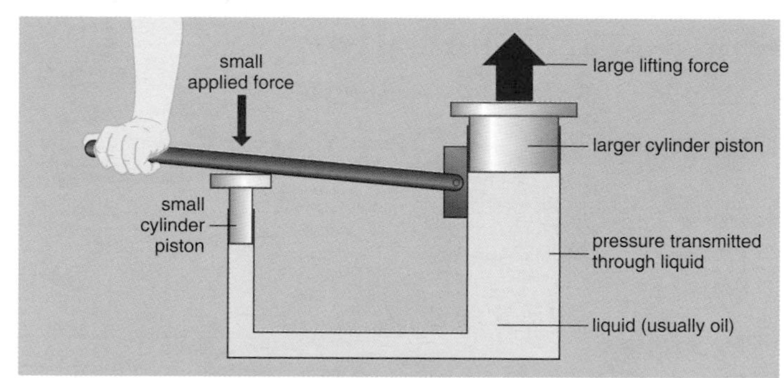

Figure 5 A hydraulic jack

When the handle is pushed down, it applies a force on a small piston which creates a pressure in the liquid. This pressure is transmitted through the liquid to the larger piston where a much larger lifting force is created, i.e. a small force has been changed into a much larger one. The hydraulic jack is a **force multiplier**.

Hydraulic brakes

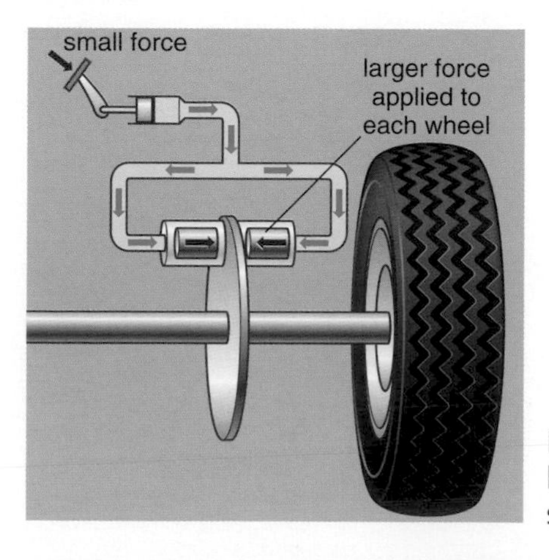

Figure 6 The hydraulic brake system of a car

The hydraulic brakes on a car work in a similar way. When the driver presses the brake pedal, his force is increased and directed through tubes of brake fluid to the wheels of the car.

Pressure in gases

Pressure in gases is created by collisions of the particles in the gas.

The study and use of pressure in gases is called **pneumatics**.

When the top of this plastic bottle is removed there is air inside and outside its walls. The pressures created by the colliding particles are equal and balanced. The bottle therefore keeps its shape. If some of the air inside the bottle is removed, the pressure inside the bottle is now less than the **air pressure** outside and the bottle is crushed.

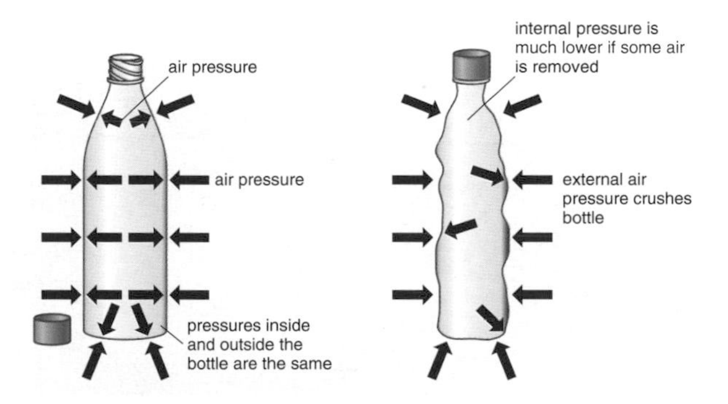

Figure 7 Demonstrating air pressure

Drinking through a straw

When you initially suck through a straw, you remove some of the air particles inside the straw. The pressure exerted by the air particles on the surface of the liquid outside the straw is larger. So liquid is pulled up the straw.

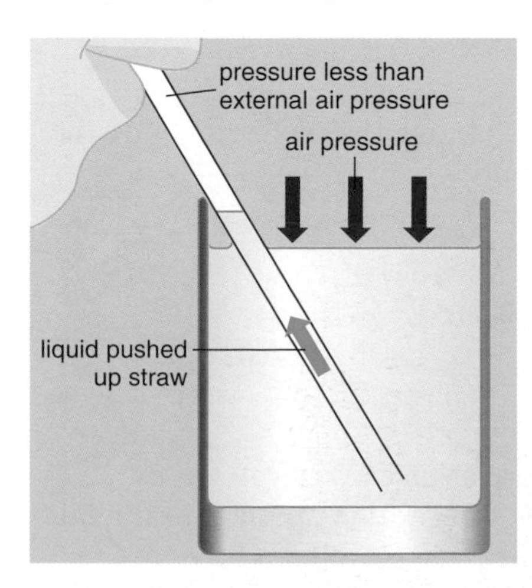

Figure 8 Drinking through a straw

Aerosols

An aerosol spray can contains a liquid such as paint or polish and a gas. The gas is at a pressure which is much higher than **atmospheric pressure**. When the nozzle is pressed, the liquid in the can is pushed up the tube by the gas and out of the nozzle opening.

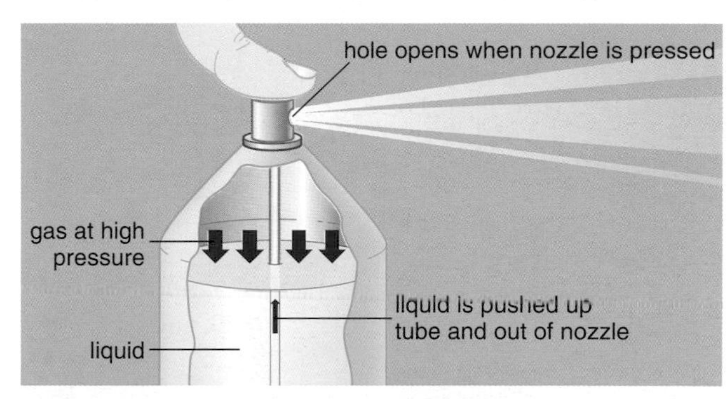

Figure 9 How an aerosol works

Key terms

Check that you understand and can explain the following terms:

- ★ incompressible
- ★ transmit
- ★ hydraulics
- ★ force multiplier
- ★ pneumatics
- ★ air pressure
- ★ atmospheric pressure

Questions

1. Explain why the base of a dam wall is always wider than any other part.
2. A spherical balloon is inflated and then taken deep below the sea to the sea bed.
   a) What happens to the size of the balloon as it goes deeper under water?
   b) What shape is the balloon when it is on the sea bed?
3. A syringe with a sealed needle is filled with water. Explain why it is impossible to push the piston into the cylinder of the syringe.
4. Why is a hydraulic jack described as being a force multiplier?
5. How do air particles create atmospheric pressure?
6. Why will a balloon explode if too much air is put into it?

19.3 Turning forces or moments

Sometimes when you apply a force to an object you can make it turn or rotate. The **turning effect** of a force is called a **moment**.

The size of the turning effect depends upon

★ the size of the force you apply

★ the place where it is applied.

The point around which the force is turning is called the **pivot**.

If you apply a force a long way from the pivot, you can create a large moment.

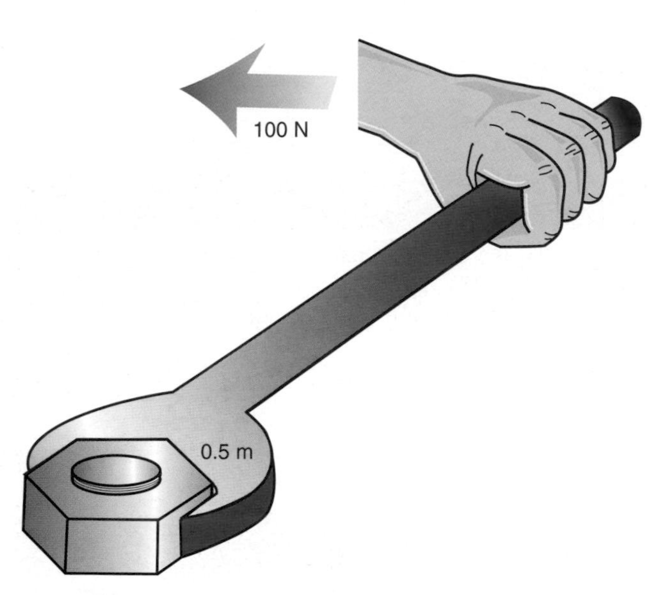

Figure 1 Creating a large moment

If you apply a similar force closer to the pivot, the moment created will be smaller.

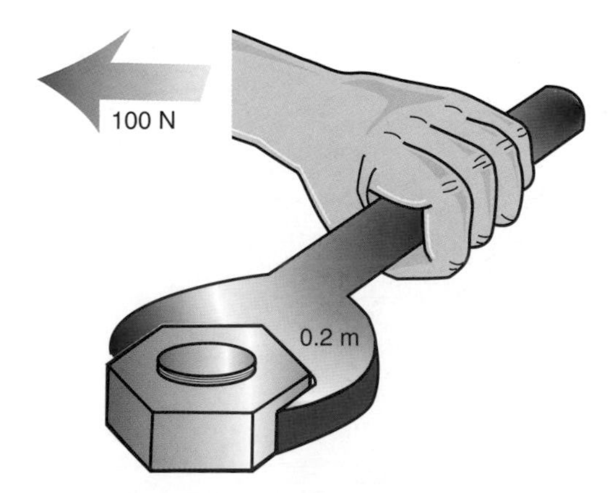

Figure 2 Creating a small moment

This explains why it is easier to undo a stiff nut using a long spanner.

The moment of a force can be calculated using the equation:

$$\text{moment of a force} = \text{force} \times \text{distance from pivot}$$

The moment created by the long spanner was:

$$100\,\text{N} \times 0.5\,\text{m} = 50\,\text{Nm}.$$

The moment created by the short spanner was:

$$100\,\text{N} \times 0.2\,\text{m} = 20\,\text{Nm}.$$

Levers

Opening a tin of paint or syrup is easy when you know how. Pushing down on the handle of the screwdriver creates a large upward force, lifting the lid of the can. The screwdriver is being used as a **lever**. A lever is a device which can be used to change the direction and/or the size of a force.

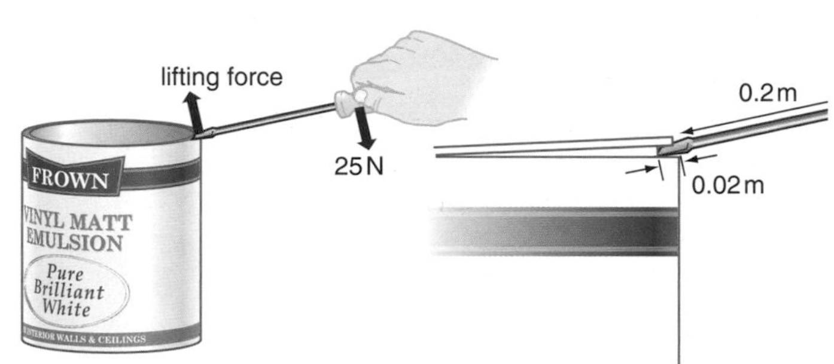

Figure 4 Using a screwdriver as a lever

When you push down on the screwdriver, you create a moment. In this case the moment is:

$$\begin{aligned}\text{moment} &= \text{force} \times \text{distance from pivot}\\ &= 25\,\text{N} \times 0.2\,\text{m} = 5\,\text{Nm}\end{aligned}$$

As the lid begins to move the moments on both sides of the pivot are equal.

Therefore:

$$\begin{aligned}\text{Lifting force} \times 0.02\,\text{m} &= 5\,\text{Nm}\\ \text{Lifting force} &= 5\,\text{Nm}\backslash 0.02\,\text{m}\\ \text{Lifting force} &= 250\,\text{N}\end{aligned}$$

Using the screwdriver as a lever you have increased the size of the force applied to the lid. A lever used in this way is called a **force multiplier**.

Balancing moments

Sometimes more than one moment can be applied to an object. The weight of the boy on the see-saw is trying to turn the see-saw anti-clockwise. The weight of the girl is trying to turn the see-saw clockwise. They can make the see-saw turn in different directions by altering their positions.

Sitting in the positions shown below, the **clockwise moment** created by the girl is larger than the **anticlockwise moment** created by the boy. The see-saw therefore turns clockwise.

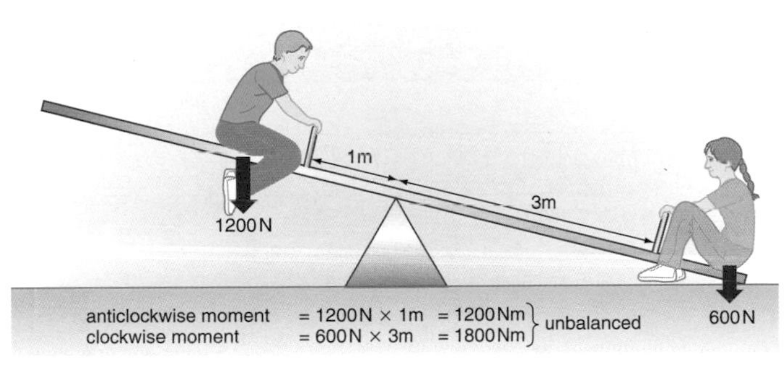

Figure 5

Sitting in the positions shown in Figure 6, the anticlockwise moment created by the boy is larger than the clockwise moment created by the girl. The see-saw therefore turns anticlockwise.

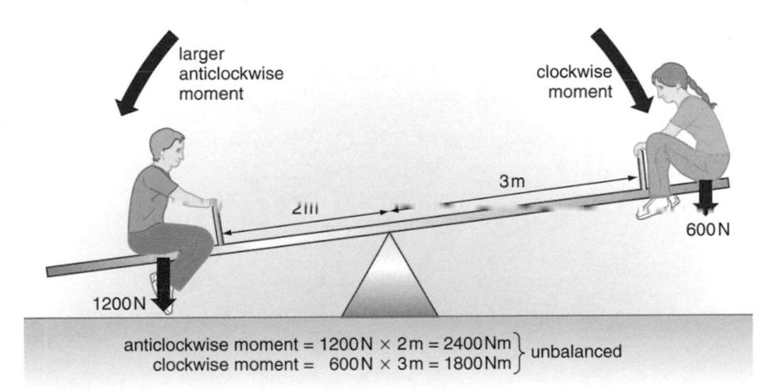

Figure 6

Sitting in the positions shown in Figure 7, the moments created by the children are equal. The seesaw is balanced.

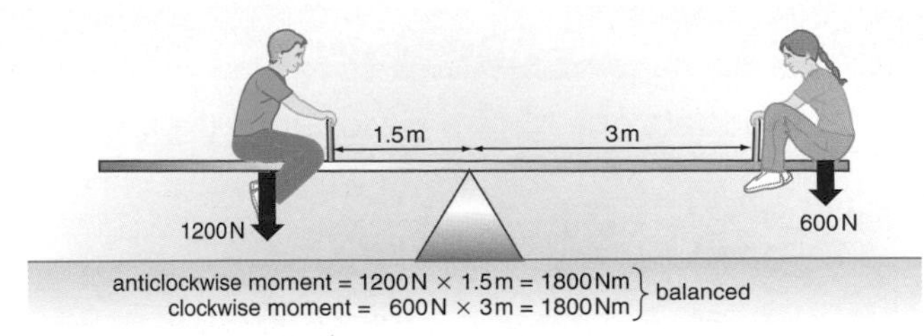

Figure 7

Key terms

Check that you understand and can explain the following terms:

- ★ turning effect
- ★ moment
- ★ pivot
- ★ lever
- ★ force multiplier
- ★ clockwise moment
- ★ anticlockwise moment

Questions

1. Which of the following is not an example of a force creating a moment? **a)** pedalling a bike, **b)** pushing a sledge, **c)** opening the page of a book, **d)** turning on a tap and **e)** striking a match.
2. Why is it easier to open a tin of paint using a long screwdriver than a short screwdriver?
3. Explain why door handles are always placed a long way from the hinges of the door.
4. A man opens a door by pushing with a force of 50 N at a point 0.5 m from the hinges. Calculate the moment he creates.
5. Give one example of two moments which are balanced.
6. Find out why tightrope walkers often carry a long pole.
7. A girl weighing 400 N sits 2.5 m from the pivot of a see-saw. How far from the pivot must a boy weighing 500 N sit in order that the see-saw balances?

Chapter 19 Pressure and moments:

What you need to know

1. The effect of a force depends upon the area over which it is applied.
2. A force concentrated over a small area will create a large pressure.
3. A force spread out over a large area will create a small pressure.
4. Pressure = force/area.
5. Pressure in a liquid acts in all directions.
6. Pressure in a liquid increases with depth.
7. Liquids are incompressible and can be used to transmit pressures and forces.
8. Gas particles collide with objects creating gas pressure.
9. Some forces cause objects to turn or rotate.
10. The turning effect of a force is called a moment.
11. The moment of a force = size of force × perpendicular distance from pivot.
12. A lever is used to change the direction or size of an applied force.
13. If a see-saw balances then the clockwise moments must equal the anticlockwise moments.

How much do you know?

1 All these questions concern forces and pressures.

a) Explain why carrier bags should have wide handles.

____

____

____

2 marks

b) Why do rescuers use ladders to reach people who have fallen through ice on frozen lakes and ponds?

____

____

____

2 marks

c) Explain how and why camels have adapted to allow them to walk more easily in desert conditions.

____

____

____

2 marks

2 Two girls of equal weight walk across a snow-covered field. One is wearing snowshoes, the other is not.

a) Explain why the girl wearing snowshoes is able to walk on top of the snow, but the girl without the snowshoes sinks deep into the snow.

____

____

____

3 marks

b) The girl wearing the snowshoes weighs 500 N. The total area of her two snowshoes is $0.5\ m^2$.

Calculate the pressure exerted by the snowshoes on the snow.

____

____

2 marks

3 The diagram below shows a can filled with water. There are three holes in the can, one at the top, one in the middle and one at the bottom.

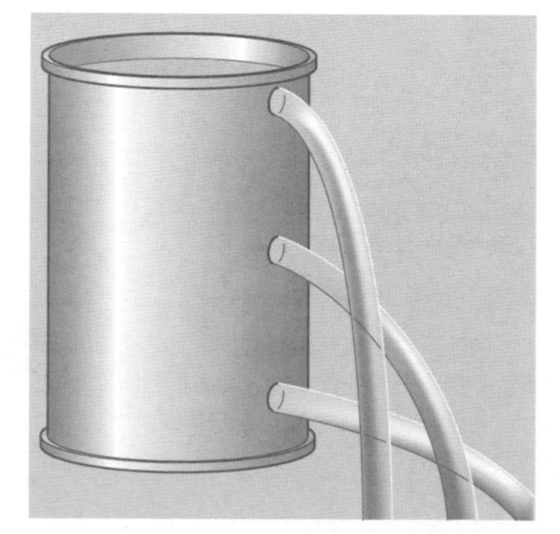

a) Out of which hole is the water flowing the fastest?

____

1 mark

How much do you know? *continued*

b) Give one reason why the water should flow out of this hole the fastest.

____

1 mark

4 The diagram below shows two sealed syringes. Syringe A is filled with water. Syringe B is filled with air.

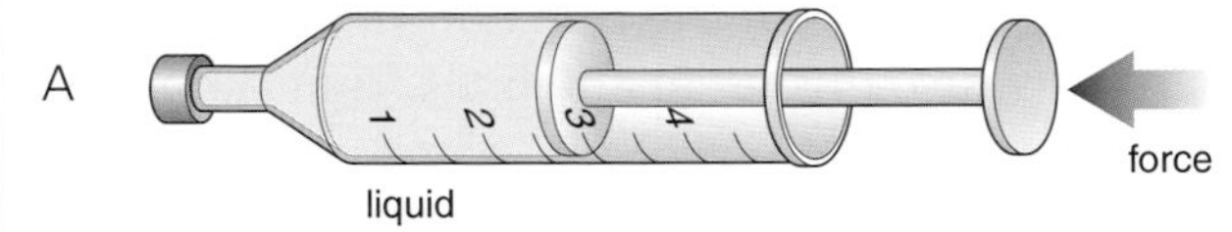

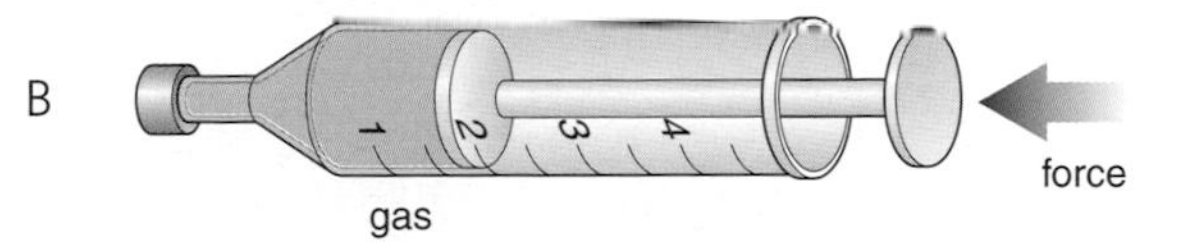

a) Describe what happens when a force is applied to syringe A.

____

1 mark

b) Describe what happens when a force is applied to syringe B.

____

1 mark

c) Explain the difference in the behaviour between syringe A and syringe B.

____

____

2 marks

5 The diagram below shows a spanner being used to undo a nut on a car wheel.

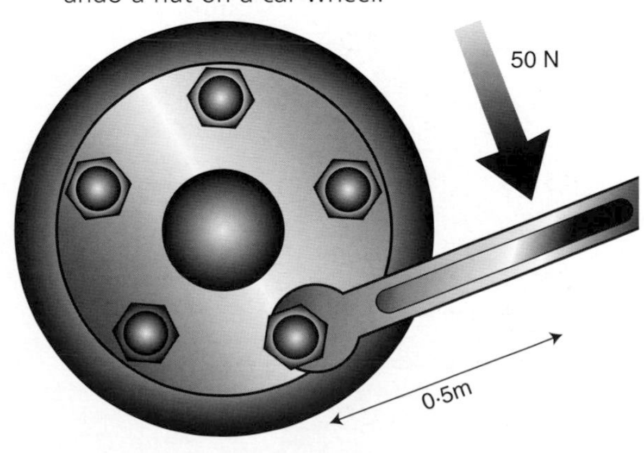

a) Calculate the size of the moment created by the 50 N force.

____

____

____

3 marks

b) If the nut was stiff and you needed to create a larger moment using the same force, what would you do?

____

1 mark

6 The diagram below shows two friends Janet and John on a see-saw. Janet and John both weigh the same.

What should John do in order to balance the see-saw?

____

____

2 marks

7 The diagram below shows a screwdriver being used as a lever to open a tin of paint. If the downward force applied to the end of the screwdriver is 50 N calculate the upward force applied to the tin lid.

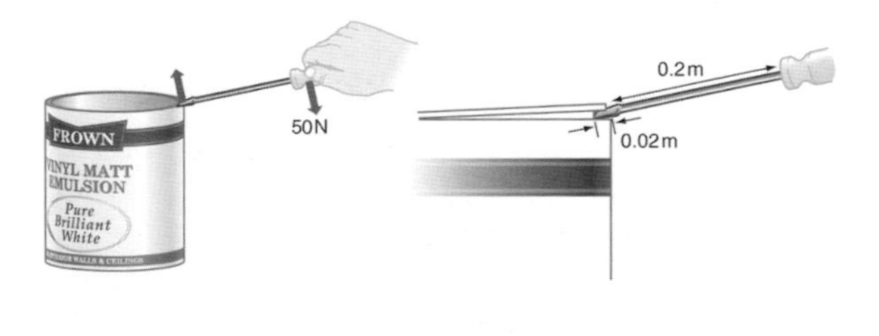

____

____

____

3 marks

20.1 The Earth in space

You, your family, all your friends, in fact all mankind live on a **planet** called Earth. If you could see the Earth from space it would look like this.

Figure 1 How the Earth looks from space

This photograph shows Europe, the Atlantic Ocean and the Mediterranean Sea. Part of the Earth's surface is covered by cloud.

The daily journey of the Sun

Although you cannot feel it, the Earth is spinning around on its **axis** like a top. This turning motion makes the Sun appear to rise in the East, travel high overhead and then set in the West.

The Earth completes one full turn or rotation every 24 hours – this is **one day**.

Day and night

As the Earth turns, part of its surface is in daylight, whilst other parts are in darkness. It is **daytime** on those parts of the Earth that are receiving light from the Sun and **night** on those parts that are not receiving light from the Sun.

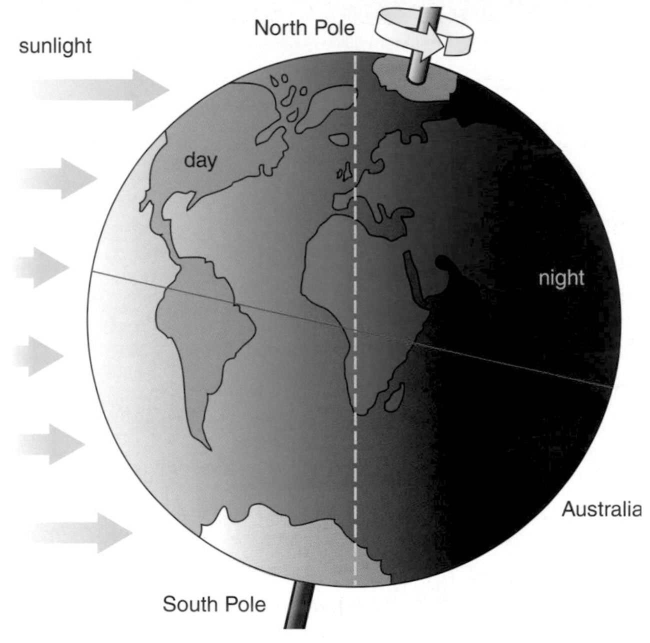

Figure 3 The explanation for day and night

What time is it here?

Different parts of the Earth's surface are at different times of their day. In Figure 4, place A is moving into the sunlight so it is early morning here. It is midday at B. It is evening at C. It is the middle of the night at D.

Figure 2 The Sun makes the same journey across the sky, but it is much lower in the sky in the winter

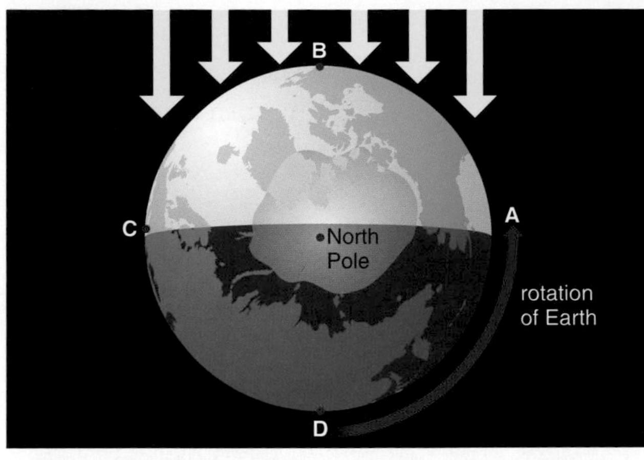

Figure 4 Different places on the surface of the Earth may be at different times of the day

A year and the seasons

The nearest **star** to the Earth is the **Sun**. It is 150 million kilometres away. The Earth **orbits** the Sun once every **year**, following a path called an **ellipse**. (An ellipse is a slightly squashed circle.)

As the Earth travels around the Sun we experience the different **seasons**. Seasonal changes to the weather and to the climate happen because the Earth's axis is slightly tilted.

When we in Britain are in **summer**, our part of the Earth (the northern part) is tilted towards the Sun. This makes our climate warmer and our days (time in daylight) longer.

When we are in **winter**, our part of the Earth is tilted away from the Sun. This makes our climate colder and our daytime shorter.

When we are in spring or autumn, our part of the Earth is tilted neither towards, nor away from the Sun.

When it is summer in the northern hemisphere it is winter in the southern hemisphere. When it is winter in the northern hemisphere it is summer in the southern hemisphere

Key terms

Check that you understand and can explain the following terms:

- ★ planet
- ★ axis
- ★ one day
- ★ daytime
- ★ night
- ★ star/Sun
- ★ orbit
- ★ year
- ★ ellipse
- ★ season
- ★ summer
- ★ winter

Questions

1 How do we know that the Earth is 'ball shaped' and not flat?

2 How long does it take for the Earth to make one complete revolution?

3 How long does it take for the Earth to orbit the Sun?

4 Explain why during the day we see the Sun rise in the East, travel across the sky and set in the West.

5 Why do we have seasons?

6 Describe two changes that take place when we move from summer, through autumn into winter.

7 If it is summer in Britain, what season is it in Australia?

8 Find out what a leap year is and why we have them.

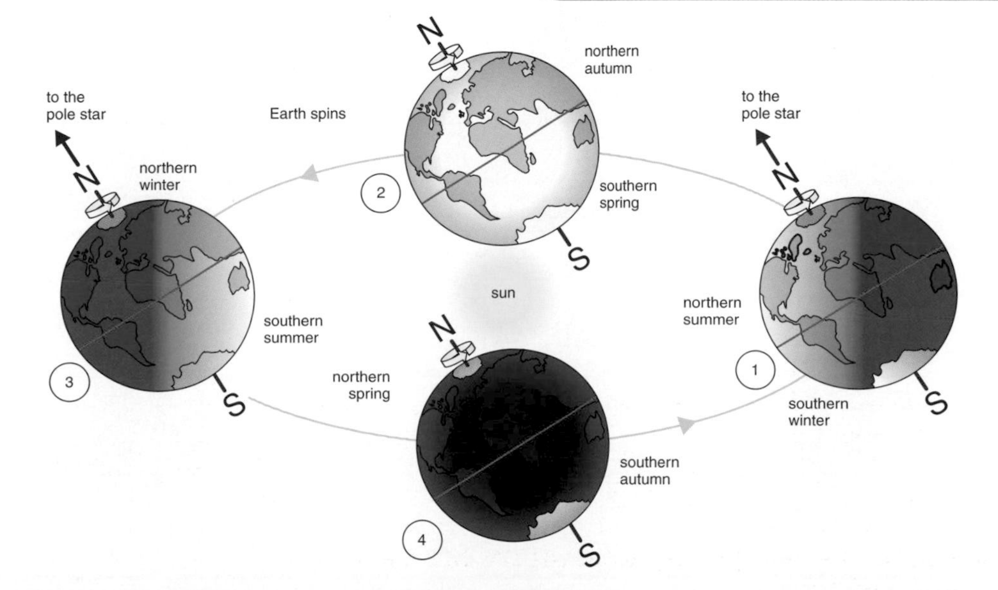

Figure 5 How the tilting of the Earth causes the seasons

20.2 The Solar System

Our **Solar System** consists of the Sun, the planets and their moons and a belt of rocks or asteroids.

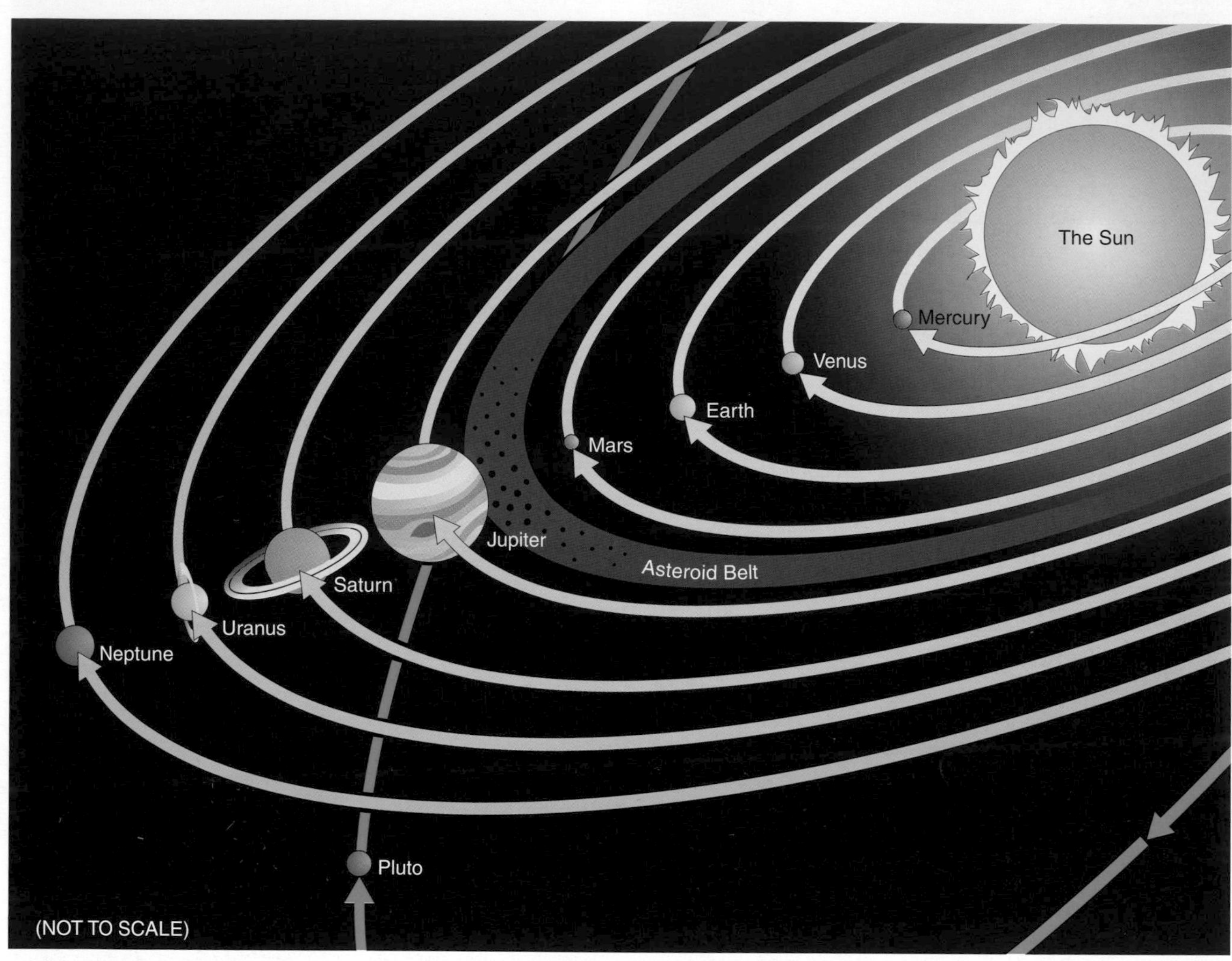

Figure 1 The Solar System

Planet	Approximate distance from Sun compared with the Earth	Approximate diameter compared with Earth	Approximate mass compared with Earth	Approximate surface gravity compared with Earth	Approximate length of one day	Approximate length of one year	Approximate surface temperature in °C
Mercury	0.5	0.5	0.05	0.4	60	0.2	120
Venus	0.75	1	0.80	0.9	240	0.6	460
Earth	1	1	1	1	1	1	15
Mars	1.5	0.5	0.1	0.4	1	2	−25
Jupiter	5	11	320	2.3	0.4	12	−73
Saturn	10	10	95	0.9	0.4	30	−140
Uranus	19	4	15	0.8	0.7	84	−200
Neptune	30	3.5	17	1.2	0.7	165	−200
Pluto	40	0.2	0.003	0.04	6.4	248	−220

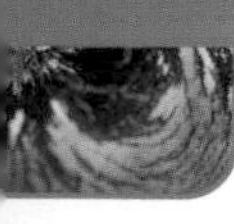

Planets

The Earth is one of nine **planets** in the Solar System which orbit the Sun. Starting with the planet nearest the Sun they are Mercury, Venus, Earth, Mars, Jupiter, Saturn, Uranus, Neptune and Pluto. Try using this sentence to help you remember the order. **M**any **V**ery **E**nergetic **M**en **J**og **S**lowly **U**pto **N**ewport **P**agnell.

It is not possible to draw this diagram to scale and fit it on this page but Figure 1 should give you some idea of the sizes of the planets and their distances from the Sun.

The table opposite gives some information about the planets in the Solar System.

Gravity in space

All the planets in the Solar System move in **elliptical orbits** with the Sun near the centre. They do this because of strong attractive forces between the planets and the Sun. These forces are **forces of gravity** and they control the positions and the movements of the planets.

The size of the gravitational attraction between two objects depends upon the masses of each object and their separation. The larger the masses, the greater the force. The larger the separation, the smaller the force. A planet such as Pluto which has a small mass and is a long way from the Sun therefore experiences small gravitational forces.

Moons

Moons are natural objects which orbit a planet. They are natural **satellites**. They are kept in orbit by the pull of gravity, i.e. their positions and motions are controlled by the gravitational forces between them and their planet. The Earth has just one moon, but Jupiter has 16 moons, Saturn has 23 moons and Mars and Venus have no moons.

The mass of our Moon is much less than the mass of the Earth. As a result its gravity is only one-sixth of that on the Earth, i.e. if an object weighs 6 N on the Earth, it will weigh just 1 N on the Moon.

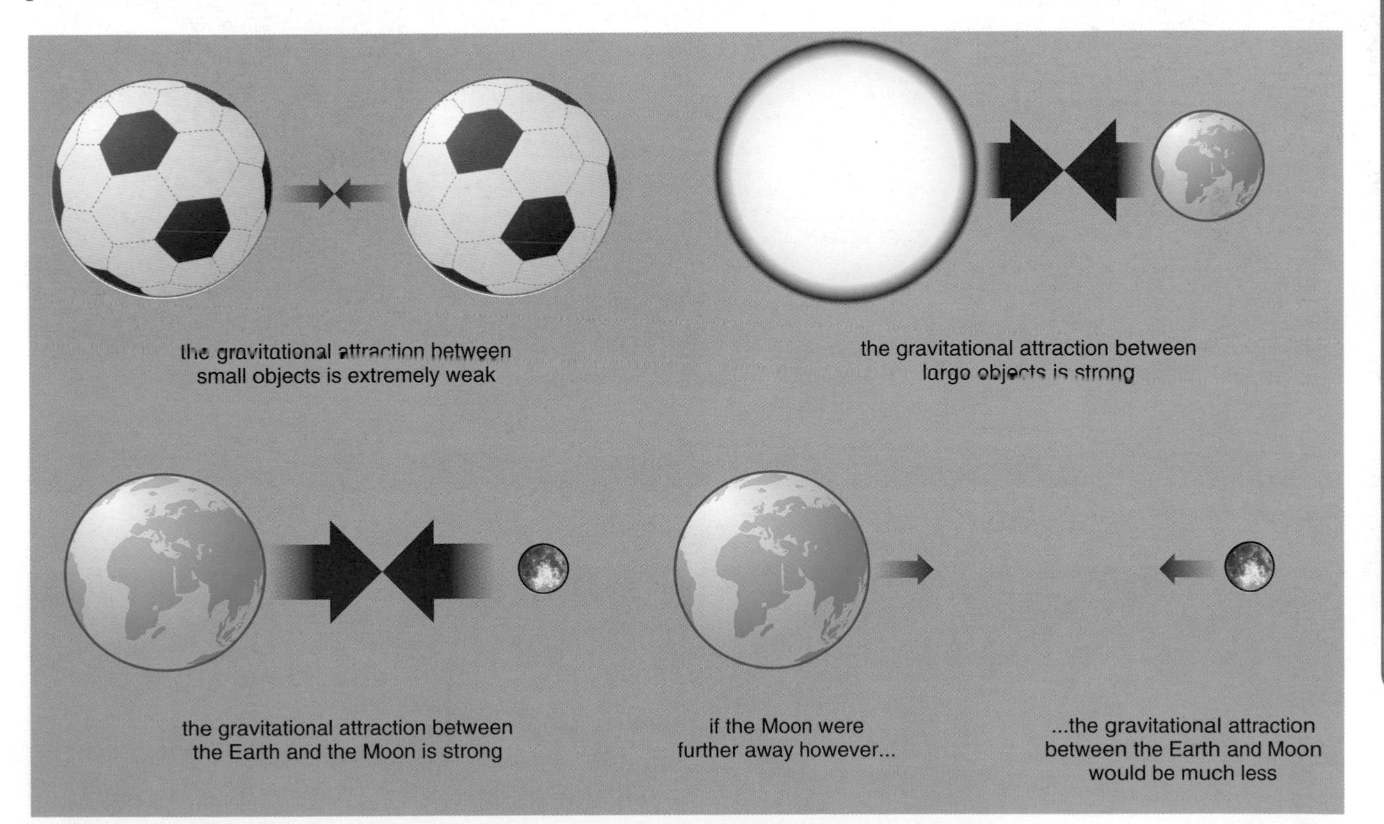

Figure 2 The gravitational attraction between two objects depends on their masses and their separation

20.2 The Earth in space (*continued*)

Phases of the Moon

The Moon is a non-luminous object. We see it because of the light it reflects from the Sun. The Moon orbits the Earth once every 28 days (a **lunar month**). When the Moon is in different parts of its orbit, it reflects different amounts of sunlight towards the Earth. As a result we see the shape of the Moon appear to change. These shapes are called the **phases of the Moon** (see Figure 3).

Solar eclipse

Sometimes as the Moon orbits the Earth it passes between the Earth and the Sun blocking off the sunlight. This is known as a **solar eclipse**. If all the sunlight is blocked off it is known as a total eclipse. If only part of the light is blocked off it is known as a partial eclipse.

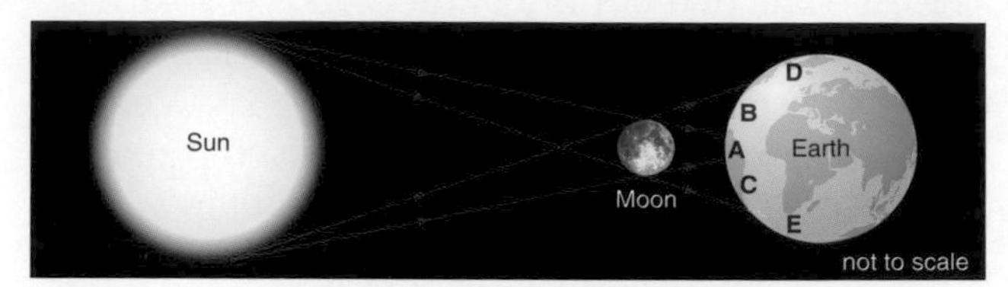

Figure 4 Solar eclipse. A – Total eclipse, B and C – Partial eclipse, D and E – no eclipse

Lunar eclipse

Sometimes the Moon will pass through the shadow created by the Earth. This is called a **lunar eclipse** (see Figure 5).

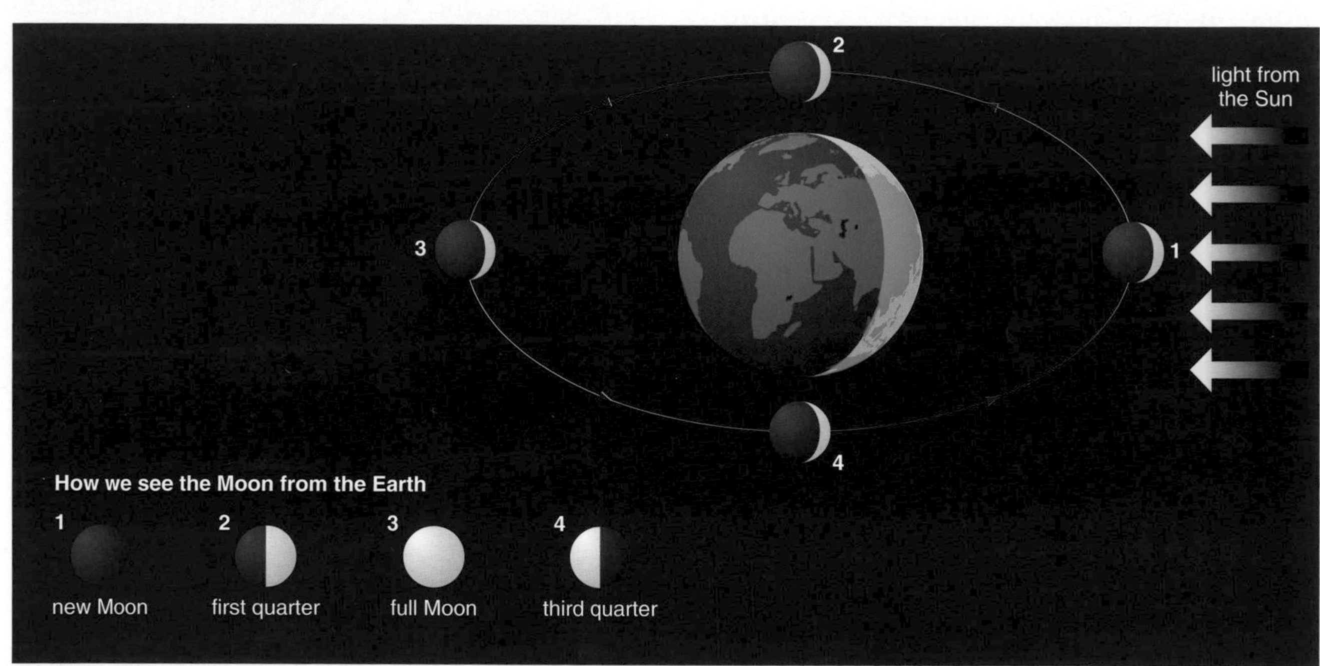

Figure 3 Phases of the Moon

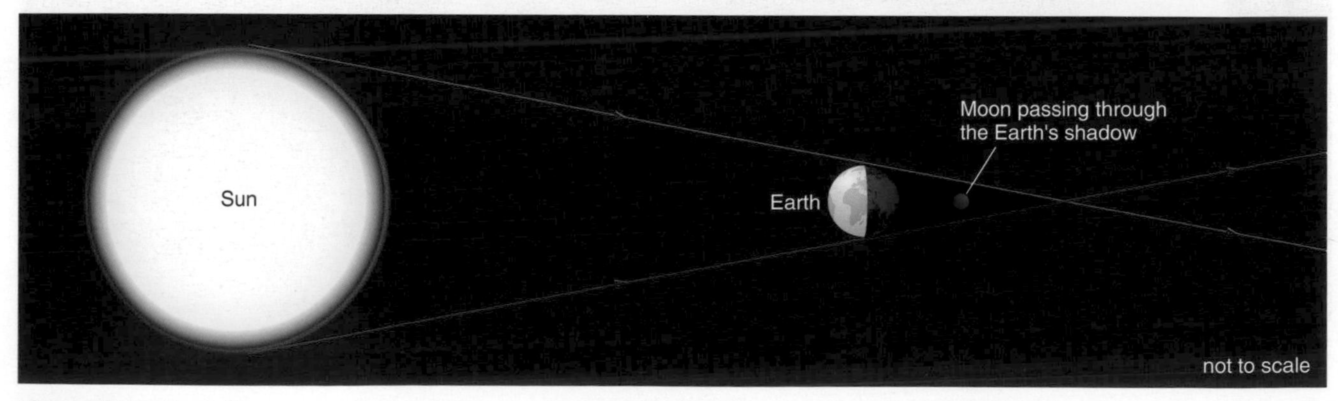

Figure 5 Lunar eclipse

Asteroids and comets

Between the planets Mars and Jupiter there is a belt of rock debris. These rocks are called **asteroids** and vary in size from just a few metres, to several hundreds of kilometres across. Asteroids are thought to be the remains of a planet pulled apart by the gravitational forces from Jupiter.

Comets are made from dust and ice. Like planets they travel around the Sun. But their orbits take them very close to the Sun and then to the outer reaches of the Solar System. As a comet approaches the Sun, some of its frozen gases evaporate creating a spectacular tail of dust and ice which can be millions of kilometres long.

Key terms

Check that you understand and can explain the following terms:

- ★ Solar System
- ★ planet
- ★ elliptical orbits
- ★ forces of gravity
- ★ moon
- ★ satellite
- ★ lunar month
- ★ phases of the Moon
- ★ solar eclipse
- ★ lunar eclipse
- ★ asteroid
- ★ comet

Questions

1. Write down the names of all the planets in order starting from the planet closest to the Sun.
2. What is the name of the forces which control the positions and movements of the planets?
3. What is the shape of a planet's orbit around the Sun?
4. Which is the largest planet in the Solar System?
5. Which is the smallest planet in the Solar System?
6. What is a moon? Which planet has 16 moons?
7. Why is gravity on the Moon much less than that on the Earth?
8. Explain the phrase 'phases of the Moon'.
9. Explain the difference between a solar eclipse and a lunar eclipse.
10. Look at the data in the table on page 184. Suggest one other planet in the Solar System where life as we know it might exist. Give one reason why you have chosen this planet.

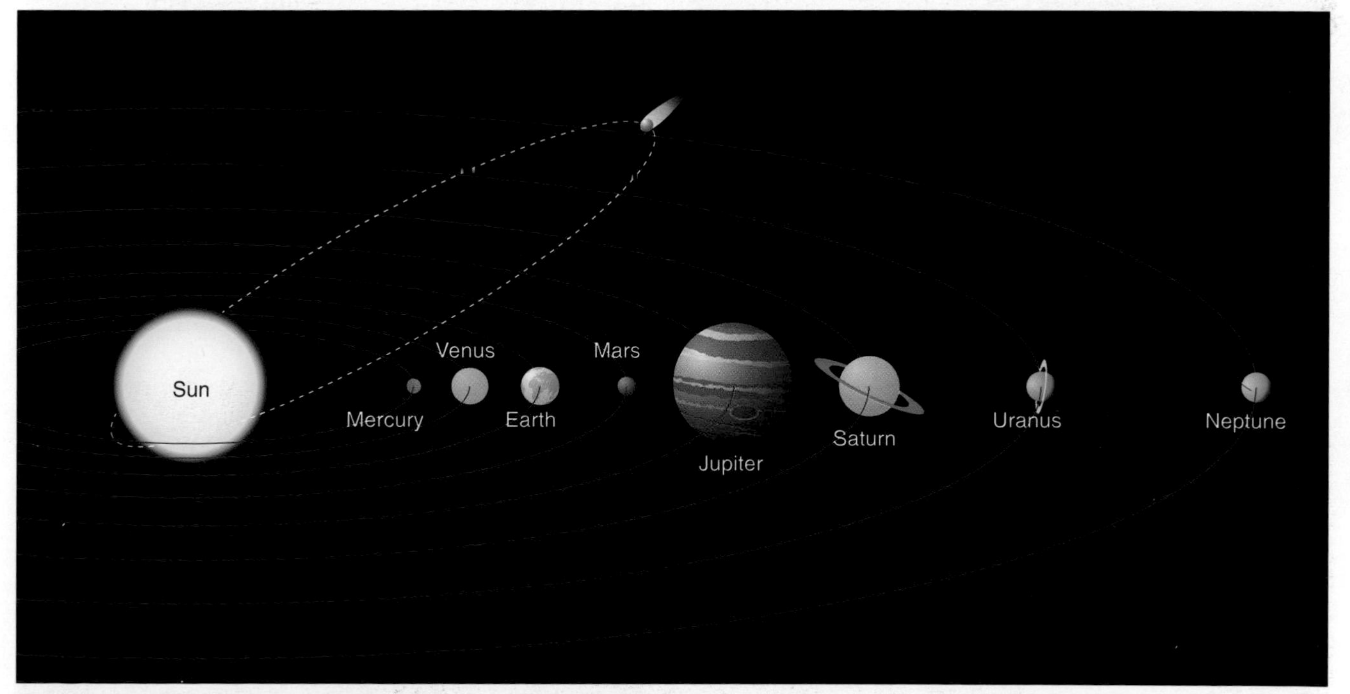

Figure 6 The orbit of a comet

20.3 Stars and satellites

Stars are luminous objects, i.e. they give out light. The star at the centre of our Solar System is the Sun. During the daytime, the light from the Sun is so bright that it is impossible to see any other stars. They only become visible as the Sun sets. These other stars seem very small and dim compared with the Sun, because they are much, much further away.

Figure 1 The sky at night

Planets and moons can also be seen in the night sky but they are not luminous objects. We see them because they reflect the light from the Sun.

Constellations

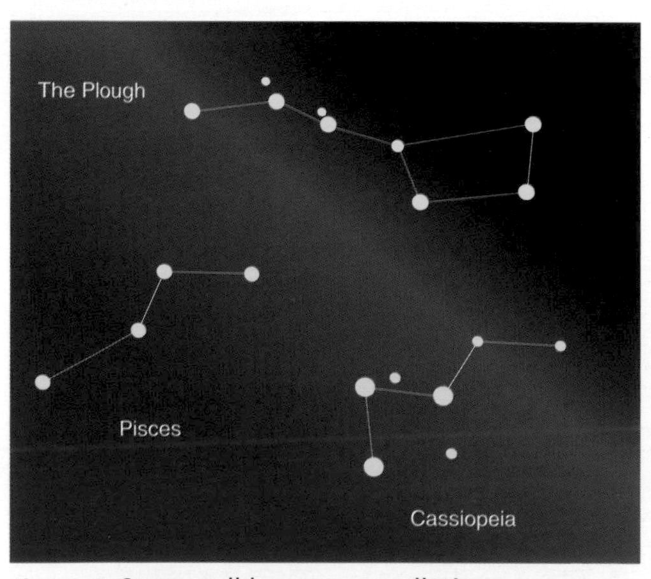

Figure 2 Some well-known constellations

Groups of suns or stars which appear close together in the sky are called **constellations**. You will probably have heard of some of them and can perhaps identify their shape. People used to use the positions of stars and constellations in the night sky to help them navigate at sea.

Galaxies

Figure 3 The Milky Way

Stars and constellations cluster together in enormous groups called **galaxies**. Our galaxy is called the Milky Way and contains about 100 billion stars. There are billions of galaxies spread throughout the Universe.

The moving night sky

If you were to take a photograph of the stars in the sky every hour, you would discover that they appear to revolve and change position. This happens because the Earth is rotating. Only one star, the Pole Star, appears not to move, as it is directly above the axis of rotation.

Figure 4 This time lapse photograph of the night sky shows how the stars appear to move

Artificial satellites

The Moon orbits the Earth – it is a **natural satellite**. There are many man-made objects which also orbit the Earth and these are called **artificial satellites**. These artificial satellites have many uses.

1. Observing the Earth from above. For example, weather satellites provide information which makes weather forecasting more reliable.
2. Allowing communication between people all over the world using radio, television, or the internet.
3. Looking away from the Earth into space, to help us learn more about the Solar System and the Universe.

   When scientists use telescopes to observe objects in space, they have to look through the Earth's atmosphere and this may prevent them from seeing objects clearly. They have recently overcome this problem by mounting a telescope on a satellite which is orbiting above the Earth's atmosphere. This telescope is called the Hubble telescope. It is orbiting 600 kilometres above the Earth's surface.

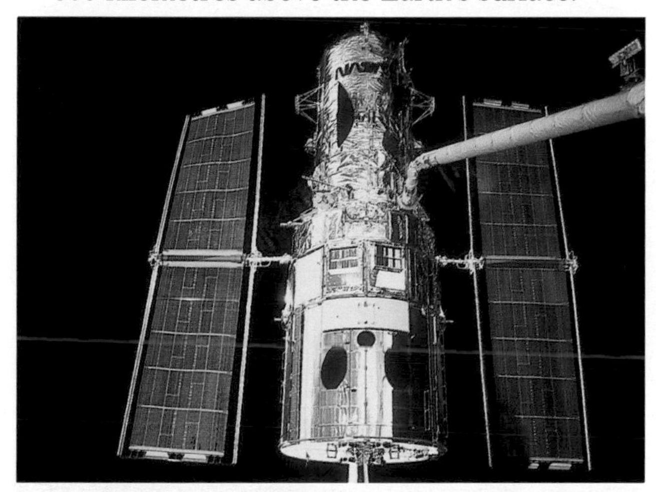

Figure 5 The Hubble telescope

Although man has shown with his visits to the Moon that he is able to travel in space, artificial satellites and probes offer a cheaper and safer way of investigating our Solar System and beyond.

Satellite orbits

Satellites can orbit the Earth at different heights and speeds. Satellites which monitor whole world conditions, e.g. the temperature of the oceans, are put into fast low orbits which take them over the North and South poles. Using these orbits they can scan the whole of the Earth's surface in a very short time.

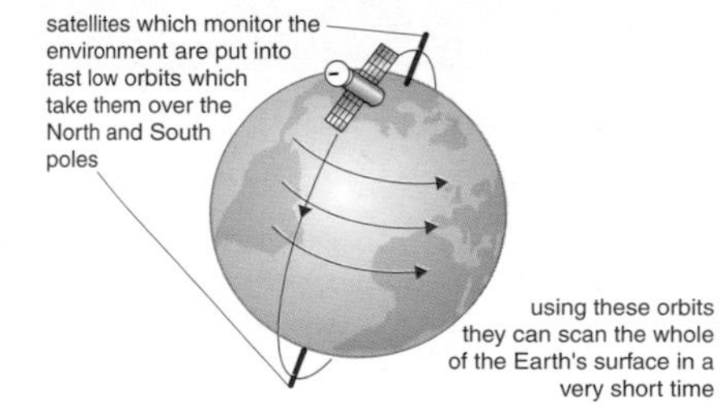

Figure 6 A satellite in low polar orbit

Some satellites are put into much higher orbits and remain above the same point on the Earth's surface the whole time. These are called **geostationary satellites** and are very useful for receiving and redirecting radio signals and observing weather conditions.

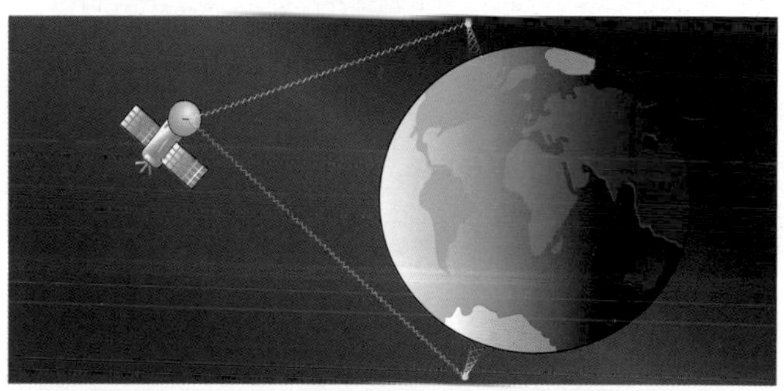

Figure 7 When the radio signal arrives at the satellite, it is redirected down to another part of the Earth where the message is received

Key terms

Check that you understand and can explain the following terms:

- ★ star
- ★ constellation
- ★ galaxy
- ★ satellite
- ★ natural satellite
- ★ artificial satellite
- ★ geostationary satellite

Questions

1. Name one luminous and one non-luminous object we can see in the night sky.
2. Why is it not possible to see stars during the day?
3. Some stars are much bigger and hotter than our Sun but in the night sky they look much smaller and dimmer. Explain why this is so.
4. Explain the difference between a galaxy and a constellation.
5. Why do stars seem to move across the sky during the night?
6. Name one natural satellite.
7. Give one use for an artifical satellite.
8. Explain the difference between a satellite which is put in a geostationary orbit above the Earth and one which is put in a low polar orbit. Give one use for each of these satellites.

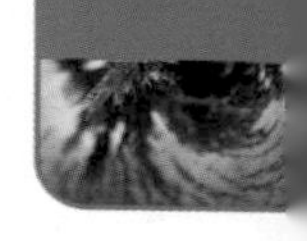

Chapter 20 The Earth in space:

What you need to know

1. Because the Earth is spinning, the position of the Sun appears to change throughout the day.
2. As a result of this spinning motion, parts of our planet are in darkness whilst other parts are in sunlight. It is night where the Earth's surface is in darkness and day where the surface is receiving light from the Sun.
3. It takes one day for the Earth to rotate once.
4. It takes one year for the Earth to orbit the Sun once.
5. The Earth's axis is tilted. As a result the seasons change as the Earth orbits the Sun.
6. The Sun is at the centre of our Solar System.
7. There are nine orbiting planets in our Solar System.
8. The Sun and the stars are sources of light. Non-luminous bodies like the planets and moons are seen because of reflected light.
9. As the Earth spins and orbits the Sun, the positions of the stars in the night sky change.
10. The Moon is a natural satellite.
11. As the Moon orbits the Earth its shape seems to change. The different shapes are called the phases of the Moon.
12. A solar eclipse takes place when the Moon passes between the Sun and the Earth.
13. A lunar eclipse takes place when the Earth passes between the Sun and the Moon.
14. The movements and positions of all the planets in our Solar System are controlled by the pull of gravity (gravitational forces).
15. The gravitational force between two objects depends on the masses of the objects and their separation.
16. Artificial satellites can be used to observe the Earth, for example weather satellites, to look away from the Earth, for example the Hubble telescope, or to travel to and explore other parts of our Solar System and beyond.

How much do you know?

1 Look at the list below:

Earth	A
Moon	B
Sun	C
Constellation	D
Universe	E
Solar System	F

Put them in order of size from smallest to largest. The smallest one has already been done for you.

B ____

5 marks

2 Use words from this list to complete the sentences below.

planets moons satellites Sun gravity galaxy orbits luminous reflect

a) Our Solar System has nine ____________ which travel around the ____________ in circular-like ____________.

3 marks

b) A large cluster of stars, such as the Milky Way, is called a ________________ .

1 mark

c) Stars give out light. They are __________ . Planets and __________ are non-luminous. We see them because of the light they__________ .

3 marks

3 The diagram below shows sunlight shining onto part of the Earth.

Sunlight

A

B

C

D

North Pole

How much do you know? *continued*

a) At which two places is it daytime?

_______________________________ *2 marks*

b) At which two places is it night?

_______________________________ *2 marks*

c) At which place is the Sun setting?

_______________________________ *1 mark*

4 The drawing below shows the positions of the Sun at different times during the day in the summer. The circle marked A shows the position of the Sun early in the morning.

a) Write the letter B in the circle which shows where the Sun will be at midday.

1 mark

b) Write the letter C in the circle which shows the position of the Sun in the late evening.

1 mark

c) Draw a circle in the above diagram and write the letter D in it to show the position of the Sun at midday in the winter.

1 mark

5 The diagram below shows a solar eclipse.

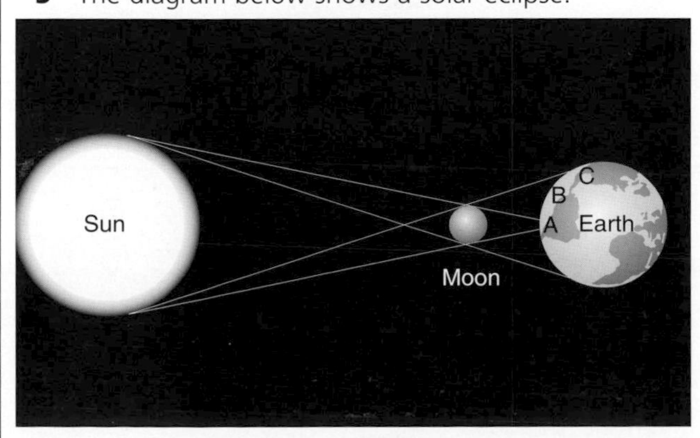

What is seen by people living on the parts of the Earth's surface marked A, B and C?

____

____

_______________________________ *3 marks*

6 The diagram below shows the positions of the Earth at different times of the year.

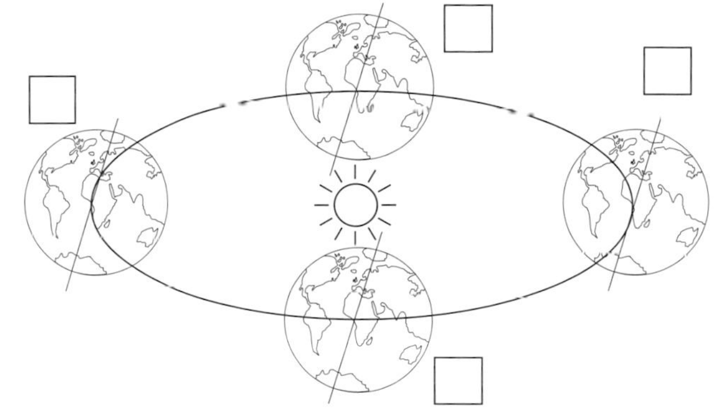

a) Put an A into the box which shows the position of the Earth when we are in summer.

1 mark

b) Put a B in the box which shows the position of the Earth when we are in winter.

1 mark

7 The diagrams below show the phases of the Moon, but apart from a the diagrams have been put into the wrong order.

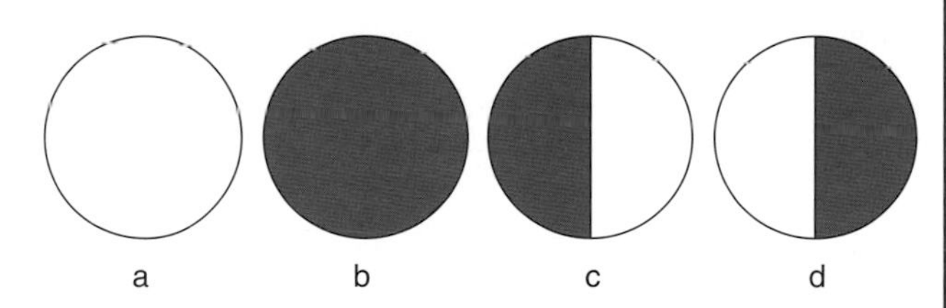

What should the correct order be?

_______________________________ *1 mark*

A
acceleration 149, 153
acid rain 81, 92, 109–110
acids 76–77, 79–81
  neutralisation 76
  reactions with metals 79, 100–101
  reactions with carbonates 79
adaptation 32–35, 39
addiction 48–49
adolescence 17
aerobic exercise 26
air pressure 177
air resistance 155
alcohol abuse 48
alimentary canal 24–25
alkalis 76–77, 80
  neutralisation 76
alternative sources of energy 122–123
aluminium electrolysis 99
alveoli 28, 29, 47
ammeters 134–135
amps 134
amplitude 171
amylase 24
anaerobic respiration 26
analysing evidence 8
angle of incidence 160–161
angle of reflection 160–161
animal cells 11
antacids 76
antagonistic pair 50
anthers 62–63
antibiotics 45
antibodies 16, 44
arteries 27
asteroids 187
atmospheric pressure 177
atoms 84–85
auditory nerve 169

B
bacteria 42
balanced and unbalanced forces 152–153
balanced diet 20, 48
balancing moments 179
ball and socket joint 50–51
bases 76–77
batteries 132
bell jar experiment 169
biceps 50
biomass 60, 123
birth 16
blast furnace 99
blood 27
  vessels 27
boiling 68–69
boiling points 68, 87
bones 50–51
breathing 28–29, 47
bronchioles 28
burning 80

C
camouflage 38
capillaries 27
carbohydrates 20
carbon dioxide
  burning 80
  photosynthesis 60
  respiration 26–29, 61
  testing for 78
carnivores 36
carpels 62–63
cartilage 51
catalytic converter 109
cathode ray oscilloscope 170
celsius scale126
cell membrane 11
cell wall 11
cell 10
  structure 11
  animal 11
  plant 11
cells and batteries 132
cell sap 11
cellulose 11, 61
cervix 16
changes of state 68–69
characteristics 56–57
chemical changes 78–81
chemical composition 86
chemical energy 119
chemical formulae 85, 87
chemical properties 98–99
chemical reactions 78, 86
  reactants 78
  products 78
chemical symbols 85
chlorophyll 11, 60
chloroplasts 11, 60
chromatography 72
cilia 43
circuits 132–135
  complete 132
  components 132
  incomplete 132
  symbols 133
  test 133
circuit diagrams 133
circulation (of the blood) 27
circulatory system 27
classification 54–55
clone 57
coal 120–121
colour 163–165
colour filters 165
colour television 165
combustion 80
  incomplete combustion 112
comets 187
community 33
compasses 142
competition 38–39
compounds 84–87
condensation 68
conduction 128
conductors 128
constellations 188
conservation of mass 115
consumers 36
  primary consumers 36
  secondary consumers 36
continuous variations 56
convection 129
convection current 129
co-ordination 49
corrosion 81, 102, 114
cross pollination 57
crystalisation 73
crystals 66
current 132–137
cytoplasm 11

D
day 182–183
deceleration 149, 153
decibel scale 171
depressants 48
diaphragm 47
diffusion 27, 67
digestion 24–25
digestive juices 24
digestive system 24–25
discontinuous variations 56
disease 43–45
dislocation 51
displacement reactions 104–105, 113
dissolving 70
distance–time graphs 149
distillation 71
dormancy 35
drag 155
drug(s) 48–49
  addiction 49
  overdose 49
  types of 48–49
dyes 77, 163

E
ear 168–169
ear drum 168
Earth 182–183
  axis of 182
  magnetic field of 142
eclipse
  lunar 186
  solar 186
egg cell (ovum) 12
elbow joint 50
electric bell 145
electric current 132–137
  symbols 133
electrical energy 118
electricity
  paying for 125
  power 125
  watts 125
electrolysis 99
electromagnets 144–145
  field strength 144
elements 84–85
ellipse 183
emphysema 47
embryo 15
energy 118–127
  from food 120
  from the Sun 121
  resources 120–123
  transfer 119
environment 32
enzymes 24–25
erosion 90
evaluating your experiment 9
evaporation 71
exam revision 5–7
excretion 25
exercise 46
exothermic reaction 80

F
fallopian tube 15
fats 20–21
  test for 22
fermentation 43
fertilisation 14–15, 63
  external 14
  internal 14
fertilisers 61

fibre 22
filtration 71
flower, structure of 62
flowering plants 62–63
fetus 16
fibre 22
food chains 36–37
food pyramids 36
food tests 22–23
food webs 37
forces 150–157, 174–179
  and moments 178–179
  and pressure 174–177
force multiplier 176, 178
fossils 92
fossil fuels 120–121
  power stations 80, 124
freezing 68
freezing point 68, 87
frequency 170
friction 154–155
fruit 63
fuels 80, 120–121
fungi 42

G
galaxies 188
gas 66–67
  as a fuel 120–121
  exchange 28, 47
  pressure of 67
generating electricity 124
generators 124
genes 57
geothermal energy 123
germs 43–45
global warming 80, 110
glucose 26, 28
gravity 150, 185
greenhouse effect 110–111
gut 24

H
habitats 32–35
herbivores 36
hallucinogens 49
hazard warning symbols 77
health 46–49
hearing range 170
heart 27
heat energy 118, 126–127
hibernation 35
hinge joint 50
hip joint 51
hormones 17
hydraulics 176–177
hydrocarbon fuels 112–113
hydroelectric energy 122, 124
hydroelectric power stations 122, 124
hydrogen . . . testing for 78

I
infrared 110–111
igneous rocks 94
  extrusive 94
  intrusive 94
illness 43–45
image 161
immune system 44–45
immunity 44
indicator organisms 110
indicators 77
inheritance 56–57
insect pollination 62–63
insulators 128
invertebrates 54
iron
  from iron ore 99
  rusting 102, 114

J
joints 50–51

K
keys 55
kilograms 150
kinetic energy 118
kingdoms 54

L
lactic acid 26
lava 94–95
leaves 60, 62
levers 178
ligaments 51
light 160–165
  absorption 163
  energy 118
  dispersion 163
  reflection of 160–161
  refraction of 162–163
  in photosynthesis 60
liquids 66–67
litmus 77
loudness 171
lubricant 154
luminous sources of light 160
lungs 28–29, 47–48

M
magma 94–95
magnetic fields 141–142
magnetic materials 140
magnetic poles 140
magnetic shielding 143
magnets 140–143
  molecular domains 140–141
  molecular magnets 140–141
malnutrition 20
mammals 14
mass 150–151
melting 68
melting point 68, 87
menstrual cycle 16
metals and acids 100–101, 103
metals and oxygen 102
metals and water 103
metals
  extracting 99
  physical properties 98–99
  reactions of 79, 100–101
metamorphic rocks 94–95
microbes 42, 44
micro-organisms 42
migration 35
minerals 21
mirrors 161
mitosis 13
mixtures 70–71, 86
molecules 85
moments 178–179
  anticlockwise 179
  clockwise 179
Moon 185–186
  phases of 186
movement 50–51
mucus 43
muscles 50

N
National Grid 124
natural defenses 43–45
natural selection 39
nerve cells 12
neutralisation 76, 101
newtonmeter 151
newtons 150–151
night 182–183
nocturnal 34
noise pollution 171
non-metals
  physical properties 99
non-renewable energy 120
nuclear energy 119
nuclear power 124
nucleus
  of a cell 11
nutrition 20–22

O
obesity 48
obtaining evidence 8
oesophagus 25
oil 120–121
omnivores 36
opaque objects 160
orbits 183, 185, 189
ores 99
organisms, characteristics of 10
organs 13
organ systems 13
ovaries 16
overdose 49
oviducts 15
ovules 62
ovum, *see* egg cell
oxidation reactions 114
oxide 114
oxygen
  debt 26
  in respiration 26–29, 61
  in photosynthesis 60
  testing for 78

P
painkillers 48
palisade cells 13
parallel circuits 134–135
particles 66
  arrangement of 66
particle theory 68
pascals 175
pathogens 43
  transmission of 43
penis 15
pepsin 24
period 16
Periodic Table 84–85
periscope 161
pH scale 77, 108
phloem 62
photosynthesis 36, 60
physical properties 98–99
physical changes 68–69
pitch 170
pivot 178–179
placenta 16
plane mirror 161
planets 182, 184–185

planning an experiment 8
plant cells 11
plants 60–63
pneumatics 177
pollen 62–63
pollination 62–63
pollutants 110
populations 33
potential energy 118
predators 38–39
pregnancy 16–17
pressure
  calculating 175
  due to a force 174–175
  in a gas 177
  in a liquid 176–177
prey 38–39
primary colours 164
producers 36
proteins 21, 25
  test for 23
puberty 17

R
radiation 129
rainforests 38
reaction time 49
reactivity series 102–105, 113
red blood cells 12
reduction reaction 115
reflection of light 160–161
refraction of light 162–163
relay switch 145
renewable energy 121, 122–123
reproduction 14–16
reproductive systems
  in humans 15
  in plants 62–63
resistance 136–137
resistors 137
  variable resistors 137
resources 38
respiration 26–30, 61
  aerobic 26
  anaerobic 26
respiratory system 47
revision timetable 5–7
rivets 127
rock cycle 95
rocks 90–95
  disintegration 90
  grains 90
  non-porous 90
  porous 90
  weathering 90–93
roots 60, 61, 62
root hair cell 13
rusting 102, 114

S
saliva 24
salts 79, 101
satellites 185, 189
  artificial 189
  geostationary 189
  natural 185
  orbits 189
saturation 72
scientific investigations 8–9
seasons 183
secondary colours 164
secondary sexual characteristics 17
sediment 92
sedimentary rocks 92–93, 94
seed dispersal 63
seeds 63
selective breeding 57
separating mixtures 71–73
series circuits 134–135
sexual intercourse 15
sexual reproduction 14
  in humans 15
  in plants 62–63
shadows 160
skeleton 50
smoking 29, 47–48
soil 108
  nutrients 108
solar energy 123
Solar System 184–185
solids 66–67
solubility 70, 72
solute 70
solutions 70–73, 86
solvent 70
solvent abuse 49
sound 168–171
  amplification 168
  compression 168
  energy 118
  rarefaction 168
  waves 168
specialised cells 12–13
species 56
spectrum 163
speed 148–149
  of light 160
  of sound 169
speed–time graphs 149, 156
sperm 15
sperm cells 12
spider diagrams 3
sprains 51
starch 24, 61
  test for 22
stars 183, 188
stem 61, 62
stigma 62–63
stimulants 49
stomata 60
stopping distances 157
streamlining 155
study skills 2–4
sugar, test for 22
Sun 182–183
  energy from121
survival of the fittest 38
switches 132
synovial fluid 51
symbols
  chemical 85
  electrical 133

T
temperature 126
tendons 51
terminal velocity 155
testes 15
thermal energy 127
thermal contraction 127
thermal expansion 127
thermit reaction 105
thermometers 126
tidal energy 122
tissues 13
titration 101
toxins 43
tranquillisers 49
translucent objects 160
transparent objects 160
transpiration 61
transpiration stream 61
triceps 50
turbines 124
turning forces 178

U
ultrasounds 170
umbilical cord 16
Universal indicator 77, 109

V
vaccination 45
vacuoles 11
vacuum 129, 169
vagina 15
variation 56–57
veins 27
vertebrates 54
vibrations
  particles 127, 128
  sound 170–171
viruses 42
vital organs 50
vitamins 22
voltmeter 135
voltage 135
voltaic cell 113

W
waste products 27
water 22
  in plants 61, 62
  reactions with metals 88
  in respiration 24
  in weathering of rocks 90–91
wave energy 123
weathering of rocks 90–93
  chemical weathering 90, 92, 109
  freeze and thaw 91
  heating and cooling 91
  physical weathering 90
weight 150–151
white blood cells 12, 44
wind energy 122
wind pollination 62–63
womb (uterus) 15
word equation 86, 99

X
xylem 62

Y
year 183

Z
zygote 15